Applied Farm Management

Jonathan Turner

Dean – School of Business
Royal Agricultural College
Cirencester, Gloucestershire

Martin Taylor

Senior Agricultural Manager
Barclays Bank – Swindon
Swindon, Wiltshire

Second Edition

Blackwell
Science

Blackwell Publishing
Editorial Offices:
Blackwell Science Ltd, 9600 Garsington Road, Oxford OX4 2DQ, UK
Tel: +44 (0)1865 776868
Iowa State Press, a Blackwell Publishing Company, 2121 State Avenue, Ames, Iowa 50014-8300, USA
Tel: +1 515 292 0140
Blackwell Science Asia Pty, 550 Swanston Street, Carlton, Victoria 3053, Australia
Tel: +61 (0)3 8359 1011

First edition published 1989 by BSP Professional Books
Second edition published 1998
Reprinted 2003

Library of Congress Cataloging-in-Publication Data
Turner, Jonathan.
Applied farm management/Jonathan Turner and Martin Taylor. – 2nd ed.
p. cm.
Includes index.
ISBN 0-632-03603-6 (pbk.)
1. Farm management. I. Taylor, Martin.
II. Title.
S561.T87 1998
630'.68–dc21 97-40286
CIP

ISBN 0-632-03603-6

A catalogue record for this title is available from the British Library

Set in 10/12pt Times
by DP Photosetting, Aylesbury, Bucks
Printed and bound in Great Britain using acid-free paper
by Marston Lindsay Ross International Ltd, Oxfordshire

For further information on Blackwell Publishing, visit our website:
www.blackwellpublishing.com

Contents

Preface to the Second Edition

The second edition has finally come to fruition with several major changes. So much has happened in the European Union that Chapter 14 has been completely rewritten and includes a section on the General Agreement on Tariffs and Trade (GATT). In other areas hedging 'green' money is dealt with, as is the BSE crisis. World grain and food markets are constantly changing and the text reflects the main trends. Other alterations relate to changes in legislation, business performance, costs and profitability.

The original concept arose because students made it clear that they required a text which would assist them in their studies of farm business management, as none existed which covered business management, finance, taxation, the European Union and marketing in one volume. In this edition we have stayed close to our original intentions and the book is produced for diploma and undergraduate students, although experience has shown that students on more advanced courses will find it useful material on which to base their studies. Students of agrifood and agribusiness will also find many parts of the book relevant to their studies as the principles involved in managing a business are transferable between sectors of the economy.

The text provides the basis for an integrated study of the subjects which present and future farm, estate and food industry managers will find necessary. The contents should prove useful for teachers and lecturers, bank managers, and advisors to farmers and food producers throughout the world, as the principles explained are relevant to many working environments.

Jonathan Turner
Martin Taylor

Acknowledgements

Our grateful thanks go to Penny, Christine, Hannah and Frances; Janette, Daniel, Anna, Rebekka, Thomas and Josef, who have done without us while we laboured. To our colleagues, past and present, who have provided encouragement and information, we are indebted. To students of 'The Royal' who have in their own way provided the motivation for producing this work and for whom much will be familiar. To Barclays Bank and other organisations for allowing us to use some of their own data, we are most appreciative. Thanks also to Bekky for word processing the updates from original 'scribblings'. Finally, we express our gratitude to John who was involved in the conception but not the 'birth' of this work.

1 Farm Management

The most important criterion in successful farm management is attention to detail in all aspects of husbandry, even though pressures on the farm business continue to increase. These pressures tend to increase costs due to greater emphasis on animal welfare, the environment and food safety and quality. This chapter describes some of the other factors that are important in seeking *profits* and *satisfaction*.

What is management?

In its simplest form 'management' means making the best possible use of available scarce resources – land, labour and capital – to achieve the objectives set. Management is a dynamic process involving responsibility for the effective operation of a business, this responsibility entailing the installation and implementation of procedures which will ensure adherence to plans and the selection, guidance, motivation and control of the people involved in the business.

The key aspects of the job of management can be identified as follows:

(1) It involves the identification and application of the best husbandry methods and techniques available
(2) It involves analysing, organising, planning and controlling the performance of the business
(3) It is a complex, dynamic process, not a single activity
(4) It is concerned with people
(5) It involves the acceptance of responsibility for achieving planned objectives

Management is a practical process, not simply a theoretical subject. The resources of the business have to be managed.

The most effective way of managing resources is to set objectives (see Fig. 1.1) which can be achieved realistically within a predetermined period of time. The objectives should be made known to all members of the management team and the work of the business can then be directed towards the fulfilment of objectives in a logical manner.

Management is effective if it enables a business to progress towards and/or to achieve its objectives. The objectives of a business may be many and varied. The simple objective (considered to be the basic one in economic theory) of maximising profit rarely reveals the true purpose of a farm business. A farmer's objective may be to continue a business which will provide an acceptable way of life for himself and his family. Expansion may be the goal of some farmers, while others may see the

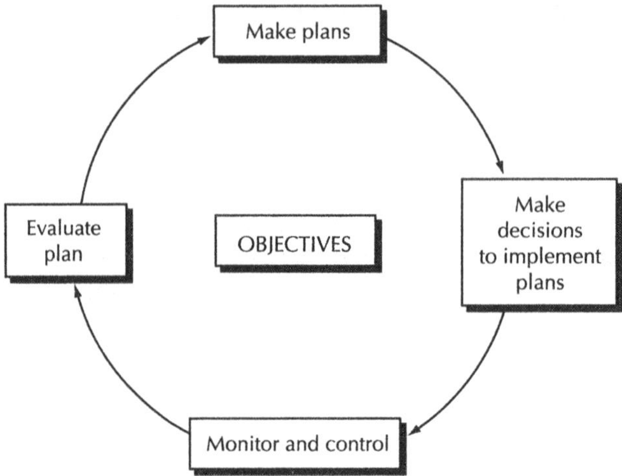

Fig. 1.1 The basic management functions.

build up of a large capital asset as being the most desirable objective to achieve. Pursuit of profit may, in these circumstances, only be a means to achieving the ultimate objective. Objectives are set within the economic, political, social and legal environment of the farm. To achieve success, plans must be made, implemented, evaluated and, if necessary, revised to ensure continued progress towards the set goals. This is illustrated in its simplest form in Fig. 1.1. Figure 1.2 shows the areas in which objectives have to be set. These are:

(1) Production
(2) Financing
(3) Marketing
(4) Staffing

Setting objectives is only one part of the process of management which involves a complex series of interrelated activities (see Fig. 1.2) which may at different times entail the following:

- Objective setting
- Innovating
- Planning
- Developing resources
- Organising
- Supervising
- Policy making
- Co-ordinating
- Communicating
- Motivating
- Leading
- Controlling

These activities do not occur in any stylised order and in the course of a working day any member of a management team may be engaged in many of them. For example, a person supervising a gang harvesting peas will be organising, co-ordinating, communicating, motivating, leading, controlling and supervising all at the same time.

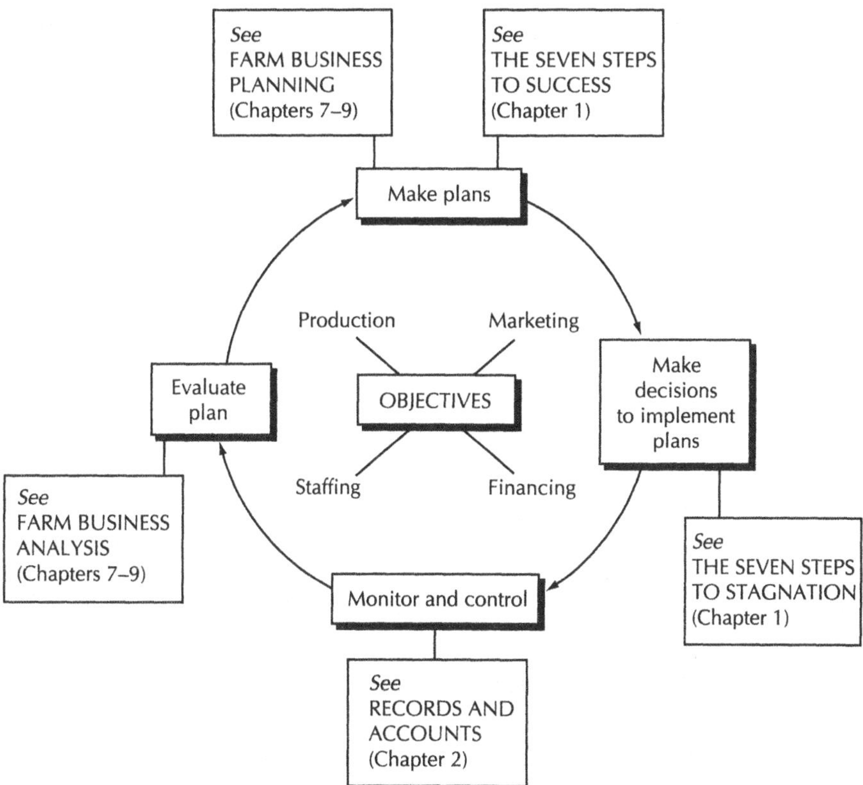

Fig. 1.2 Decision areas in management.

This chapter explains the areas identified in Figs 1.1 and 1.2 in greater detail.

Objectives

These can be strategic or tactical. *Strategic objectives* can be thought of as those which involve the whole business in the longer term. They determine the ultimate destination of the business and relate to productivity (what, where, when, how and how much to produce), business structure and organisation (sole trader, partnership, limited company) and capital (source and disposition of funds). *Tactical objectives* relate more to daily, often routine, activities. Once the objectives are set, plans have to be made and implemented.

Planning

The details of farm planning are dealt with in Chapters 7 to 12. In general, it is necessary to plan ahead in order to:

(1) Make the most effective use of time available, especially at peak periods of demand for men and machines when weather and soil conditions are good
(2) Optimise use of machines and physical effort
(3) Avoid loss and wastage
(4) Reduce costs and maximise profits.

Planning is not just to save time, however, but also to improve output; that is, to make use of any time saved. It is also aimed at not only getting more work done but also making it easier, safer, more organised and therefore more enjoyable and effective.

Decision making

When implementing plans, there are many factors and decision areas to consider. Decision making can be facilitated using various techniques, such as decision trees (see Fig. 6.1), game theory, linear programming and programme planning, network analysis or just plain intuition based upon experience or 'animal cunning'. The process of decision making can be divided into five steps:

(1) *Observation.* Find and define the problem that determines the critical factor – the element which has to be changed before anything else will alter. The rules within which the solution must be made have also to be identified – for example, quota restrictions, climatic constraints, or company policy not to borrow more than half of any capital required.
(2) *Analysing.* It must be determined who is to make the decision, who must be informed and who will convert the decision into action. The manager must know what facts are available, how relevant they are and how accurate. Priorities needed to analyse a situation are:
    (a) How long will a decision affect the business?
    (b) The effect of a decision on the rest of the business
    (c) How often does the decision occur?
(3) *Developing alternative solutions.* Each alternative should be considered. Even no change is an alternative which must not be overlooked. Indeed, there are *seven steps to stagnation* which must be avoided:
    (a) We have never done it that way
    (b) We are not ready for that
    (c) We are doing all right without it
    (d) We tried that once
    (e) It costs too much
    (f) That is not our responsibility
    (g) It would not work for us
(4) *Finding the best solution*
    (a) The costs and benefits of each alternative must be considered
    (b) Determine which will give the best results for the least effort

   (c) Is a long consistent effort required or does the solution demand urgent action?
   (d) Who will carry out the course of action – will it suit them? No decision can be better than the person who carried it out
(5) *Making the decision effective.* In many situations, the decision-making manager is not the one who will carry out the effective action demanded by a solution. He can, however, communicate to others what they ought to be doing and, by regular review and reporting back sessions, encourage and motivate them.
(6) *Monitoring and control.* The records and the accounts necessary to carry out the monitoring of performance, both physical and financial, are described in Chapter 2. *Excellence in husbandry is the most vital aspect of any farm business* and is only neglected at the risk of business failure. Decisions on monitoring the financing of the business are explained in Chapters 8, 9 and 11, while the principles of agricultural marketing are considered in Chapter 13.

Having outlined the process of management, let us concentrate on the steps that must be taken to ensure the success of the farm business.

In its simplest form, we believe there are *seven* basic steps towards successful farm management. These are:

(1) Husbandry: technical excellence
(2) Records and accounts: appropriate for the business
(3) Cost control
(4) Review and analysis of the business
(5) Budgeting for change
(6) Re-organising the resources of the business
(7) Investment of capital to expand or intensify

Husbandry

It is important to recognise at the earliest possible date that attention to detail at all levels of husbandry and stockmanship is the most important criterion of profitable farming. It is possible to be successful in farming if high standards of husbandry are achieved with little regard to the sophisticated farm business management techniques that are available. The converse is impossible. In this book, therefore, we do not intend to dwell upon the most important aspect of farming for we expect farmers and students to acknowledge this truth, regularly up-date themselves on current developments in husbandry, and apply any relevant new technology, with diligence, to their enterprise(s).

On the other hand, we also accept that farming is no longer just a nice way of life. It is big business and must be treated as such. For that reason, we believe that some of the management techniques which we incorporate in this book have a *real* application to most farm situations and, provided the husbandry is right in the first place, can help farmers strive towards greater profits.

Records and accounts

There are two reasons for keeping accurate records and accounts. The first is that farmers must obviously obey the laws of the land and accommodate the Inland Revenue and Customs and Excise. Taxation matters are briefly discussed in Chapter 12. In addition to taxation, there are certain statutory records that must be kept; these are explained in Chapter 2. In this book, however, the records and accounts referred to will be 'management accounts'. Compiling management accounts is the second and main reason for record keeping. We need to see 'real' accounts that can help analyse past trends and performances and also assist in planning the future policy of the business. These 'management accounts' cover physical and financial data of the whole farm, and individual enterprises. They are explored in Chapter 2 (and in Chapter 3), where traditional trading, profit and loss accounts are converted to a more meaningful form, and in Chapter 6, where individual enterprises are studied. The use of computers for recording purposes is also considered in Chapter 2 and it is emphasised that, when properly applied, good records can help a farmer to improve his performance, even though it may already be of a high standard.

Remember, attention to detail in husbandry is vital, even in situations where there are cost pressures due to greater emphasis on animal welfare, the environment and food safety quality.

Critical analysis of a farm or an enterprise can only be achieved if good farm management records are available. In turn, this leads to small changes which may lead to increased profits. If husbandry standards are already good and are maintained at high levels through the application of monitoring devices, then the next logical step towards greater profits must be *to keep costs under control*!

Cost control

There is no longer just a matter of 'pulling in the belt' another notch. The whole procedure is an essential part of today's farming business and not just in the UK. Data required refers to the records and accounts available, but three aspects are considered in depth within this section of the book:

(1) The cash-flow in Chapter 8.
(2) The balance sheet and its interpretation in Chapter 5
(3) Enterprise costings and control measures in Chapter 6

As cost pressure increases, farmers must run faster to stand still. It means that they must improve management output by better husbandry but also try to keep costs under control.

Review and analysis of the business

Having considered husbandry records and accounts and cost control, it is essential from time to time to sit back and review the farm business as a whole, and individual enterprises in particular. This is sometimes best done with a friend or outside advisor who can help look at the business with a 'fresh pair of eyes'. Frequently an outsider can identify problem areas which appear to stick out like a sore thumb. However, in most cases, it will be necessary to collect and examine current levels of physical and financial performance. This data should preferably cover three to five years. If good records have been kept, then this is no problem. If records have not been kept, the difficulties encountered at this stage can be numerous, for on the one hand, farmers' estimates of performance standards frequently tend to be optimistic, whilst most accounts which are available under these circumstances are so historic that they are of no great value. Also, as accounts prepared for the Inland Revenue follow a particular convention, they are less useful for management analysis.

Chapter 4 attempts to discuss the difficulties that arise in data collection for farm analysis and explains the procedure of adjusting accounts for future planning.

The whole purpose of this review of the farm's present position is to look at the current performance levels in the following areas:

(1) Production – husbandry!
(2) Marketing
(3) Financing
(4) Staffing

Our review of these functions is to determine whether changes should be made to the business objectives and leads into the methods that can be used to implement these changes.

Budgeting for change

The techniques available for evaluating the average annual effect on the farm profits are explained, with examples, in Chapters 7 to 10. Budgeting to estimate the effect of minor changes to the farm is called 'partial budgeting'. Partial budgeting involves calculating changes in farm profits that do not entail major changes to the structure of the farm. It is essentially a method of action. It is not the means whereby a change of plan was conceived, but more likely derived from an intuitive idea by the farmer or by a new cropping or stocking opportunity which has appeared. The application and deficiencies of partial budgeting are discussed in Chapter 7.

Business planning

This stage involves reorganising the resources of the business and the investment of capital. The 'tools' available for assessing the effects of major changes in the business or for planning a new business are explained in Chapters 7 to 10. Examples of these, of course, may not apply to every farm business or planning situation. The emphasis is placed on the relationship between *profit, cash* and *capital* and how the deployment of farm resources affects and is affected by these three facets of the business. It is imperative that they are not considered in isolation or the picture may be misleading and poor decisions made. Thus we explain in Chapter 8 how profit, cash and capital fit together.

2 Records and Accounts

Record keeping is often seen as a necessary evil by farmers, indeed some may even say it is an unnecessary evil! They prefer 'the day's work' on the farm to the office work of 'paper pushing'. Chapter 1 outlines the steps necessary for successful farming. Good records are a vital part of this process and must be *simple* and *accurate* to be of any use. Like medicine, to do any good, records should be appropriate and digestible both for the user and the person doing the recording.

It is sometimes thought that the historical nature of recorded information renders it unusable. This criticism, often used as an excuse for not keeping good records or any records at all, is not acceptable.

Chapter 1 shows that successful management depends upon being able to analyse the present state of the business. This analysis forms the basis for future farm planning and it can only be assessed using the most up to date physical and financial information available from the farm business.

There are two reasons for keeping records and accounts on the farm:

(1) *Mandatory (legal requirements)*
(2) *Management purposes*
    (a) Analysis
    (b) Planning

Naturally, some of the mandatory records can be used for both analysis and planning purposes but this is not always the case. For example, the statement of profitability presented for tax purposes to the Inland Revenue may not represent the 'real' performance of the business in the past year because of:

- Capital allowances (now substantially lower than they were)
- Profit averaging
- Records may be two years out of date
- Treatment of repairs and renewals of buildings
- Valuation policy for stock and crops

No matter what the purpose for keeping records, they must:

(1) *Serve a definite purpose* (which can be demonstrated to the arable worker or stockman who is keeping the record)
(2) *Be accurate* (the person keeping the record must also be motivated to record the required information accurately)

(3) *Be easy to complete* (not too time-consuming or complicated)
(4) *Be up to date*
(5) *Be easily stored and readily accessible* (hence computer recording systems are becoming more widely used in farm business management)

Mandatory records

Mandatory records serve the purpose of fulfilling legal requirements, they cover the following areas.

Health and safety at work

The main Act of 1974 has been augmented by regulations on:

- Health and Safety (First Aid, 1981)
- Food and Environment Protection Act (FEPA, 1985)
- Reporting of Injuries, Diseases and Dangerous Occurrences (1985)
- Control of Pesticides (1986)
- Control of Substances Hazardous to Health (COSHH, 1988)
- Noise at Work (1989)
- Health and Safety Information for Employees (1989)
- Electricity at Work (1989)
- Pressure Systems and Transportable Gas Containers (1989)
- Management of Health and Safety at Work (1992)
- Provision and Use of Work Equipment (1992)
- Personal Protective Equipment at Work (1992)
- Manual Handling Operations (1992)
- Health and Safety (Display Screen Equipment 1992)

More emphasis has been focused on health and safety at work and the main requirements are as follows:

- Assess the risks of the health and safety of employees and anyone else who may be affected by the work activity
- Record the findings of the assessment (for employers with five or more employees) and prepare a written health and safety policy
- Make arrangements for implementing the measures highlighted in the assessment (and record progress if five or more employees are involved)
- Set up emergency procedures
- Appoint competent people to help devise and apply measures to comply with legal duties
- Provide understandable information to employees
- Ensure employees are adequately trained and competent at their job in order to avoid risks.

In general, employers have to:

- Try and make the workplace safe and free from risks to health
- Keep dust, fumes and noise under control
- Ensure plant and machinery are safe
- Ensure articles and substances are moved, stored and used safely
- Provide free protective clothing, or equipment required by law
- Report certain injuries, diseases and dangerous occurrences to the authorities

The records that have to be kept are:

- Accident Record Book
- Written Scheme of Examination for Pressure Systems
- Assessment of Risk Report
- Implementation of action necessary to reduce risks
- Logbook of pesticide applications, stock control and disposal. Poisonous substances must be kept in a locked and clearly marked chemical store
- Health and maintenance details

Comprehensive details of all these regulations and mandatory records can be obtained from the Health and Safety Executive.

Movement of animals

Under the Animal Health Act (and relating to notifiable diseases), restrictions are placed upon animal movements in accordance with various orders made since 1928, specifically the Movement of Animals (Records) Order 1 1960 and Amendment Order 1961. Briefly, anyone moving or permitting the movement of animals shall keep a record of the movements. EU directive 92/102 EEC has the objective of establishing an EU-wide Animal Registration and Identification Scheme so that animals can be traced to their farm of origin. The Bovine Animals (Records, Identification and Movement) Order 1995 means that since April 1995 the use of new official tags in the right ear of new born cattle is compulsory. The new regulations create four different record-keeping requirements;

(1) An on-farm record with individual details of all cattle
(2) The Premium Record, detailing premium claims by ear tag number
(3) The Movement Document which travels with cattle to market or slaughterhouse
(4) An Annual Stocktake, which can be kept in the on-farm record

A Continuous Flock record must be kept for sheep and goats and when they are moved off the farm it must be possible to identify them from a 'temporary' mark (inside the UK) and a permanent tag or tattoo if they go outside the UK.

The police and local authorities are responsible for enforcing orders made under the various Animal Health Acts and records on livestock movement must be retained on the farm for three years.

Value Added Tax (VAT) records

It is essential that the farm business be registered so that any recoverable VAT can be claimed from Customs and Excise. Since many purchased items used to produce food are standard rated, farmers can claim significant sums, especially on fertiliser and machinery. Most farm inputs are taxed at the standard rate while most outputs are zero-rated; cash flow may be improved by submitting monthly, rather than quarterly, returns to reclaim the net VAT being paid. The following list indicates items which are exceptions to this general rule and which relate to the farm business:

(1) *Exempt outputs and inputs*
- Purchase and renting of land (land includes buildings, walls, trees, plants and other structures and natural objects attached to the land so long as they remain attached)
- Granting of mineral rights, whether by licence or by lease
- Wages
- Postal services, bank charges and bank interest
- Insurances
- Local authority and water rates (although the latter are standard rated where water is used in any industrial process)
- Tithe redemption annuities; sales of land.

(2) *Zero-rated inputs*
- Livestock (of a food-producing kind)
- Animal foodstuffs, hay and straw
- Erection of new buildings (but not maintenance and repairs which are chargeable at the standard rate)
- Seeds, tubers of edible vegetables, fruits and herbs
- Plants producing edible fruit
- Grass and other herbage seeds
- Propane, methane and butane, petrol, diesel and paraffin
- Purchasing of grazing or mowing rights
- Hatching eggs
- Creosote
- Emptying cesspools and septic tanks
- All sales to overseas customers are only zero rated if the correct procedures are followed. There are separate rules for sales inside and outside the EU.

(3) *Lower rated inputs*
All *domestic* fuel (coal, oil, gas, electricity, wood, etc.) carries a VAT charge of 5%. This is not a deductible input, as it applies to private, rather than business use.

(4) *Output taxable at the standard rate*
- Wool, cider and perry
- Flowers, flower seeds, cuttings, bulbs, rootstock
- Ornamental nursery stock, trees and hedgerow timber
- Leasing of quota
- Canary and sunflower seed and seed for similar purposes

- Farmyard manure
- Semen
- Secondhand machinery, hiring machinery, contracting
- Peat, turf and topsoil
- Horses (a special sales scheme applies to second-hand horses) and dogs
- Agistment (charging a fee to look after livestock)
- Granting of rights to take game or fish
- Provision of camping facilities and car-parking
- Holiday accommodation, meals and farmhouse teas

It is imperative that the farmer charges VAT on any items sold or any work done which is liable to VAT. If VAT is not charged on these items then the farmer will have to pay the VAT due out of any net income, since the VAT will be collected by the Customs and Excise whether the farmer has charged it or not.

VAT records can be included in the cash analysis book or kept in an entirely separate VAT sales and purchases book. All VAT records must be retained for six years and can be inspected by Customs and Excise at any time.

The farmer can recover VAT by setting it against the VAT he has charged on his own sales and only sending the balance to the Customs. If he has made no taxable sales, a refund can be claimed, normally on a quarterly basis but possibly monthly by special arrangement with Customs.

Wages book

Records must be kept of wages paid to employees. The Ministry of Agriculture, Fisheries and Food (MAFF) appoints inspectors to check the records on behalf of the Agricultural Wages Board. Records that will be inspected are such items as PAYE, income tax deductions, National Insurance and any pension contributions which have been stopped from the gross wage of the worker. The inspector will also check the records to ensure that employees who worked more than the minimum weekly hours are correctly paid at the prescribed overtime rate.

Statement of profitability (income tax returns)

For tax purposes, the Inland Revenue must receive a statement regarding annual profit made by the business. This normally takes the form of a trading and profit and loss account. Sales and transfers of assets must also be recorded and the return made to the revenue for inheritance (formerly capital transfer) tax and capital gains tax.

Integrated Administration and Control System (IACS)

These are for European Union purposes so that set-aside and stocking rate calculations can be done and area compensation payments claimed under the Common Agricultural Policy reform operative since 1993.

Agriculture census returns

These are required from farmers in June for all registered holdings (150 000 holdings) and in December (a sample of approximately 50 000 holdings) each year. This information is used to compile some of the information which is published in *The UK Farm* in February of each year.

Records required by landlords

Many tenancy agreements require that the tenant keeps a record of field cropping, yields, stocking rates and all produce sold off the farm. Naturally, the farmer should have all this information anyway for his own management purposes, so the maintaining of these records for the landlord is not so onerous as may be thought at first sight.

Other records

Other records which need to be kept by farmers include:

(1) Details of any journeys relating to heavy goods vehicles
(2) A rent book is required for cottages let to non-farm workers
(3) Licences will be required for water extraction from rivers or bore-holes and the amount of water which is used in each prescribed period must be recorded
(4) Licences are required for petrol storage. Usage should also be recorded
(5) Certificates are required to show that all sheep have been dipped during the two prescribed periods in the year. The Ministry send notices of intention to dip to each flock owner, and these forms must be sent to the local authority three days before dipping is to take place. After dipping another form should be sent to the local authority within seven days of the dipping taking place to notify them that all the sheep have been dipped in accordance with the regulations. Further details of all these records can be obtained from the MAFF

Records for management purposes

(1) To assess the physical and financial performance
(2) To analyse the performance
(3) To assess progress over the past year
(4) To reveal strengths and weaknesses in the business
(5) To establish a base for planning changes in the business and to determine the likely effect of any changes made
(6) To facilitate monitoring of performance and instil discipline in moving towards the fulfilment of targets, which have been set as a result of the recorded information from the past three to five years

What are the main physical and financial records that need to be kept for management purposes?

Physical records

The farm map

An accurate farm map is an essential and valuable asset to the running of the farm business. It is remarkable how many of those in use were not up to date before the IACS forms of 1993, because fences have been removed, new buildings erected or roadworks have taken place. It is useful to colour-code the particular cropping of the farm over previous years and some farmers would even record details of fertiliser, seed and spray use on the map itself. This is not recommended as it is preferable to maintain a detailed individual record of each field and there is not normally enough room to do so on a map.

A map should be kept of any gas pipes or other public utilities which cross the land and it is vital that records of drainage are kept on the farm map for future reference should any drainage problems occur on any particular area of the farm.

The daily diary

This is a most useful and a very simple way to record the events on the farm. It is simple to write down the sales of arable products or the movement of animals off the farm, and even to record input costs on grazing land when particular animals were using it, so that forage costs can be properly allocated to each enterprise. This diary can then be used as a basic record for filling out more detailed physical records which the farmer may wish to keep.

Barn records

These are mainly used to record the amount of feed which certain classes of livestock have used. It is essential that feed is accurately recorded as it is usually the main cost involved in livestock production. As the recording of feed use can be time-consuming, it is necessary to keep these records down to a minimum. It is recommended that the person mixing the feed in the barn simply records the type of mix (which should be given a number), the tonnage fed, the class of stock which have received it, and the date. This can be done on a daily basis and, as only four items are being recorded, should not prove to be too time-consuming or cause any loss of efficiency. The daily slips recorded by the men on the farm should be reconciled with the delivery notes of feed brought onto the farm.

Livestock numbers

To claim the livestock payments from the CAP and to analyse performance, each farmer must know his precise stock numbers. For cattle this means individual ear tags, for sheep a continuous stock record is necessary. For this purpose, it is usual to record livestock numbers on a monthly basis, although some farmers may do it weekly (particularly with pigs and poultry) and others only do it at the annual farm valuation. A simple monthly stock summary and reconciliation chart (see Fig. 2.1) will supply the necessary information for the farmer to calculate average livestock numbers for each enterprise.

Fig. 2.1 A livestock inventory chart

Livestock	Stock at start	Births	Purchases	Transfers in	Transfers out	Deaths	Sales	Stock at end
Sheep:								
commercial rams	15		5			1	3	16
vas. rams	5							5
ewes	450		50			4	40	456
gimmers								
ewe lambs	50					1		49
hoggets								
lambs	200	230				3	197	230
Sub-total	720	230	55			9	240	756
Beef:								
stock bulls								
intensive bulls								
intensive steers								
over 12 mths	75			69			75	69
under 12 mths	71			70	69	2		70
calves 0 to 3 mths	20	68			70			18
Sub-total	166	68		139	139	2	75	157
Dairy cattle:								
bulls								
cows in milk	81			44	19	1	15	90
heifers in milk	20			18	20			18
dry cows	22			19	22			19
in-calf heifers	20			17	18		2	17
bulling heifers	17			18	17			18
calves 3 to 15 mths	19		8	12	18	1		20
calves 0 to 3 mths	12	17			12			17
Sub-total	191	17	8	128	126	2	17	199
Grand total	1077	315	63	267	265	13	332	1112

Information for the livestock inventory chart can be gained from the daily dairy of events or the sales day book and purchases book. Should the numbers not reconcile, then stock must have been lost or stolen!

Field records

The individual field records should show the action which is taken on that particular field during each production year. The example (Fig. 2.2) shows a cereal crop and the items and events recorded include:

- ❑ Cultivations
- ❑ Animal movements
- ❑ Crop inputs
- ❑ Crops harvested

This particular kind of form allows for the costings to be done at the time that fertiliser, seed or spray are applied and this will facilitate the keeping of the financial records on the farm. The cost of fertiliser particularly is now so great a factor in crop production that many farmers take special care in ensuring that staff apply the correct amount. To aid in this process, a slip such as that demonstrated in Fig. 2.3 may be used.

These slips are produced in duplicate or triplicate: one copy is given to the person applying the input, one is kept by the farmer or manager (and the other one, if in triplicate, stays in the book). When the fertiliser has been applied to the particular field, the driver fills in the relevant section and returns it to the manager or to the farm office where the amount he has applied can be checked against budget to determine how accurate the application has been. The cost of fertiliser applied can then be calculated. By putting such detail in the field records, it is possible to calculate a gross margin per field. This allows a more comprehensive analysis of performance to be made on the farm.

Financial records

Sales and purchases

Often these are recorded with VAT in the cash analysis book. It is a simple procedure to complete the cash analysis book for each transaction as it occurs, then to strike a total (called ruling off the cash book) at each month end for each item of expenditure.

(1) *The cash analysis pages.* Set out the cash analysis book using one page for payments and another page for receipts
(2) *The page headings.* Set out the column headings (if the cash analysis book is not already set out for you) using (from the left of the page) separate columns,

Fig. 2.2 A field record

Field records						
Farm: Manor Farm				Harvest year: 1997		
Field: 17 Paddimore	Hectares: 23.22		Current crop: bronze winter barley Previous crop: winter wheat			
Date	Detail	Product	Total quantity used	Rate per ha	Unit price	Cost per ha
1/9/93	Shallow vibroflex × 1					
3/9/93	Cambridge rolled × 1					
15-18/9/93	Ploughed (8″ deep)					
23/9/93	Disced × 1					
24/9/93	Triple K × 1					
25/9/93	Crumbler × 1					
26/9/93	Drilled	Bronze	88 × 50 kg	189 kg	£342/t	£64.81
29/9/93	Drilled headlands	Bronze				
2/11/93	Residual herbicide	Harlequin Ardent	90 l 22.5 l	4 l 1 l	£7.962 £11.20	£30.86 £10.85
22/2/94	Topdress (N: 49 kg/ha)	Olympic (33.5 N)	3390 kg	146 kg	£90/t	£13.14
18/4/94	Topdress (N: 65 kg/ha)	Nitram	4388 kg	189 kg	£90/t	£17.01
18/4/94	Fungicide spray (Rhynco and Netblotch)	Cosmic Punch C	33 l 6.5 l	1.5 l 0.3 l	£4.76 £37.30	£6.76 £10.44
20/4/94	Compound topdress (47 kg P and 47 kg K/ha)	Olympic (0:24:24)	4500 kg	194 kg	£99/t	£19.19
28/4/94	Final topdress (total N = 161 kg/ha)	Nitram	3163 kg	135 kg	£90/t	£12.16
		Income				
		Variable costs				
		Field gross margin				

Fig. 2.3 Input use on crops and grass

Field: Barn	Field Type	Weight required	Date: 15 Oct Weight applied
Seed Fertiliser Spray	Olympic (0:24:24)	1350 kg	1375 kg

for the 'date', 'details' 'cheque (or voucher) number', 'bank' and 'contra'. Head the other columns as required by your business leaving the last two columns on the right of the page for 'VAT claimable' and 'inputs at standard and zero rate' on the payments page, and 'VAT charged' and 'outputs at standard and zero rate' on the receipts page

(3) *Making entries*

(a) Entries are only made in the bank column when money flows out of the bank account by cheque, standing order, direct debit or cash payment for a particular transaction [see Section (5): Checking bank statements], or when money flows into the bank account by cheque or cash receipts. Entries are made in the contra column when money is owed to (or due from) a client but is 'paid' (or received) by deducting it from the value of goods which you have delivered to them. In this way, you do not need to write a cheque for the amount you owe but you will only receive, for your goods (e.g. milk or barley), a cheque reflecting the net value of your sales less the amount you pay by contra.

NB:

- Contra values *must* be shown on both the receipts and payments page of the cash analysis book
- Any payments made by contra which include VAT at the standard rate are entered for VAT purposes in the same way as payments made by cash or cheque which incur VAT at the standard rate
- The bank and contra columns show the total value of transactions occurring in the business
- Only money paid into the bank and money paid out of the bank will appear on the bank statement

(b) On 1 January 1993 farmers were given an alternative to VAT registration if they so desired. The alternative is known as the Flat Rate Scheme for Agriculture, which allows farmers to apply to their local VAT office for a certificate as a flat rate farmer. If granted this exempts the farmer from submitting VAT returns and therefore accounting for or reclaiming VAT. Before applying to join the scheme farmers should consult their accountants. Farmers have to pay VAT on certain items (see page 11). This amount, however, can be claimed from Customs and Excise provided the farmer is registered for VAT. Hence, the amount of VAT actually paid on any inputs is recorded in the VAT claimable column. To facilitate VAT returns, the total amount paid out on standard and zero-rated inputs (which accounts for almost all farm inputs) is recorded as well

Farmers also have to charge VAT on what are mostly non-food items. Thus farmers have to collect the VAT and pay it to the Customs and Excise. This is recorded in the 'VAT charged' column and the total of standard and zero-rated outputs is entered in a separate column

(4) *Checking totals; ruling off the cash book.* At the month end or year end, add up all the entries in each column. The sum of 'bank' and 'contra' should be equal to the sum of all the other columns (excluding the one used to show the total of standard and zero rated inputs and outputs). If these columns do not reconcile, then there must be a mistake in the addition or in the entries

(5) *Checking bank statements and the bank reconciliation.* The bank statement is checked against the bank receipts and bank payment columns. The cash analysis book entries should be the same as statement entries. If they are not, then either the bank or the cash analysis book is wrong, *or* some monies have not been cleared and are still in transit

N.B. Interest and bank charges shown on the statement need to be entered on the cash analysis book payments page.

Bank statements, paying-in book, cheque book stubs, invoices and delivery notes

These should be kept so that *all* transactions can be traced and the cash book reconciled with the bank statement at the end of each month.

Petty cash book

Records all cash payments but it is easier in most farm businesses if these are kept to a minimum. In any event, all cash receipts should go into the bank and wages or private drawings should be drawn from the bank (and not from petty cash), if the system is to remain accurate and simple.

Valuations

Valuations are dealt with in greater detail in Chapter 5. The value of capital invested in live and dead stock should be calculated annually. This allows a full and accurate picture of the year's profit to be made.

These financial records will provide enough detailed information to construct a trading and profit and loss account (TPL) and, with the addition of a statement of debtors and creditors, enough information to construct a balance sheet (balance sheets are dealt with in Chapter 5). The items required to construct the trading profit and loss account are interrelated, as shown in Fig. 2.4.

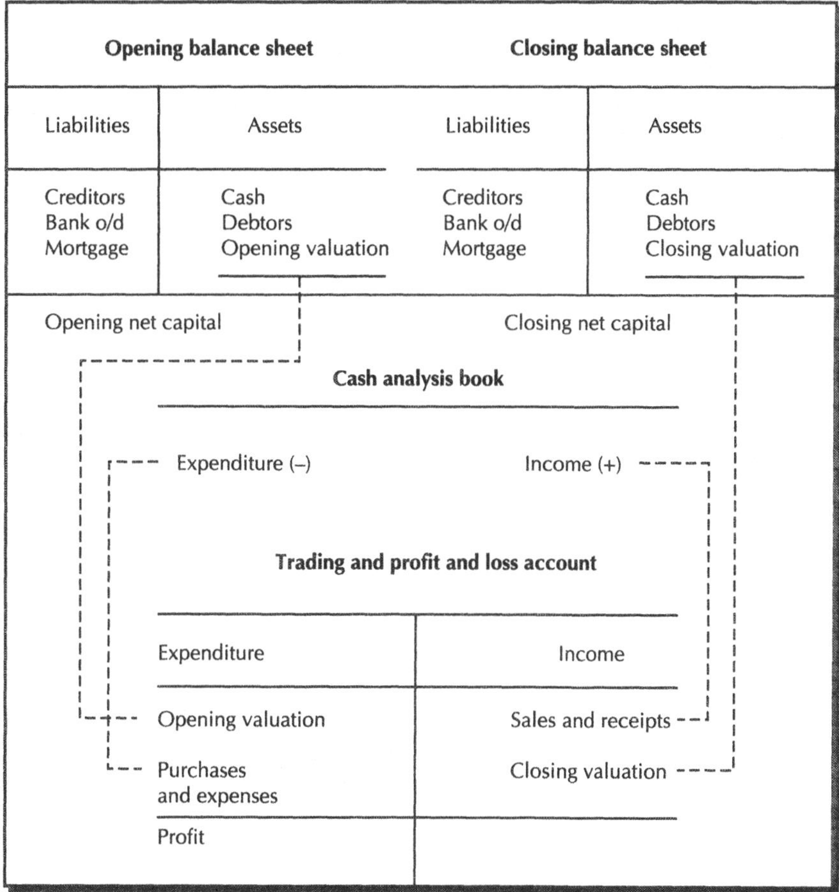

Fig. 2.4 The relationship between TPL and balance sheet

The profit and loss account is normally laid out with expenditure on the left-hand side and income on the right-hand side. Valuation figures are included, with opening valuations and details of expenditure on the left-hand side and closing valuations on the right-hand side, along with income.

If the income plus closing valuation exceeds the opening valuation plus expenditure then the business is making a profit!

The following information can be obtained from the profit and loss account:

(1) *Profit or loss*. This figure is recorded before drawings or capital expenditure have been taken into account
(2) *Direct or variable costs*
(3) *Indirect or fixed costs*

Closer analysis of the account, however, will reveal that there are costs which are associated with particular enterprises [marked (1) on the TPL account shown in

Fig. 2.5] and there will also be costs which are common to all the enterprises [marked (2) on the TPL account shown in Fig. 2.5].

In many profit and loss accounts, the breakdown of information is minimal (especially when they are presented for tax purposes as in Fig. 2.5) and it can be quite difficult to determine what proportion of each item of expenditure has been incurred by each enterprise on the farm during the year. It is much more useful for analysis purposes, and is easier to gain an understanding of the activities in the business, if all items such as fertilisers, seeds, sprays, feeding stuffs, AI, veterinary and medicine costs are set down individually rather than including them in much larger groupings (as shown in Fig. 2.6). Even if the entries in the trading and profit and loss account are very carefully broken down, it can still be difficult to gain sufficient detailed information to make an accurate assessment of the performance of individual enterprises on the farm which have given rise to the costs shown in the profit and loss account. To achieve a complete breakdown necessitates the use of a reconciliation chart as illustrated in Example 2.1(a).

Fig. 2.5 Trading and profit and loss account year ending 30 September 1997 (for taxation purposes)

Opening valuation	£	*Receipts*	£
Livestock	176 750	Livestock	143 060
Crops	—	Livestock products	138 600
Crop products	2 500	Crops (including area payments)	25 663
Stores	3 000	Miscellaneous (including set-aside)	3 817
Sub-total (a)	182 250	Sub-total	311 140
Expenses		*Closing valuation*	
Fertiliser	10 415 (1)	Livestock	156 420
Seeds	2 760 (1)	Crop products	21 176
Sprays	2 920 (1)	Stores	5 000
Feeds	136 150 (1)	Sub-total	182 596
Vet and medicine	6 800 (1)		
Other costs	5 195 (1)		
Wages	26 069 (2)		
Power and machinery	16 750 (2)		
Rent	10 868 (2)		
Bank	1 270 (2)		
Sundries	8 365 (2)		
Total expenses (b)	227 562		
Sub-total (a) + (b)	409 812		
Profit	83 924		
	493 736*		493 736*

* Note both sides balance.

Fig. 2.6 Trading and profit and loss account year ending 30 September 1997 (for management purposes)

Opening valuation	£	*Sales and receipts*	£
Cows (93)	74 400	Wheat (55 t)	5 225
Heifers (69)	34 500	Barley (12 t)	10 953
Sows (120)	12 600	Area payments	9 485
Pigs (850)	55 250	Cull cows (20)	7 800
Silage, hay, etc.	2 500	Calves (64)	5 760
Stores	3 000	Milk (6300 l/cow)	138 600
		Cull sows (60)	6 000
		Bacon pigs (1900)	123 500
		Set-aside	2 837
		Contract	980
Sub-total	182 250	Sub-total	311 140
Variable costs		*Closing valuation*	
Fertilisers	10 415	Wheat (160 t)	14 240
Seeds	2 760	Cows (102)	81 600
Sprays	2 920	Heifers (75)	37 500
Concentrates	114 670	Sows (70)	7 350
Vet and medicine	6 800	Pigs (666)	29 970
Brewers grain	3 000	Stores	5 000
Creep feed for pigs	15 680	Silage @ £22/t	6 936
Straw	2 800	Sub-total	182 596
Other costs	5 195		
Total variables	164 240		
Fixed costs			
Wages	26 069		
Power and machinery	16 750		
Sundries	8 365		
Rent and rates	10 868		
Bank	1 270		
Total fixed costs	63 322		
Profit	83 924		
	493 736		493 736

Example 2.1 (a) Reconciliation of fertiliser purchase and use (fertiliser purchased and used in the same year)

Opening valuation:	£	Used on:	£
20 tonnes @ £120/t	2 400	wheat	
		barley	
		grassland	10 415
Purchases:		Closing valuation:	
(from TPL a/c)	10 415	20 tonnes @ £120/t	2 400
	12 815		12 815

In this instance, all the fertiliser purchased was, in effect, used during the account year but it is often the case that this does not happen. This is illustrated in Example 2.1(b).

Example 2.1(b) Reconciliation of fertiliser purchase and use (purchased fertiliser not all used in same year)

	£		£
Opening valuation:		Used on:	
20 tonnes @ £120/t	2400	wheat, barley and grass (ex store 20 t) in trading year	10415
Purchases:		Closing valuation:	
95 tonnes @ £115/t	10925	25.3 tonnes @ £115/t	2910
	13325		13325

To determine the total usage of fertiliser on the farm during the year, therefore, requires not only a record of what is purchased but also of what was on the farm at the start and end of the year. To allocate the cost to each enterprise requires a very accurate record of where each amount of fertiliser taken from the barn was used during the year. These physical barn records, together with valuation records, are necessary as the basis upon which to accurately assess the true financial position of the business.

The other point which must be borne in mind is that the trading and profit and loss account may not reflect the true profit for the calendar period which it covers. This is because the inputs (fertiliser, etc.) could be purchased in the 1995–1996 trading year, for use on the 1996–1997 crops, while the crop (and livestock) products are sold in the 1997–1998 trading period. Thus, to get the true profit in the 1996–1997 year means extracting the cost of fertiliser and other inputs from the 1995–1996 trading and profit and loss account, and the income (or at least part of it) from the 1996–1997 trading and profit and loss account. These are some of the weaknesses of the trading and profit and loss account and it is true to say that its use is somewhat limited for management purposes.

To overcome this problem, and to aid management in assessing more specifically which enterprises are contributing to the profitability of the business, the gross margin system of accounting was devised and is now almost universally accepted in agricultural circles. This system is described in Chapter 3.

Having briefly outlined the records which need to be kept it is necessary to consider how to organise a system which will operate in a farm office to enable the recording of the vital daily statistics, and how computers can fit into a farm recording system.

Office procedure: purchases and sales

Purchases

The farmer fills out an order for a particular item which he requires on the farm. For example, he may order ten tonnes of animal feed. When this is delivered to the farm, he should check that the correct goods, and amount of goods, have been delivered against the delivery note. If he is really well organised, he will check the delivery note against his original order. Several days later, he will receive an invoice from the firm who supplied the feed. This invoice should be filed until such time as payment is made. When the invoice is paid, usually by cheque, the farmer should record details of the VAT part of the payment in his VAT record book and also ensure that he enters the correct cost amount under animal feed in his cash analysis book.

Separate files should be kept for each of the steps in the purchasing procedure as follows.

(1) Orders
(2) Deliveries
(3) Invoices outstanding
(4) Invoices paid
(5) VAT
(6) Cash analysis

Six files are required, essentially to ensure that the farmer's purchase records are carefully and logically organised, which in turn will facilitate the accurate recording of all information relating to purchases. The procedure is illustrated in Fig. 2.7.

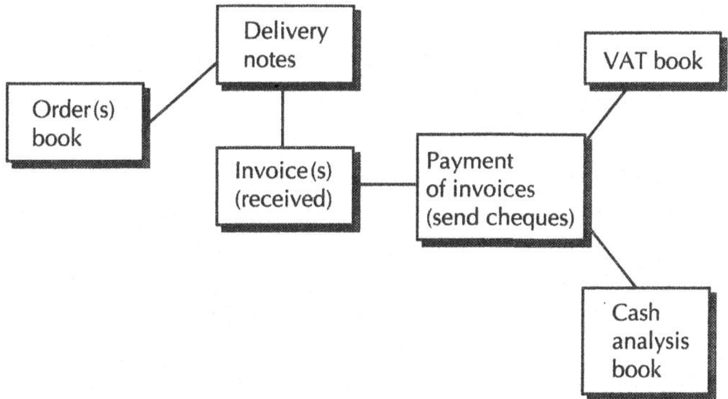

Fig. 2.7 Office procedure: purchases.

Sales

When items leave the farm, the farmer should make out a sales note. He may then issue an invoice claiming payment for the goods or, if he has been issued with a statement for the goods from the purchaser, await payment on the due date, either in the form of a cheque or as a direct credit into his bank account. If the latter occurs, he must ensure that the credit paid into the bank account is for the correct sum.

Any cheques received as payment for goods should be put into the bank as soon as possible (to minimise potential interest charges) using a bank paying-in book. The information from the paying-in book should then be transferred into the appropriate column in the cash analysis book. Credits paid directly into the bank account by a buyer should be taken from the bank statement and recorded in the cash analysis book. Details of all VAT which is paid to the farm should be recorded in the VAT sales book. The procedure is illustrated in Fig. 2.8.

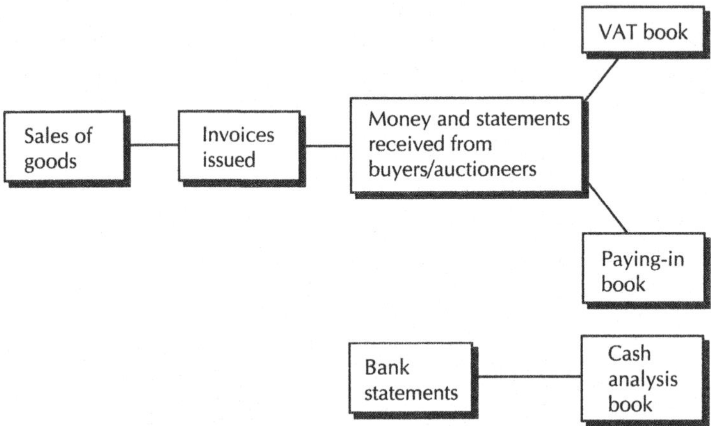

Fig. 2.8 Office procedure: sales.

Computers in agriculture

Computers are good at storing, analysing and retrieving information. Thus, a computer-based system generally allows large amounts of information to be safely stored and quickly retrieved without taking up a great deal of office space. The information can be analysed extremely quickly ('number crunching') and in a wider variety of ways than could be done by hand or using a desk-top calculator. The human element involved is still quite strong, however, as the computer has to have the raw information fed in, and this task can be as tedious and time consuming as filling out ordinary ledgers and files. There are, however, some aids to raw information collection (data capture) on the farm which can allow the stockman to record data which can be automatically transferred into the computer for processing.

At the other end of the system there is again a human function: once the computer has produced an answer, the recipient has to decide what it means. Computers, therefore, are an aid to management, especially in processing information. The provision of data and the decision making still largely have to be done by people.

Users of computers in agriculture

(1) *Linear programming.* This is widely used for determining least-cost rations by feed manufactures. Farmers who mix their own feed could equally well apply the principles of linear programming for this purpose on their personal computers (PCs). For a small cost a linear programming program can be added using a spreadsheet package such as Lotus 123 or Microsoft Excel.

(2) *Farm accounts.* There are programs available which enable the keeping of a whole variety of records ranging from simple records, such as the cash analysis book, to full farm financial packages which will produce the cash analysis book, cash flows and balance sheets from one set of input data.

(3) *Budgeting.* There are many applications for computer-aided planning ranging from partial budgets looking at the effect on the business profit of small changes to enterprise mix or capital investment, to whole farm budgets which may be used when major changes are being considered to an existing business or when rent tenders are being prepared. It is also possible to analyse labour utilisation [labour profiles and the potential efficiency of capital investment using discounted cash flows (DCF)] by computer.

The programs mentioned above can be purchased 'off the shelf' as packages (from Farmplan, Hylton Nomis, Sage or Sum It) or can be 'written' by the individual user on a spreadsheet (Works, Lotus, Excel). Spreadsheet programs have several advantages over the specialised packages available:

(a) They are written to include items which will relate to the individual business, i.e. you do not pay for what you may not use.
(b) They are simple to use and can be altered easily to suit any changes in the business
(c) They are written by the user and he can therefore understand them!

(4) *Enterprise records.* Physical and financial records can be kept using herd management programs. These can either be used by the farmer on his own PC, or by one of the agencies, such as Genus, ADAS or the Meat and Livestock Commission (MLC) who operate a bureau service. This means that the farmer provides the basic herd information and Genus analyses it and provides action lists for the farmer, on a regular basis, for the herd as a whole and for individual animals. Items such as bulling (service) dates, calving dates and drying-off periods are dealt with as in any herd management recording scheme. There may be a slight drawback with these bureau services as there is a delay in receiving some of the data and, of course, a monthly rent is required, but the task of putting the information into the computer is avoided for those who do not relish such work. Packages are available for dairy, pigs, arable, sheep, beef, horses and estates.

(5) *Payroll.* These will calculate and, if required, even pay wages. The computer is eminently suited to this, as indeed it is to all regular routine tasks. Insurance, pension and tax contributions are automatically calculated, once the program is set up, but one should be careful to ensure that updates are possible when, for example, tax rates change.

(6) Word processing. There are now many word processing programs available (notably Word and WordPerfect). The word processor has a large memory, almost instant recall and uses only a small space in the office allowing correction, updating and rewriting of text to be carried out without 'magic white stuff' (correction fluid!). The small, less complicated business may not justify the purchase of a word-processing program but certainly the larger businesses and estate offices will find it useful to have one. Many of the programs come as integrated packages having spreadsheet, word processor, graphics and database facilities all linked together, if required, while some of the most sophisticated have the ability to produce high-quality documents and are called Desktop Publishing (DTP) packages (e.g. Lotus Smartsuite).

(7) *Aids to mechanisation.* There are a wide range of applications in this area:
   (a) Environmental control in glasshouse cultivation
   (b) Spray application systems
   (c) Storage and drying of grain
   (d) Dairy parlour automation
   (e) Animal feeding systems
   (f) Design of machinery and equipment
   (g) Satellite-tracking recording systems (remote sensing)

(8) *Information provision.* Prestel, Telecom Gold and E-Mail are examples of computer based information systems which farmers can use. At present, a great deal of the information they provide can be obtained more cheaply from the farming press or the ancillary industry direct. The advantages provided by these systems are:
   (a) Speed of obtaining information
   (b) The ability to communicate with buyers or sellers in the market place is not limited to a period in the day between 0900 and 1700, when the farmer is most likely to be out on the farm.
   (c) Ease of use – sit at home and communicate with the outside world at your leisure

   At present, these systems are not well used by many farmers or members of the ancillary industries. As more join these services, they will become more useful and cost effective for the farming community

Some thoughts on computers

Computers need information to be put into them. For the majority of users this process will be a very frustrating procedure as at present the method most used is 'direct data entry' (typing). Indeed touch-tying, or keyboard skill, is a valuable life-skill to acquire, and it is imperative that users do not condemn computers because of their own inadequacies in keyboard skills! In future this difficulty may be overcome by increased use of 'touch screen', 'mouse' and voice recognition facilities which minimise (or may eradicate) keyboard operations.

(1) *Do not expect the computer to solve all your problems.* It may, in fact, create more if care is not taken to ensure that the correct program is selected which

will do all you want it to now, and have a capability to be expanded and updated in the future

(2) *Determine who will operate the computer and involve them in the decision-making process*
(3) *Spend as much time as possible getting experience with the program(s) and machine(s) before buying, i.e. talk to a current user*
(4) *Select your computer ('the hardware') after finding what you consider to be the correct program 'the software')*
(5) *Ascertain the level of after-sales service which will be available as it is unlikely that it will be possible to become fully operational without assistance*
(6) *Determine the cost of a system and decide whether your business can justify its purchase.*

N.B. If you operate a keyboard relatively well before starting on a PC, a great deal of frustration, which has nothing to do with computer programs or hardware, will be avoided.

Summary

Records can be:

(1) Mandatory
(2) Physical
(3) Financial

They are used for:

(1) Legal purposes – tax liability and planning
(2) Business analysis
(3) Business planning and control

The value of accurate records will be demonstrated in subsequent chapters as attention is focused on the construction of management accounts (Chapter 3) and farm business analysis and planning (Chapters 4–11). The usefulness of the results of analysis and planning is determined by the accuracy and adaptability of the information upon which they are based and how well they relate to the practical potential of the farm and the farmer.

3 Gross Margin Analysis

The gross margin system

Chapter 2 has described the records that need to be kept on a farm in order to fulfil legal requirements and for management purposes. The management purposes are to enable the analysis of performance of the whole business, both physically and financially.

One way of doing this is to construct, from the records kept by the farmer, a trading and profit and loss account. Most farmers do this in order to make a statement of profitability, for tax purposes, to the Inland Revenue. Armed with this annual profit/loss figure the farmer felt either comfortably happy or suicidal depending upon whether it appeared good or bad. The biggest difficulty with this method is that it does not identify the enterprises, in a multi-enterprise business, which are contributing to the profit of the business. This was readily recognised and a system of accounting was developed which constructed individual profit centres within a business. To do this all the income and costs associated with each enterprise had to be accounted for and it was then possible to get a profit per hectare for each enterprise.

It is relatively easy to determine the income (or output) of each enterprise from the cash analysis book. Calculating the costs generated by each enterprise is more difficult. Variable costs, such as feed, seed, fertilisers and sprays, all of which are directly related to the level of output from the enterprise, are relatively easy to allocate (with the possible exception of forage costs) from field and barn records being reconciled with cash analysis entries and valuation figures. However, overhead costs such as depreciation and repairs to machinery, labour costs, miscellaneous costs, fuel and power costs, interest or leasing charges, are very difficult to allocate to individual enterprises. A system which aims to allocate all costs to individual enterprises in order to get a cost per tonne of wheat or per litre of milk is usually referred to as a full cost accounting system. However this result cannot easily be achieved because:

(1) There are too many joint costs on UK farms, for example, farm workers who use their time on more than one enterprise
(2) The records needed to share out all costs are too complicated and time-consuming, e.g. fuel costs allocation would need information on soil type, engine revs, field conditions, and time period for each different enterprise on which machinery was used, and every time it was used
(3) It ignores the complementary relationships which exist between enterprises, so that no value is attached to use of labour in slack times, improvements to soil

fertility or weed and disease control or crop management as a result of rotational cropping or mixed grazing

The fact is that although full cost accounting has its advantages, for most farm businesses it is impractical. As a result of this situation a system has been developed which overcomes the problem of cost allocation largely by ignoring the costs which are difficult to allocate (regarding them as overhead costs) and concentrating solely on the performance of enterprises relating to output and variable costs.

This system is called the gross margin system and has the advantage of being a quick and simple means of assessing enterprise performance provided that it is noted that no overhead costs are taken into account when gross margins are constructed.

The construction of gross margins

The simplest gross margins to construct are for cereal enterprises. Thus for a winter wheat enterprise it would be as shown in Fig. 3.1.

For livestock enterprises the gross margin is more complex, especially since there are likely to be internal transfers of animals from breeding to rearing and fattening enterprises on the same farm; replacements have to be included, and because of this valuation figures for animals have to be included.

Fig. 3.1 Winter wheat gross margin

				£/ha
Sales: wheat	6.5 t @ £107/t =			695.50
straw				—
area payment*				193.53
Gross output			(a)	889.03
Variable costs:		£/ha		
Seed	190 kg/ha @ 22 p/kg	41.80		
Fertiliser N	190 kg/ha @ 30 p/kg	57.00		
P and K	60 kg/ha @ 58 p/kg	34.80		
Sprays: herbicide		25.00		
fungicides		20.00		
insecticides		2.00		
growth regulators		12.00		
Total variable costs			(b)	192.60
Gross margin = gross output (a) – variable costs (b)				
£889.03 – £192.60			=	696.43

* Area payments can and do vary from year to year based on the e.c.u./£ exchange rate and the annual price agreements.

There are several points to note in the dairy gross margin. First to calculate the gross output figure the opening and closing valuation for the herd must be included as shown in Fig. 3.2. It must be emphasised that in this example the closing valuation figure is higher than the opening valuation because there were four cows more. The value per cow remained the same at the end, being £800 per cow. If the closing value per cow was increased, the effect would be to raise the gross output figure without contributing anything more to sales. Thus a 'paper' or false increase would have been made to profitability. Great care must be exercised when constructing and analysing gross margins to check for changes in valuation figures.

The receipts of the dairy cow enterprise include all items sold off the farm and any items consumed by farm workers, or the farm owner or manager (in this example, milk). Also calves kept for rearing as replacements for dairy cows in two to three years time must be shown, as indeed should any beef calves which are transferred to the beef unit on a farm. The value of such transfers must be at current market prices otherwise an unrealistic level of performance will be shown.

Fig. 3.2 Dairy cow gross margin (100 cow herd on 55 ha)

Expenses	£	*Receipts*	£
Opening valuation:		Milk 540 000 @ 20 p/l	108 000
100 cows at £800/head	80 000	Calves 66 @ £80	5 280
Purchased heifers:		Transfers out 30 @ £80	2 400
5 @ £800	4 000	Cull cows 19 @ £420	7 980
Transfers in: 18 @ £750	13 500	Deaths 0	0
		Closing valuation:	
		104 @ £800	83 200
Sub-total (b)	97 500	Sub-total (a)	206 860
Gross output (a – b) = c	109 360		
	206 860		

(Gross output per cow = £1093.60)

Variable costs:			£	£/cow
Feed:concentrates	1.56t/cow	@ £150/t	23 400	234
sugar beet pulp	30t	@ £85/t	2 550	25.50
brewers grains	100t	@ £20/t	2 000	20.00
AI		@ £20/cow	2000	20.00
Vet and medicine			1 800	18.00
Transport			250	2.50
Straw			700	7.00
Minerals			700	7.00
Sundries			350	3.50
Forage costs			5 500	55.00
Total variable costs (d)			39 250	392.50

Total gross margin £70110 (c – d)
gross margin £/head 701.10
gross margin £/ha 1274.72

If any deaths occur, of either calves or adult animals, then these must be shown whether any money is received for them or not, otherwise the numbers of animals in the herd and on the farm at the start and end of the year will not balance properly. On the expenses side, along with opening valuation, the transfers in of in-calf heifers from the dairy replacement enterprise are shown at realistic market values. If any cattle are purchased, these too will be included here at actual cost. The gross output figure for the enterprise is then calculated at £109 360 as in Fig. 3.2.

The relationship between cull cows sold and replacements transferred or bought in annually is referred to as replacement cost or herd depreciation and can be expressed as a total figure or on a per cow basis. In Fig. 3.2 the total cost is £17 500 – £7980 = £9520 or £95.20 per cow. However, this figure contains an element of herd expansion, that is, 19 replacements were needed to replace 19 cull cows, but 4 others were purchased to expand the herd in this example. It would therefore be more accurate to say that replacement cost was (£13 500 + £800) – £7980 = £6320 or £63.20 per cow. Care must again be taken when constructing or analysing gross margins in relation to replacement costs.

The gross margin for the enterprise is gross output minus variable costs, in Fig. 3.2 being:

£109 360 – £39 250 = £70 110, or £701.10/head,
or £1274.72/ha (£70 110 divided by 55 ha)

Use of the gross margin system

Gross margins are used to identify individual enterprise performance. Thus it is possible to convert a trading and profit and loss account (tax account) into individual enterprise gross margin figures which can be analysed for management purposes. Consider the trading and profit and loss account in Example 3.1.

With the assistance of some further details the gross margins can be constructed.

Additional information to enable the construction of gross margins from a TPL account

(1) Farm size:

Total	119 ha
Wheat	28 ha
Barley	21 ha
Grass	61 ha
Set-aside	9 ha

(2) Average number of livestock:
Cattle: 100 dairy cows
72 followers
Pigs: 112 sows
1900 bacon pigs

Example 3.1 Trading and profit and loss account year ending 30 September 1997 (for management purposes)

Opening valuation	£	*Sales and receipts*	£
Cows (93)	74 400	Wheat (55 t)	5 225
Heifers (69)	34 500	Barley (121 t)	10 953
Sows (120)	12 600	Area payments	9 485
Pigs (850)	55 250	Cull cows (20)	7 800
Silage, hay, etc.	2 500	Calves (64)	5 760
Stores (consumable)	3 000	Milk (6300 l/cow)	138 600
		Cull sows (60)	6 000
		Bacon pigs (1900)	123 500
		Set-aside	2 837
		Contract	980
Sub-total	182 250	Sub-total	311 140
Variable costs		*Closing valuation*	
Fertilisers	10 415	Wheat (160 t)	14 240
Seeds	2 760	Cows (102)	81 600
Sprays	2 920	Heifers (75)	37 500
Concentrates	114 670	Sows (70)	7 350
Vet and medicine	6 800	Pigs (666)	29 970
Brewers grain	3 000	Stores (consumable)	5 000
Creep feed pigs	15 680	Silage @ £22/t	6 936
Straw	2 800	Sub-total	182 596
Other costs	5 195		
Total variables	164 240		
Fixed costs			
Wages	26 069		
Power and machinery	16 750		
Sundries	8 365		
Rent	10 868		
Bank	1 270		
Total fixed costs	63 322		
Profit	83 924		
	493 736		493 736

(3) Thirty-five calves transferred out to the follower unit at £90 per head
(4) Twenty-nine in-calf heifers transferred to the dairy unit at £750 per head
(5) One thousand eight hundred and six weaners transferred out to the pig fattening unit at £27 per head
(6) Ten in-pig gilts transferred to the sow unit at £120 per head
(7) Expenses (£):

		Total	Wheat	Barley	Grass
(a)	Fertilisers	10 415	2120	1275	7020
(b)	Seeds	2760	1120	840	800
(c)	Sprays	2920	1680	840	400

(d) Bought feed 136 150:

Cows:	concentrates	24 250
	brewers grains	3 000
Heifers:	concentrates	5 600
	feed straw	2 800
Sows:	concentrates	23 520
	creep feed	15 680
Bacon pigs:	concentrates	61 300

(e) Vet and medicine 6 800:

Cows	2 000
Heifers	750
Sows	2 100
Pigs	1 950

(f) Other costs 5 195:

Cows	1 800
Heifers	695
Sows	1 800
Pigs	900

Using this information the individual enterprise gross margins can be constructed. The allocation of forage costs presents one of the difficult areas in construction. To enable this to be done forage costs are shared between the grazing livestock using the grazing livestock unit (GLU) system. This system ascribes a value to animals in relation to their grazing requirements. A mature Friesian cow, weighing 650 kg and producing 5000 l of milk plus a calf each year, represents the basic unit to which all other animals are related. Thus the main GLU values are shown in Table 3.1.

Table 3.1 Grazing livestock units

Dairy cows	1.0
Beef cows	0.8
Cattle (over 2 years)	0.8
Cattle (1–2 years)	0.6
Cattle (under 1 year)	0.4
Ewes (and lambs)	0.2
Other sheep (over 6 months)	0.1

Using this system in Example 3.1 there are:

Livestock:	Livestock units:
100 cows ($\times 1$)	100
72 followers ($\times 0.5$)	36
Total	136

The forage costs in Example 3.1 are:

	£
Fertiliser	7020
Seed	800
Spray	400
Total	8220

These costs can be allocated as follows:

$$\text{Cows use } \frac{100}{136} \times 61\,\text{ha} = 45\,\text{ha}$$

$$\text{Followers use } \frac{36}{136} \times 61\,\text{ha} = 16\,\text{ha}$$

$$\text{Therefore: cows } \frac{45}{61} \times £8220 = £6064$$

$$\text{Followers } \frac{16}{61} \times £8220 = £2156$$

The individual enterprise gross margins are shown in Fig. 3.3

Fig. 3.3 Individual enterprise gross margins

Dairy cow gross margin (45 ha)

	£		£
Opening valuation:		Sales: milk (6300 l/cow)	138 600
93 cows @ £800	74 400	calves (64)	5 760
		culls (20)	7 800
Transfers in:		Transfers out:	
29 heifers @ £750	21 750	35 calves @ £90	3 150
		Closing valuation:	
Enterprise output:	140 760	102 cows @ £800	81 600
	236 910		236 910

Variable costs:	
Feed: concentrates	24 250
brewers grain	3 000
Vet and medicine	2 000
Other costs	1 800
Forage costs*	6 064
Total	37 114

Enterprise gross margin £103 646 (140 760 – 37 114) or £2303/ha

Fig. 3.3 *(continued)*

Young stock gross margin (16 ha)

	£		£
Opening valuation:		Sales	
69 stock @ £500	34 500	Transfers out	
Transfers in		29 heifers @ £750	21 750
39 calves @ £90	3 150	4 deaths	—
Enterprise output:	21 600	Closing valuation:	
	59 250	75 stock @ £500	37 500
			59 250

Variable costs:

Feed: concentrates	5 600
straw	2 800
Vet and medicine	750
Other costs	695
Forage costs*	2 156
Total	12 001

Enterprise gross margin £9599 or £699.94/ha

Breeding sow gross margin

	£		£
Opening valuation:		Sales:	
120 sows @ £105	12 600	Cull sows: 60 @ £100	6 000
Transfers in:		Transfers out:	
10 gilts @ £120	1 200	1806 @ £27	48 762
		Closing valuation:	
Enterprise output:	48 312	70 sows @ £105	7 350
	62 112		62 112

Variable costs:

Sow concentrate	23 520
Creep feed	15 680
Vet and medicine	2 100
Other costs	1 800
Total	43 100

Enterprise gross margin £5212 or £46.54/sow

Young pigs gross margin

	£		£
Opening valuation:		Sales:	
850 pigs @ £65	55 250	1900 baconers @ £65	123 500
Transfers in:		Transfers out:	
1806 weaners @ £27	48 762	10 gilts @ £120	1 200
		Deaths: 80 @ £0	—
		Closing valuation:	
Enterprise output	50 658	666 @ £45	29 970
	154 670		154 670

Fig. 3.3 *(continued)*

Variables costs:			
Feed	61 300		
Vet and medicine	1 950		
Other costs	900		
	64 150		

Enterprise gross margin – £13 492 or – £7.10/pig

Wheat gross margin (28 ha)

	£		£
Opening valuation:	—	Sales:	
		55 tonnes @ £95	5 225
		Area payments	5 419
		Closing valuation:	
Enterprise output:	24 884	160 tonnes @ £89	14 240
	24 884		24 884

Variable costs:	
Seed	1 120
Fertiliser	2 120
Sprays	1 680
Total	4 920

Enterprise gross margin £19964 or £713/ha

Barley gross margin (21 ha)

	£		£
Opening valuation:		Sales:	
		121 tonnes @ £90.52	10 953
		Area payments	4 066
Enterprise output:	15 019	Closing valuation:	—
	15 019		15 019

Variable costs:	
Seed	840
Fertiliser	1 275
Sprays	840
Total	2 955

Enterprise gross margin £12064 or £574.48/ha

Gross margin account

	£	£
Cows	103 646	
Young stock	9 599	
Breeding sows	5 212	
Young pigs	–13 492	
Wheat	19 964	
Barley	12 064	
Total gross margin	136 993	

Fig. 3.3 *(continued)*

Add miscellaneous income:	Set-aside	2 837	
	Contract	980	
	Increase in stores value	2 000†	
	Increase in silage value	4 436†	
	Total	10 253	
Grand total of income			147 246
Deduct overhead costs:	Wages	26 069	
	Power and machinery	16 750	
	Rent	10 868	
	Bank	1 270	
	Sundries	8 365	
	Total overhead costs	63 322	
Profit			83 924

* Adjusted for changes in stocks (see p. 37) any increase in silage valuation should be deducted from cows and young stock forage costs because they have not used the silage for which all the forage costs were incurred.
† Difference between opening and closing valuation.

Having established the individual gross margins, it is possible to get an appreciation of performance by comparing them with some standard data. This will be dealt with in Chapter 4, but for the meantime Table 3.2 shows some likely gross margins.

With the gross margins calculated it is possible to construct some whole farm profitability figures. These are net farm income (NFI) and management and investment income (MII) which are used for comparative analysis purposes and as an indication of whole farm performance when using the gross margin system.

Derivation of NFI and MII

To facilitate the comparison of profitability and performance, farms are put onto the same cost basis, that is, a debt-free tenant farmer without salaried management. This means that some items in the overhead costs will be adjusted.

(1) Interest payments are excluded altogether
(2) If the rent is not average for the area or the farm is owner-occupied an average rent for the farm is imputed. This is normally enough to get a reliable comparison but some managers could take further steps
(3) Exclude what are normally landlord's expenses, such as buildings, insurance, depreciation of fixed capital improvements and part of maintenance and repair costs
(4) Exclude the managerial element of a manager's salary, that is, only include the part which relates to his physical work on the farm
(5) Make a charge for physical labour which is supplied on a 'free' basis by the family

The overhead costs will thus be as shown in Table 3.3.

Table 3.2 Range of gross margins for selected enterprises (March 1997)

Enterprise	Price	Production standards					Implied gross margins (£/head)				
		1	2	3	4	5	1	2	3	4	5
Milk production	per litre	litres per annum									
Dairy cows non-C.I.											
AYR calving	22 p	4700	5300	5900	6600	7300	659	751	843	951	1059
Spring calving		4500	5000	5500	6300	7100	629	706	777	907	1030
Autumn calving		5000	5500	6000	6800	7600	708	792	862	985	1108
Dairy cows C.I.											
AYR calving	28 p	3300	3700	4100	4500	4900	547	627	708	788	868
Cattle rearing	per head	kg concentrate fed per beast									
Dairy replacements, non-C.I.											
Autumn calving 24/26 months	£680	985	935	905	840	785	214	236	263	293	317
Dairy replacements, C.I.											
Autumn calving 24/26 months	£390	806	756	722	660	606	48	63	84	107	123
Calf rearing (sold 12 weeks)	£203	145	145	165	150	145	0	2	10	20	23
Beef breeding	per kg live weight	calf sale, kg live weight									
Beef cows, lowland, autumn	£1.08	305	320	335	350	365	141	159	177	196	215
Beef cows, upland, autumn	£1.08	295	310	325	340	355	148	166	184	203	222
Store cattle with premium, spring born		kg concentrate fed per beast									
0–18 months, sold as store	£1.06	765	715	690	627	578	153	166	184	204	218
0–18 months, sold finished	£1.04	1018	945	901	819	751	184	200	221	244	260
Store cattle with premium autumn born											
0–18 months, sold as store	£1.08	700	663	653	592	546	186	197	213	234	247
0–18 months, sold finished	£1.06	1222	1130	1069	971	888	183	201	224	250	268
Store cattle without premium autumn born											
0–18 months, sold as store	£1.06	615	585	580	527	488	95	104	119	137	149
0–18 months, sold finished	£1.04	1055	980	933	849	779	89	104	125	147	163

Beef finishing	per kg d.c.w.	kg D.C.W.									
Veal calves	£3.50	85	90	95	100	105	–27	–9	10	28	46
Sheep	per kg d.c.w.	lambs reared per ewe per annum									
Lowland ewes	222 p	1.25	1.4	1.5	1.6	1.75	10	16	20	24	30
Lowland, early fat lamb	280 p	1.25	1.4	1.5	1.6	1.75	6	13	18	23	30
Hill ewes, self-maintained	234 p	0.8	0.9	1.0	1.1	1.2	11	14	17	21	24
Upland ewes	262 p	1.1	1.2	1.3	1.4	1.5	12	16	20	24	28
	per litre	litres per annum									
Milk sheep	72 p	200	250	300	350	400	91	136	181	227	272
Pigs: breeding only	per weaner	weaners per sow per annum									
Weaner production, indoors	£32	19.0	20.0	21.5	23.0	24.5	159	186	227	263	296
Weaner production, outdoors	£32	18.5	19.5	20.5	22.0	23.0	82	115	149	196	232
Pigs: feeding only	per kg dead weight	F.C.R. from transfer (25 kg)									
Pork pigs (53 kg dead weight)	£1.14	2.6:1	2.5:1	2.4:1	2.2:1	2.1:1	4.01	5.20	6.56	7.75	8.94
Cutter pigs (64 kg dead weight)	£1.12	2.7:1	2.6:1	2.5:1	2.3:1	2.2:1	9.29	10.78	12.26	13.91	15.40
Bacon pigs (70 kg dead weight)	£1.05	2.8:1	2.7:1	2.5:1	2.4:1	2.2:1	7.19	8.84	10.49	12.14	13.79
Heavy pigs (83 kg dead weight)	£0.92	3.1:1	3.0:1	2.9:1	2.7:1	2.6:1	0.18	2.04	4.06	5.92	7.93
Poultry	per dozen	dozen eggs per bird per annum									
Laying hens, intensive	46 p	20	21	22	23	24	–0.92	–0.21	0.41	1.04	1.74
Laying hens, free range	69 p	18.5	20	21.5	22	22.5	0.39	1.87	3.27	3.98	4.77
	per kg live weight	kg live weight at sale									
Broilers	64 p	2.2	2.2	2.2	2.2	2.2	0.19	0.24	0.28	0.30	0.32

(*Continued*)

Table 3.2 *Continued*

Enterprise	Price	Production standards					Implied gross margins (£/head)				
		1	2	3	4	5	1	2	3	4	5
Other stock	per litre	litres per annum									
Goats	40 p	750	800	850	900	950	116	136	156	176	196
	per kg dead weight	kg dead weight at sale									
Red deer	£3.50	23	24	25	26	27	48	52	55	59	62
Rabbits	£1.12	95	100	105	110	115	20	23	27	31	35
	per kg	kg of fish									
Rainbow trout	£2.00	1000	1000	1000	1000	1000	390	450	510	570	630
Winter wheat	per tonne	tonnes per acre									
Feed, sold Oct–Dec	£90	2.00	2.50	3.00	3.50	4.00	172	217	262	299	335
Feed, organic	£180	1.00	1.25	1.50	1.75	2.00	252	297	342	387	432
Feed, seed	£110	2.00	2.50	3.00	3.50	4.00	195	250	305	351	397
Milling, sold Oct–Dec	£105	2.00	2.40	2.80	3.20	3.60	192	234	276	310	343
Milling, organic	£200	1.00	1.25	1.50	1.75	2.00	271	321	503	371	421
Milling, seed	£125	2.00	2.40	2.80	3.20	3.60	216	266	316	357	398
Spring wheat											
Feed, sold Oct–Dec	£90	1.80	2.00	2.20	2.40	2.60	175	193	211	222	233
Feed, organic	£180	1.00	1.20	1.40	1.60	1.80	248	284	320	356	392
Feed, seed	£110	1.80	2.00	2.20	2.40	2.60	194	216	238	253	268
Milling, sold Oct–Dec	£105	1.60	1.80	2.00	2.20	2.40	176	197	218	232	247
Milling, organic	£200	1.00	1.20	1.40	1.60	1.80	268	308	348	388	428
Milling, seed	£125	1.60	1.80	2.00	2.20	2.40	193	218	243	261	279
Winter barley											
Feed, sold Oct–Dec	£88	2.00	2.30	2.60	2.90	3.20	187	213	240	259	278
Feed, organic	£168	1.00	1.25	1.50	1.75	2.00	242	284	326	368	410
Malting, sold Oct–Nov	£113	1.80	2.10	2.40	2.70	3.00	210	244	278	304	331

Spring barley											
Feed, sold Oct–Nov	£88	1.60	1.80	2.00	2.20	2.40	170	187	205	217	228
Feed, organic	£168	1.28	1.44	1.60	1.76	1.92	285	312	338	365	392
Feed, seed	£108	1.60	1.80	2.00	2.20	2.40	186	208	230	245	260
Malting, sold Oct–Dec	£113	1.40	1.60	1.80	2.00	2.20	187	210	233	249	266
Malting, seed	£133	1.40	1.60	1.80	2.00	2.20	210	236	263	283	303
Winter oats											
Feed, sold Oct–Dec	£86	2.10	2.30	2.50	2.70	2.90	208	225	242	253	264
Feed, organic	£170	1.44	1.52	1.68	1.76	2.00	316	329	357	370	411
Spring oats											
Feed, sold Oct–Dec	£86	1.60	1.80	2.00	2.20	2.40	174	192	209	221	233
Feed, organic	£170	1.12	1.28	1.44	1.60	1.76	259	286	314	341	368
Oilseed rape											
Winter, sold Oct–Dec	£170	1.0	1.2	1.4	1.6	1.8	243	277	311	331	352
Spring, sold Oct–Dec	£170	0.6	0.75	0.9	1.1	1.3	201	227	252	278	304
Cereals											
Durum wheat	£145	1.20	1.50	1.80	2.00	2.20	170	214	257	280	303
Triticale	£90	1.5	1.8	2.3	2.6	3.0	150	177	222	244	275
Rye	£95	1.9	2.1	2.3	2.5	2.7	190	209	228	247	266
Straw sales (including home use)	£25	1.1	1.1	1.1	1.1	1.1	26	26	26	26	26

AYR – all year round.
D.C.W. – dressed carcass weight.
F.C.R. – food conversion ratio.
Source: Barclays Bank.

Table 3.2 *Continued*

Enterprise	Price	Production standards					Implied gross margins (£/acre)				
		1	2	3	4	5	1	2	3	4	5
Arable	per tonne			tonnes per acre							
Maincrop potatoes	£85	13.0	15.0	17.0	19.0	21.0	445	568	690	784	889
Early potatoes	£150	6.0	7.0	8.0	9.0	10.0	288	418	547	677	807
Sugar beet	£41.50	12.5	15.0	17.5	19.0	20.5	288	379	470	525	580
Dried peas	£120	1.0	1.3	1.5	1.65	1.8	180	216	240	258	276
Winter field beans	£116	1.2	1.4	1.6	1.7	1.8	228	251	274	286	297
Spring field beans	£116	1.1	1.3	1.5	1.6	1.7	212	235	258	270	282
Vining peas	£250	1.5	1.7	1.9	2.1	2.3	112	157	202	247	291
Navy beans	£350	0.6	0.7	0.8	0.9	1.0	122	157	192	227	262
Lupins	£130	0.8	1	1.2	1.4	1.6	194	220	246	272	298
Borage	£2200	0.08	0.10	0.12	0.14	0.16	74	118	162	206	250
Evening primrose	£1700	0.12	0.14	0.16	0.18	0.20	97	131	165	199	233
Linseed	£165	0.3	0.5	0.7	0.8	0.9	132	165	198	215	231
Sunflowers	£205	0.6	0.7	0.8	0.9	1.0	197	218	238	259	279
	per kilo			kilos per acre							
Herbage seed – Italian ryegrass	80 p	350	400	450	500	550	140	180	220	260	300
Herbage seed – perennial ryegrass	82 p	350	400	450	500	550	146	187	228	269	310
Herbage seed – timothy	140 p	152	172	192	202	212	122	150	178	192	206
	per tonne			tonnes per acre							
Flax	£26	0.6	0.8	1.0	1.2	1.4	193	206	220	233	247
Fodder crops	per tonne			tonnes per acre							
Hay for sale	£65	1.6	1.8	2.0	2.5	3.0	63	76	89	121	154
Fodder beet for sale	£18	22	25	28	31	34	246	300	354	408	462
Bag silage for sale	£25	4	5	6	7	8	16	41	66	91	116

Vegetables	per tonne		tonnes per acre								
Carrots – maincrop	£100	11.0	12.5	14.0	15.5	17.0	185	309	434	558	683
Parsnips	£195	7.0	8.0	9.0	10.0	11.0	663	790	917	1044	1171
Swedes – fresh market	£100	10.0	12.0	14.0	16.0	18.0	415	527	638	750	861
Brussels sprouts	£265	3.2	4.5	5.7	6.5	7.3	289	569	827	999	1171
	per crate		crates per acre								
Cauliflower – summer	£2.61	500	550	600	650	700	188	251	314	377	440
Cauliflower – winter	£2.90	380	430	480	530	580	112	180	247	315	382
	per tonne		tonnes per acre								
Cabbage – spring greens	£200	4	6	8	9	10	103	313	523	628	733
Cabbage – winter	£127	4	6	10	11	12	–21	166	540	634	727
Dry bulb onions	£112	10	12	14	16	18	234	350	466	582	698
Leeks	£564	5.0	6.0	7.0	8.0	9.0	739	978	1217	1456	1695
Asparagus	£1590	0.5	0.75	1.0	1.25	1.5	240	420	600	780	960
	per dozen		dozen per acre								
Lettuce – outdoors	£2.75	800	1000	1200	1400	1600	612	924	1236	1548	1860
	per tonne		tonnes per acre								
Runner beans	£387	5.5	7.0	8.5	10.0	11.5	481	682	883	1084	1285
French/dwarf beans	£185	1.7	2.2	2.7	3.2	3.7	76	169	261	354	446
Fruit	per tonne		tonnes per acre								
Dessert apples	£325	3.2	4.0	4.9	7.3	9.7	326	458	606	1002	1398
Culinary apples	£270	3.6	4.9	6.1	10.1	14.2	312	494	662	1222	1796
Pears	£400	2.8	3.6	4.5	5.3	6.5	588	805	1050	1268	1594
Plums	£800	1.21	2.43	3.64	6.10	8.10	410	1014	1613	2831	3821
Blackcurrants – processing	£575	2.00	2.50	3.00	3.50	4.00	509	766	1024	1281	1539
Strawberries – fresh market	£1300	1.62	2.43	3.24	4.05	4.86	327	764	1200	1637	2074
Strawberries – pick-your-own	£1450	1.62	2.43	3.24	4.05	4.86	1163	1911	2660	3408	4157
Raspberries – fresh market	£1600	1.35	1.85	2.35	2.85	3.35	389	674	959	1244	1529
Raspberries – pick-your-own	£1650	1.35	1.85	2.35	2.85	3.35	1355	1933	2510	3088	3665
Set-aside											
Set-aside – rotational		—	—	—	—	—	123	123	123	123	123
Set-aside – permanent		—	—	—	—	—	123	123	123	123	123

Table 3.3 Overhead costs used to calculate NFI and MII

	NFI	MII	Profit
Regular labour and National Insurance	×	×	×
Physical labour of farmer and wife		×	
Unpaid family labour	×	×	
Paid management	×		×
Rent (actual if average)	×	×	×
Notional rent (if owner-occupier)	×	×	
Machinery depreciation	×	×	×
Machinery running and repairs	×	×	×
Interest			×
Miscellaneous (including: office expenses, accountant fees, telephone, insurance, membership fees)	×	×	×

Calculation of depreciation

The amount deducted for depreciation represents a charge for the value of machinery or buildings used in the year.

Although individuals can choose the rate at which they depreciate machinery and buildings, for management purposes, it is generally accepted that machinery is depreciated at 20% per annum and buildings at 10%. Thus on average it is expected that machinery will last five years and buildings ten years. Some farmers go a stage further than these simple rules and use a different rate of depreciation for each item of machinery; this is perfectly acceptable and reflects the expected life of each machine. It is also good practice to record repair costs for each machine separately.

Methods of calculating depreciation

There are several methods of calculating depreciation. Two of them are described here.

(1) *The straightline method.* This is normally used for buildings. If a building costs £100 000 new, then the depreciation would be:

$$100\,000 \times \frac{10}{100} = £10\,000 \text{ per annum}$$

So the depreciation would be £1000 for each of the ten years.

(2) *The diminishing balance method.* This is normally used for machinery. The cost of all machinery on the farm is (usually though not necessarily) added together to form the 'pool' value. This is then depreciated annually at the rate of 20%. If

the opening value of machinery was £20 000 in year one, the depreciation would be:

$$\text{Year one: } £20\,000 \times \frac{20}{100} = £4000$$

$$\text{closing value } (£20\,000 - £4000) = £16\,000$$

$$\text{Year two: } £16\,000 \times \frac{20}{100} = £3200$$

$$\text{closing value } (£16\,000 - £3200) = 12\,800$$

$$\text{Year three: } £12\,800 \times \frac{20}{100} = £2560$$

$$\text{closing value } (£12\,800 - £2560) = 10\,240$$

If the sale value of a depreciated asset exceeds (or is less than) its net book value, the profit (or loss) will appear in the trading, profit and loss account, as a profit (or loss) on sale of assets.

Net farm income (NFI) represents the reward to the farmer and his wife for their own manual labour, management and interest on tenant-type capital invested in the farm whether it is borrowed or not.

Management and investment income (MII) represents the reward to management, both paid and unpaid, and the return on tenant-type capital invested in the farm whether it is borrowed or not.

Both NFI and MII are used for the purposes of comparing the performance of farms; this is why the overhead costs are standardised so that the comparison is made on a like-for-like basis. However, this does mean that the figures are unrealistic in relation to the actual performance of the business; hence the profit figure is much more useful as a measure for the farmer to determine his own real business performance compared to last year's.

The strengths and weaknesses of the gross margin system

Weaknesses of the gross margin system

(1) It does not produce a profit figure but provides a useful figure for comparing enterprises from farm to farm, or year to year on the same farm so long as the figures are compared on a like-for-like basis
(2) It does not take account of overhead costs and since enterprises create different levels of overhead costs it can be misleading to compare enterprises on a farm purely on gross margins
For example:
Two farms (identical) growing sugar beet.
A: Uses own labour and harvester; thus these costs will be shown in the farm overhead costs and will not appear in the enterprise gross margin

B: Uses casual labour and a contractor; thus these costs will be shown as variable costs in the enterprise gross margin

Thus the gross margin for farm A will be higher than for farm B but the real performance may be no different

(3) The gross margin must therefore be interpreted only in relation to overhead costs and not between enterprises without prior knowledge of the whole farm cost picture
(4) Profit is not proportional to gross margin. Increasing the intensity of the enterprises on a farm may well increase the gross margin but not necessarily the farm profit because the overhead costs may have risen faster
(5) It does not make any allowance for complementary or detrimental relationships between enterprises: for example, the effect which sheep have on soil nutrients for the following crops; the lateness of drilling an autumn cereal following a potato or sugar beet crop; slug infestation after oilseed rape

The strengths of the gross margin system

(1) It is a simple method of assessing efficiency
(2) The records required and the accounting are easy and quick
(3) Farm planning can be carried out relatively easily using gross margins provided it is remembered that overhead costs are not included
(4) It does indicate areas of strength and weakness in the business

The weaknesses of the gross margin system have led to it being referred to as the 'gross illusion' and have caused a move to the use of the net margin.

The calculation and use of net margins

The net margin of an enterprise is its gross margin less all or a proportion of the overhead costs. Care should be taken in interpreting a net margin figure, especially when no indication is given of the overhead costs which have been deducted. If all costs are deducted then the net margin is equivalent to a profit per enterprise figure which is most useful for analysis of the business. However, it is difficult to allocate all costs accurately to each enterprise and so in some cases the figure referred to as the net margin is the gross margin less a proportion of the overhead costs which are easy to allocate to an enterprise. In Chapter 6 the term 'attributable' is used to describe the overhead costs deducted before arriving at the net margin for an enterprise. There is nothing particularly significant in this usage and indeed attributable and allocatable can be used synonymously.

4 Comparative Analysis

Now we have described the gross margin system, it can be used to assess the performance of the business and to identify areas of strength and weakness.

The starting point is to get a measure of the profitability of the farm using NFI, MII or profit and then to analyse the overhead costs as these are what we have to farm against. Certainly in the short term it is not easy to alter overhead costs as they are a result of historical decisions to invest capital or use a particular production system, and of course it is necessary to generate enough output to cover overheads and produce a margin for profit, living expenses, taxation, expansion and repayment of borrowing.

Steps to analyse a business

Two methods can be adopted:

(1) To compare the figures with some standard data from, for example, one of the farm management surveys around the country (Exeter, Reading, Aberystwyth, Wye, Cambridge, Nottingham, Askham Bryan, Manchester and Newcastle), or other farm management data (Genus, ADAS). This is done by choosing comparative data from the category which is similar in size and type to the farm being analysed. In order to do this type of comparison, farm figures have to be standardised (as described in Chapter 3) on the basis of a debt-free tenant, which creates a certain unrealism and is a source of criticism of this method of analysis. For this reason it is *not* possible to get an exact and fair comparison between farms. It does, however, provide an indication of, and guidelines for, the level of performance currently being achieved, and for the potential performance too.

(2) To compare this year's performance with previous years', on the same farm. This provides a direct relationship which immediately shows improvement (or not) and is often more useful as an indicator of ongoing performance for an established business.

Some measures of performance

Farm comparisons are usually made by calculating the following figures.

Profitability

- Net farm income per hectare
- Management and investment income per hectare
- Profit per hectare
- Percentage return on tenant's capital invested

However, care must be exercised in looking at these figures in isolation since they may not show the full picture, as demonstrated in Example 4.1 where two farms have the same MII but different profit figures.

Example 4.1 Trading profit and MII in two farm businesses

High Farm
250 hectare, managed with interest payments of £60 per hectare. There is an average overdraft of £75 000 during the year at 10% interest, and management fees are £7500

	£/ha	£/ha
TPL account (loss)		(15)
Add: overdraft interest	30	
mortgage interest	60	
management fee	30	120
Deduct: notional rent		(80)
MII		25

Low Farm
150 hectare, owner-occupied with no overdraft. The farmer uses approximately one-third of his time physically working, and there is no paid management

	£/ha
TPL account	130
Deduct: notional rent	(80)
farmer's physical labour	(25)
MII	25

For this reason overhead (common) costs must be analysed.

Overheads

Overheads are analysed under the following headings:

- Total overheads per hectare
- Rent and rates per hectare
- Labour costs per hectare
- Machinery costs per hectare
- Miscellaneous costs per hectare

Output

After overheads have been considered, output is analysed (including physical and financial performance) under the following headings:

- Total gross output per hectare
- Gross output per £100 labour costs
- Gross output per £100 machinery costs
- Gross output per £100 labour and machinery costs

Inputs

Finally the variable inputs are analysed (including physical and financial performance).

The steps shown in Figure 4.1 provide a logical sequence through which aspects of trading performance can be investigated.

The likely range of figures which can be expected in whole farm analysis are shown in Table 4.1.

Of course the performance of individual farms may differ quite markedly from those in Table 4.1, these serve only as a broad guide to the sort of figures which could be expected.

Table 4.1 Likely range of performance figures

	£/ha	
	Low	High
Net farm income	80	250
Management and investment income	60	230
Overhead costs	300	700
Rent	85	160
Labour and NI	100	300
Machinery	75	225
Miscellaneous	40	80
Tenant's capital	650	1300
Percentage return on tenant's capital	5%	20%
Gross margin	350	900
Gross output	450	1050
Gross output per £100 labour	300	400
Gross output per £100 machinery	450	525

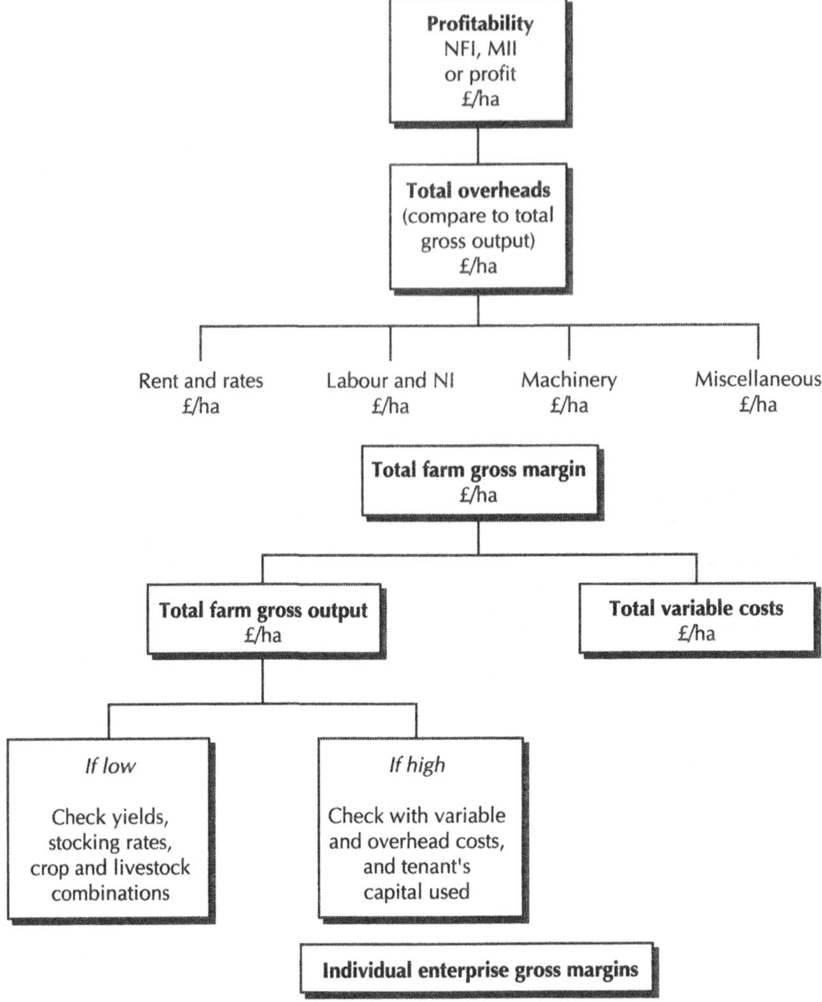

Fig. 4.1 Steps in the analysis of a farm business.

The significance of analysed figures

In terms of the all-round performance of the business there are some ratios which financiers, agents and consultants use as a guide to the soundness and stability of a business; these are shown in Table 4.2.

It would be imprudent to condemn a business because it does not meet one or two of the guidelines in Table 4.2. However, where three or more are out of line, warning lights are being shown and careful detailed analysis of the weak areas must be carried out in order to determine whether plans can be introduced to prevent further profit, cash or capital problems occurring.

Table 4.2 Some guidelines for performance

		£/ha
Output		
Gross output for cereals, more than		700
Gross output: labour costs	5:1	
Margin		
Gross margin for cereals, more than		500
Costs		
Overhead costs, less than		550
Leasing and depreciation, less than		100
Tax and personal drawings, less than		125
Capital		
Current assets: current liabilities	1.5:1	
Borrowing		
Short-term borrowing, less than		120
Long-term borrowing, less than		1400
Gross output: rental equivalent	7:1	

5 Balance Sheet Analysis

Chapter 2 discussed the trading and profit and loss account (TPL), which is produced annually for a business. This chapter will look at the way in which the profit and loss account is linked to the balance sheet of a business, and how the balance sheet in combination with the profit and loss account can be used to assess the current and future health of a business.

How are balance sheets and profit and loss accounts linked?

The trading and profit and loss account for a year reflects the income for the year (sales and receipts) less the expenditure (purchases and expenses). This net figure is then adjusted for the farm valuation at the beginning of the year and the farm valuation at the end of the year to arrive at the net profit.

The closing balance sheet will link to the trading and profit and loss account by showing the closing valuation as an asset and the net profit/loss as an increase or decrease in the proprietor's capital as illustrated in Fig. 5.1.

Fig. 5.1 TPL account for the year ending 31 March 1997

Expenditure	£	Income	£
Opening valuation	400	Sales and receipts	540
Purchases and expenses	500	Closing valuation	410
Net profit	50		
	950		950

Opening balance sheet 1/4/1996				*Closing balance sheet 31/3/1997*			
Liabilities	£	Assets	£	Liabilities	£	Assets	£
Creditors	10	Cash	5	Creditors	5	Cash	5
Bank overdraft	15	Debtors	20	Bank overdraft	10	Debtors	45
Mortgage	125	Valuation	400	Mortgage	120	Valuation	410
Proprietor's capital	275			Proprietor's capital	325		
	425		425		460		460

What is a balance sheet?

A balance sheet is a statement, which can be likened to a 'snap shot', at a particular point in time, of the assets utilised by a business and the sources of funds used to finance these assets. It is a record of the assets of the business and its liabilities on the day the balance sheet is constructed.

Unlike the profit and loss account, which records the profit generated over a period of time, the balance sheet is a picture of the assets and liabilities of the business at a specific point in time and is drawn up showing the assets and liabilities on the last day of the profit and loss period.

It is important to note that the composition of the balance sheet could vary dramatically over a very short period and a balance sheet does not measure the average capital employed.

Example 5.1 illustrates how, over a period of a few weeks, the balance sheet of a cereal farmer might change from having an overdraft and a high trade creditor position (merchant's bill) to no overdraft and cash in the bank.

It is assumed for purposes of simplicity that the stock is valued at sale price, which would not be the case in practice. Stock would normally be valued at cost or net realisable value, whichever was lower.

Example 5.1(a) Balance sheet of J. Oats at 31 August 1997

Liabilities	£	Assets	£
Trade creditors		Cash	—
(e.g. merchant's bill)	10 000	Debtors	4 000
Bank overdraft	25 000	Stock: cereals	50 000
AMC loan	35 000	fertilisers	20 000
Proprietor's capital		Machinery	40 000
account (net worth)	144 000	Land	100 000
	214 000		214 000

Example 5.1(b) Balance sheet of J. Oats at 1 September 1997

Liabilities	£	Assets	£
Trade creditors	10 000	Cash	—
(e.g. merchant's bill)			
Bank overdraft	25 000	Debtors	40 000
AMC loan	35 000	Stock: cereals	14 000
		fertiliser	20 000
		machinery	40 000
Proprietor's capital		land	100 000
account (net worth)	144 000		
	214 000		214 000

At this point [Example 5.1(a)] cereals have been harvested and are shown in the balance sheet as an asset. Mr Oats agrees a contract with his merchant to sell £36 000 of the cereals which results in a change in the balance sheet as shown in Example 5.1(b).

Now that Mr Oats has sold some of his cereals they are recorded in the balance sheet as debtors. Payment is due and received on 1 October. The position on that date is shown in Example 5.1(c).

The cash for the cereals which were sold has now been received and used to pay some of the trade creditors and clear the bank overdraft. There is cash on deposit at the bank.

Although the composition of the balance sheet has changed the farmer's capital has remained the same. However, if the stock had been valued at cost and then sold at a profit the later balance sheets would have shown an increase in the farmer's net worth which would represent the net profit on the transaction.

From the example it is seen that the composition of the assets and liabilities making up the balance sheet are fluid. This is particularly so in the short term where the current assets (cash, debtors, stock) and current liabilities (creditors, bank overdraft) are concerned. The fixed assets and long-term liabilities will, generally, alter over a longer time scale.

Example 5.1(c) Balance sheet of J. Oats at 1 October 1997

Liabilities	£	Assets	£
Trade creditors	5 000	Cash	6 000
(e.g. merchant's bill)		Debtors	4 000
Bank overdraft	—	Stock: cereals	14 000
AMC loan	35 000	fertilisers	20 000
		Machinery	40 000
		Land	100 000
Proprietor's capital account (net worth)	144 000		
	184 000		184 000

Definition of the terms used in balance sheets

Assets

The assets of a business are what the business owns. These assets fall into three categories:

(1) *Fixed assets.* Fixed assets are those which are used by the business to generate income over a number of years. Generally speaking if a fixed asset is sold it is immediately replaced to ensure that the means of generating medium and longer-term profits continue to be available to the business. Examples of fixed assets are breeding livestock, plant and machinery, land and buildings.

(2) *Intangible assets*. These are assets where no tangible market value can be given to the asset unless a total sale of the business occurs. Goodwill or company formation expenses are examples of intangible assets.
(3) *Current assets*. The current assets of a business are the more liquid assets which tend to change on a day to day basis. They are the assets which are a basic part of the production cycle. Current assets can generally be turned into cash without fundamentally changing the nature of the business. Figure 5.2(a) demonstrates how the current assets, in the form of growing crops, mature to become the harvested crops in store which are then sold to a corn merchant (debtor) and finally cash is received, after which the cycle starts again. Similarly Fig. 5.2(b) demonstrates the current asset cycle for a livestock enterprise. Debtors are the individuals and firms who owe the business money for goods received.

Current assets fall into two classes: *liquid assets* are current assets which are in the form of cash or can easily be converted into cash, e.g. debtors, harvested crops or fat cattle; *working assets* are current assets which have not yet reached the stage of being readily convertible to cash without some potential loss, e.g. growing crops in the ground or immature cattle.

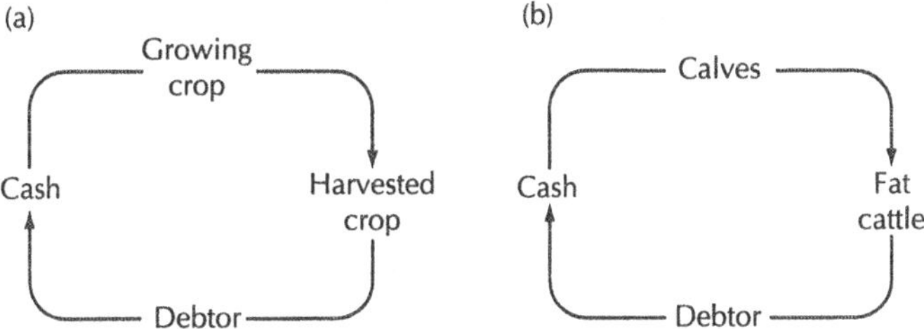

Fig. 5.2 Current asset cycle for livestock and crop enterprises.

Liabilities

The liabilities of a business are what the business owes. They represent all the outside sources of funds the business has available. These creditor funds, with the proprietor's own capital, fund the assets of the business. Like the assets of a business they can be divided into categories:

(1) Current liabilities. Current liabilities are those which have to be paid, or may have to be paid, in the short term, e.g. trade creditors (less than one year), bank overdraft, current taxation. Creditors are the individuals or firms to whom the business owes money for goods and services received
(2) *Long-term or deferred liabilities*. These are the liabilities which are repayable over longer periods of time, normally one year or more, e.g. bank loans, loans from relatives, Agricultural Mortgage Corporation loans, future or deferred taxation

Proprietor's capital (proprietor's equity, proprietor's stake, net worth, surplus)

If all the assets of a business are sold and all the liabilities repaid, then the net proceeds of the sale, subject to tax, belong to the proprietor and represent his capital in the business. There are a variety of terms in common use (e.g. net worth, proprietor's equity, proprietor's stake, surplus) which are used to describe proprietor's capital. In the balance sheet of a business the proprietor's capital appears as a liability. This is logical if proprietor's capital is thought of as a source of finance used to fund the assets of the business. In a company the owners of the business are shareholders and the shares and the accumulated profit represent shareholders' funds and equate to proprietor's capital.

The relationship of assets, liabilities and proprietor's capital

It is important that the relationship between assets, liabilities and proprietor's capital is clearly understood. Assets have been defined as the amount which a business owns, with liabilities and proprietor's capital as the sources of funds to finance these assets. Thus, by definition, the balance sheet value of the assets will be equal to the balance sheet value of the liabilities including proprietor's capital. The amount owed to the outside creditors of the business will vary from time to time and can be readily ascertained from the books of the business. Proprietor's capital will vary by retained profits (if any) and any injection of new capital into the business.

The difference between the value of the assets and the outside liabilities in the balance sheet will represent the proprietor's capital as illustrated in Fig. 5.3.

Fig. 5.3 Balance sheet as at 31 March 1996

	£		£
Liabilities	50	Assets	100
Proprietor's capital	50		
	100		100

Presentation of balance sheets

Historically, balance sheets were presented on a 'side by side' basis with liabilities on the left ranked in order of their 'life' (i.e. short term, medium term, long term) and assets on the right ranked in order of their liquidity (see Fig. 5.4). This method is used in this book for ease of demonstration.

Balance sheets are now presented in a 'columnar' form starting with fixed assets at the top followed by net current assets (current assets less current liabilities).

Fig. 5.4 Balance sheet presented on a 'side by side' basis (£'000)

Liabilities	£				£	
Trade creditors	20			Cash	—	
Bank overdraft	30			Debtors	25	
				Stock and growing crops	75	
Current liabilities		50		Current assets		100
Hire purchase	—			Breeding stock	—	
Bank loan	50			Machinery and equipment	100	
AMC loan	250			Land and buildings	900	
Deferred and long-term liabilities		300		Fixed assets		1 000
Proprietor's capital/ Net worth/surplus		750		Intangible assets	—	
Current liabilities Long-term liabilities Proprietor's capital		1 100	=	Current assets and fixed assets and intangible assets		1 100

Fixed assets and net current assets are then added together to give the net assets which are then said to be financed by the long-term liabilities and the proprietor's capital. Figure 5.5 illustrates this style of presentation.

Balance sheet analysis

A business may be shown to be making a profit but an evaluation of performance cannot be solely measured in profit terms. It is equally important to ensure that a business avoids damaging future profit potential. This is best achieved by ensuring that the financial structure of the business is healthy and will remain so in the future.

There are two objectives in balance sheet analysis:

(1) To establish the long-term stability and viability of a business
(2) To establish the ability of a business to meet its short-term commitments

It is important to understand that successful interpretation of balance sheets requires the study of the trends shown over a series of balance sheets of a business as well as considering the absolute figures and ratios.

Establishing the long-term stability and viability of a business

There are five basic points to consider:

Fig. 5.5 Balance sheet presented on a columnar basis (£'000)

	£	£	£	£
Fixed assets:				
machinery and equipment			100	
land and buildings			900	
				1 000
Current assets				
stock and growing crops	75			
debtors	25			
cash	—			
		100		
Less current liabilities:				
trade creditors	20			
bank overdraft	30			
		50		
Net current assets			50	
				1 050
Financed by:				
bank loan			50	
AMC loan			250	
proprietor's capital			750	
				1 050

(1) *The potential within the business to survive adverse trading conditions*. In a loss situation the first person to lose money is the proprietor of the business. It is therefore essential that a business has sufficient proprietor's capital to be able to withstand any short-term losses. There is no 'magical' figure for the minimum level of proprietor's capital in a business but experience suggests that in general terms a farmer should be able to fund at least 60% of the total assets of the business from his own capital.

In Example 5.2(a) it can be seen that where the business has 60% of proprietor's capital it is able to survive the setback of a substantial loss, compared to the business with 20% proprietor's capital shown in Example 5.2(b) where the business becomes insolvent because its liabilities exceed its assets

It is the end of the road for the business shown in Example 5.2(b) with not only the proprietor losing his money but also some creditors.

(2) *The potential which the business has to raise creditor finance*. All businesses use creditor funds to a greater or lesser extent to help finance their activities. The ability of a business to call on creditor finance to fund the purchase of profit-producing assets or in adverse times to support temporary loss-making situations is important to the long-term stability of a business. Apart from measuring potential profitability, and thus the capital servicing capabilities of a

Example 5.2(a) Balance sheet with initial proprietor's capital 60% (the business makes a loss of £22 000)

Balance sheet before loss				Balance sheet after loss	
Liabilities	£	Assets	£	Liabilities £	Assets £
Trade creditors	15 000	Cash	—	20 000	—
Bank overdraft	25 000	Debtor	8 000	25 000	5 000
		Stock	40 000		30 000
		Plant and machinery	52 000		48 000
Proprietor's capital	60 000			38 000	
	100 000		100 000	83 000	83 000

Example 5.2(b) Balance sheet with initial proprietor's capital 20% (the business makes a loss of £22 000)

Balance sheet before loss				Balance sheet after loss	
Liabilities	£	Assets	£	Liabilities £	Assets £
Trade creditors	45 000	Cash	—		—
Bank overdraft	35 000	Debtor	8 000	50 000	5 000
		Stock	40 000	35 000	30 000
		Plant and machinery	52 000		48 000
				Capital deficiency	2000
Proprietor's capital	20 000				
	100 000		100 000	85 000	85 000

business, the creditor knows that the safety of his investment depends, to a large extent, upon the proprietor's capital stake in the business because the proprietor must lose his money before the creditors start to lose theirs! As a result the higher the percentage of proprietor's capital in a business (as a general rule) the greater will be the chance of creditors providing funds. They are interested in the solvency of the business. The relationship of proprietor's capital to creditor capital is monitored by means of equity ratios which are discussed later in this chapter

(3) *The ability of the business to meet its trading commitments on time (liquidity)*. A farmer's capital could represent 90% of the total asset value of the business but because it is all in the form of fixed assets (e.g. land and buildings) there might be inadequate liquid resources available to the business for meeting short-term

commitments, such as merchants' bills, electricity or hire purchase payments. Clearly if this is the situation then the stability of the business will be jeopardised and it may require a complete restructuring of the business (e.g. sale of land) to rectify the position

(4) *The capability of the business to make an adequate return on the capital it is employing.* One of the components determining the viability of a business is return on capital. Unless an acceptable return is made on the capital employed then in the medium and long term the proprietor's capital base will become inadequate to support the requirements of a thriving business

(5) *The ability of the business to generate sufficient profits to provide adequate cash to fund personal drawings, loan repayment, reinvestment and taxation*

Establishing the ability of a business to meet its short-term commitments

The ability of a business to meet its day to day commitments was discussed as a factor which affects the long-term viability of a business. Clearly, if the day to day business transactions cannot be met on time a business will collapse very quickly, and to gauge the likelihood of such a situation developing the liquidity of the business is investigated.

How is the health of a business assessed from reviewing balance sheets?

The study of trends in balance sheet ratios

It is very difficult (if not impossible), from the study of a single balance sheet, to arrive at any conclusion about the health of a business. No balance sheet ratio is absolute and what is acceptable in one business might be disastrous for another. A wealth of information, however, can be gleaned by:

(1) Studying a series of balance sheets for a business
(2) Investigating the trends which occur in the relationships between the different components of a balance sheet

Balance sheet ratios are discussed later in this chapter with the objective of establishing the following points.

The solvency of a business

A business is solvent if the assets are greater than the liabilities. This means that the proprietor of the business has some of his own money in the business. If the assets of the business are less than the liabilities then the business is insolvent. Not only will there be no money for the proprietor, but some of the creditors will lose money too, as shown in Example 5.2(b).

Because a business is solvent and profitable does not, in itself, mean that it is healthy. The proprietor's capital might be at such a low level that a short-term change in trading profits can drain away what capital there is in the business. Where a business has a high sales turnover on a very low capital base it is said to be overtrading. Clearly it is important to establish the capital base of a business and to investigate any significant change in that base.

The liquidity of a business

Liquidity is probably the key to a successful business. The liquidity of a business reflects the proportion of assets which are in the form of cash or assets that can be readily converted into cash. There are many interacting factors which influence the liquidity of a business and these are illustrated in Table 5.1.

Table 5.1 Factors influencing the liquidity of a business

	External factors	Internal factors
Increase liquidity	Capital injection Increase in credit Tax refund Government grants	Sale of fixed assets Profit
	LIQUIDITY	
Reduce liquidity	Reduction in credit facilities Losses Re-investment in fixed assets Inflation	Tax payments Credit repayment Dividends Drawings Bonuses

The performance of a business

A business can be solvent and liquid but over the medium term it has to be profitable to survive. It is also essential that the terms of trade are such that the assets and liabilities of the business are managed sensibly. This can be determined by considering various operating and performance ratios.

Balance sheet ratios

The health of a business is influenced by the financial proportions of the assets and liabilities used by the business rather than by the absolute amounts involved. Balance sheet analysts work almost exclusively with ratios. There are no perfect ratios, although the history of business failures suggests that there are some entirely unacceptable ratios. Whether a ratio is acceptable or not will depend on its composition and on the financial management ability of the proprietor of the business. There are often cases where businesses will survive, even though they have apparently disastrous balance sheet ratios, because of the financial genius of the proprietor. However, a business with poor ratios is vulnerable and may collapse. It is often poorly structured because of the proprietor's ignorance of his predicament!

If only one balance sheet is available for analysis then it will only be possible to probe for poor financial health. The ratios which are revealed can only be looked at against ratios seen in other farming balance sheets. Hopefully there will always be a series of balance sheets available to the analyst for his interpretation so that the trends in ratios can be investigated.

The interpretation of balance sheets should be based on a combination of sound financial experience and commonsense. Whenever analysing balance sheets it is important to ensure that the valuations in the balance sheet are sensible. It is common knowledge that the valuations in some balance sheets are unrealistic and these must be adjusted.

Liquidity ratios

Liquidity ratios reveal the extent to which a business is able to meet its short-term commitments.

Current ratio

The current ratio is probably the best known and most used of all balance sheet ratios. It measures the ability of the business to meet its short-term commitments: current liabilities (creditors, bank overdraft) from its current assets (cash, debtors, stock), without having to resort to sale of fixed assets which would jeopardise future profit potential and change the structure of the business.

$$\text{Current ratio} = \frac{\text{current assets}}{\text{current liabilities}}$$

Classically the current ratio is given as 2:1, i.e. current assets should be twice current liabilities, but in practice this is rarely the case. The ratio which is acceptable will depend on the composition of the current assets and liabilities. If there are a high proportion of liquid assets (i.e. cash, debtors) in the current asset figure then a lower ratio might be acceptable than if there is a high proportion of less liquid assets, i.e. growing crops or store cattle. A low ratio may be acceptable because the creditors are prepared to wait for their money until the production cycle is complete!

Whilst a low ratio may mean lack of liquidity a high ratio might suggest inefficient use of resources.

Acid test ratio

The acid test ratio helps to highlight the liquidity of the current assets and might explain why a low current ratio exists. It identifies the ratio of very liquid assets to current liabilities.

$$\text{Acid test} = \frac{\text{cash} + \text{debtors} + \text{harvested crops and mature stock}}{\text{current liabilities}}$$

The problem is to decide which are the liquid assets. Normally cash and debtors are regarded as liquid assets but some debtors may be excluded if the terms of trade gave the debtor extended credit. Where a business has an agreed overdraft facility only partly utilised at the balance sheet date then some authorities would include the overdraft limit in the current liabilities and show the unused portion as part of the liquid assets available to pay creditors. A guide ratio is 1:1.

Working capital ratio

Working capital is defined as the capital required to fund the production cycle and can be measured by deducting current liabilities from current assets.

The working capital ratio is designed to give some indication of the amount of additional working capital that would be required for a given increase in sales, as illustrated in Example 5.3.

$$\begin{array}{c}\text{Working}\\ \text{capital}\\ \text{ratio}\end{array} = \frac{\text{current assets} - \text{current liabilities (working capital)}}{\text{sales}}$$

The level of working capital required will depend to some extent on the length of the production cycle of the enterprises on the farm. The higher the working capital the easier it should be for a business to meet its commitments on time.

If it is assumed that the terms of trade of the business will remain the same then using the working capital ratio it is possible to estimate the likely increase in working capital requirements to support the increase in sales. In Example 5.3:

$$\begin{aligned}\frac{\text{Working capital}}{\text{sales}} &= \frac{\text{current assets} - \text{current liabilities}}{\text{sales}}\\ &= \frac{80\,000 - 50\,000}{300\,000}\\ &= 0.1,\ \text{i.e. } 10\%\end{aligned}$$

Example 5.3 An illustration of the potential increased working capital requirement to fund an increase in sales

A farm business has annual sales of £300 000 and having carried out a planning exercise which included the purchase of a machine to improve throughput now anticipates sales will increase to £400 000 per annum. The farmer is concerned that whilst he has been able to fund the machine by way of bank loan he has not quantified the effect of the increase in sales turnover on his working capital requirements. His latest balance sheet shows the following:

Liabilities	£	Assets	£
Creditors	10 000	Cash	—
Bank overdraft	40 000	Debtors	15 000
		Harvested stock	35 000
		Mature animals	30 000
	50 000		80 000
AMC loan	160 000	Dairy herd	50 000
		Plant and machinery	32 000
Proprietor's capital	702 000	Land	750 000
	912 000		912 000

This means that every £100 of additional sales will require £10 of additional working capital. Therefore by increasing sales by £100 000 (£300 000 to £400 000) the working capital requirement will increase by £10 000.

Average credit given and average credit taken

The control of the amount and duration of credit given to debtors and how this is balanced by the credit that can be sensibly taken from suppliers is important to the smooth running of a business. Utopia is to be able to safely take twice as long credit from suppliers compared to the credit given to debtors! In practice it is more likely that the time taken from creditors will equal the time given to debtors and average around four to six weeks. If the time taken by debtors to pay increases and creditors are still paid on their same credit terms then clearly another source of funding will be required, normally increased overdraft facilities which may increase costs.

Earlier in this chapter the current asset cycle was shown (Fig. 5.2) and in Fig. 5.6 this is linked with the 'current liability cycle' to show how the two interact and form the working capital cycle.

Average credit given

This is assessed as follows:

$$\frac{\text{Average debtors}}{\text{Sales}}$$

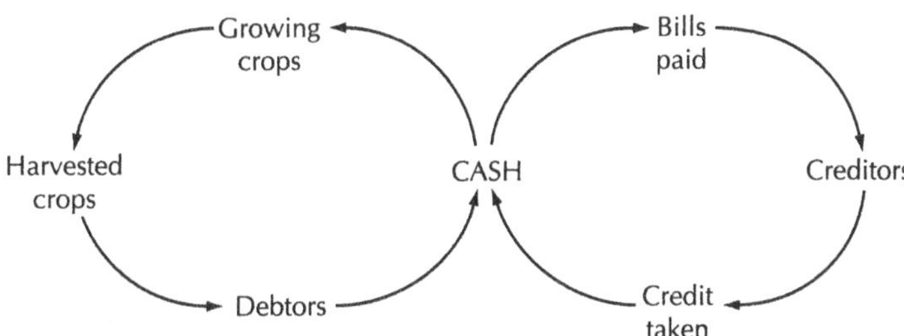

Fig. 5.6 The working capital cycle.

The average debtor figure is normally calculated by taking the debtor figures from the balance sheets of two consecutive years and dividing this by two. The sales figure will be the sales in the intervening 12 months.

To make the resulting figure more meaningful it is useful to have the answer expressed in days, weeks or months and this is achieved by multiplying by 365, 52 or 12 as illustrated in Example 5.4.

Example 5.4 The calculation of the period of credit given by a business

The debtor figure in the balance sheet of Cotswold Farms as at 31 March 1996 was £26 000 and as at 31 March 1997 £30 000. The sales for the year ended 31 March 1997 were £300 000.

$$\text{Average debtors} = \frac{£26\,000 + £30\,000}{2}$$

$$= £28\,000$$

$$\text{Average credit given (in days)} = \frac{\text{Average debtors}}{\text{Sales}} \times 365$$

$$= \frac{£28\,000}{£300\,000} \times 365$$

$$= 34 \text{ days}$$

If over a series of balance sheets the period of credit given was increasing significantly then it would be sensible to investigate why. It could be that there were some potentially bad debts building up, resulting in lack of cash to pay creditors.

Bad debts are very significant to a business and it should always be remembered that to earn a profit equal to the value of a bad debt means a very significant 'multiplier' in terms of sales. For example, if a business was making a net profit of say 5% then if it had a bad debt of £1000 it would need to sell £20 000 of goods to earn another £1000 to replace the bad debt!

On the other hand, the increased time taken by debtors to pay could be poor debtor control or simply an agreed change in credit terms.

Average credit taken

This is assessed as follows:

$$\frac{\text{Average creditors}}{\text{Purchases}} \times 365 = \text{Average credit taken in days}$$

The calculation follows the same principle as shown in Example 5.4 for average credit given, except that the creditor figure is taken from the balance sheets and the purchases figure from the profit and loss account.

Management of creditors is just as important as the management of debtors because at the end of the day they are helping to finance the assets of a business. If a creditor were, for example, to go bankrupt this could have severe repercussions for a business which was relying on it for its products and the source of short-term credit it provided.

Credit from suppliers can also be expensive, particularly when discounts are not taken.

If time taken from creditors is increasing it could mean that new terms of trade have been agreed, or that the farmer is good at creditor management, or simply that the cash was not available within the business to pay the creditor on time!

Equity/solvency ratios

Equity ratios measure the proportion of proprietor's equity in a business in relation to a variety of other forms of capital and also in relation to categories of assets. These ratios give a measure of the exposure of the business to outside creditors. There are no 'correct' ratios but in general terms the farm business should be funding 60% of total assets.

(1) *Equity to total capital employed.* This ratio shows how much of the capital used by the business is supplied by the proprietors of the business. There are a number of definitions of capital. The authors regard total capital as the balance sheet total less intangible assets (i.e. goodwill, etc.). Whichever definition is used one must always be consistent
(2) *Equity to equity plus long-term debt.* This ratio indicates the proportion of the long-term capital that is provided by the owners of the business
(3) *Equity to fixed assets.* This ratio measures the proportion of the fixed assets which are financed by the proprietors
(4) *Equity to long-term debt (gearing)* Example 5.2 illustrates the problems which can arise if a business has a low capital base. To measure this relationship the 'gearing' of a business is calculated. Gearing compares the proportion of the finance house capital in the business with the owner's capital. Normally the owner would be expected to have the higher proportion. Example 5.5 demonstrates this calculation.

    Long-term debt is represented by the AMC loan and the bank loan; therefore the gearing ratio is:

$$\frac{340\,000}{565\,000} = 0.6$$

or the long-term debt in the business only represents 60% of the capital provided by the proprietor.

A business is highly geared if there is a high proportion of borrowed money. Alternatively a low gearing is when the borrowing is low compared to proprietor's equity. Care must be exercised that the bank overdraft is not representing a significant tranche of long-term funding. If it is, then it should be treated as such in the analysis. Remember a consistent approach is important

Example 5.5 Calculating the 'gearing' of a business

Liabilities	£	Assets	£
Creditors	15 000	Cash	—
Bank overdraft	10 000	Debtors	20 000
Bank loan	40 000	Stock	50 000
AMC	300 000	Dairy herd	50 000
		Plant and machinery	60 000
Proprietor's capital equity or net worth	565 000	Land	750 000
	930 000		930 000

(5) *Total borrowing to equity*. This ratio gives an alternative calculation of gearing but includes all the borrowed money and therefore overcomes the problem outlined with the equity to long-term debt ratio

There are no magical ratios which will guarantee a healthy business from the equity standpoint but in general terms a farm business should have at least 60% assets financed by equity capital with outside capital possibly supporting the current assets. The thinking behind this statement is that the profitability in agriculture is such that it is almost impossible to generate sufficient return to support creditor borrowing of a more substantial nature. Clearly the position might vary depending on the quality of management in the individual business and on the proportion of low-yielding assets in the business (i.e. land).

It follows from a discussion of the level of creditor funds which a business can usefully use that it would be helpful to know how much leeway there is in the business to cover the payments to investors, in the form of interest and capital.

Times covered

Times covered measures how many times the servicing and repayment charges of creditors could be made out of the profits of the business.

$$\text{Times covered} = \frac{\text{Profit before interest and tax}}{\text{Amount of interest and repayment charges}}$$

The amount of the payment to investors for use of their funds might vary depending on the vulnerability of the form of finance to movements in interest rates. The more times a payment is covered, the greater the likelihood that future payments can be met. This calculation shows whether the business is able to stand a drop in profits and still cover its financial charges.

Performance ratios

Analysts have many ratios which they use to assess profitability. There are three fundamental ratios which could be helpful to the analysis of the farm business. The three ratios form an equation:

$$\text{Return on capital} = \frac{\text{Sales}}{\text{Capital employed}} \times \frac{\text{Profit}}{\text{Sales}} = \frac{\text{Profit}}{\text{Capital employed}}$$

or put another way:

Rate of turnover on capital employed		Rate of return on sales		Total return on capital employed
	×		=	

Thus it is possible to look at the two key components of return on capital and see if there are weaknesses in either area.

It is again important to be consistent in the use of definitions. In this case profit should be net profit before taxation and interest charges. It should also exclude profit from exceptional transactions such as profit or loss on sale of machinery. The capital employed in the business is the total liabilities less intangible assets.

Later in this chapter performance is measured against capital employed with the capital divided into tenant's capital and landlord's capital. The suggested performance analysis here is complementary to the later formulae.

Industrial and commercial balance sheet analysts have hundreds of different ratios they might examine, depending on the particular type of business they are examining, but within agriculture the most useful ones have been discussed in this chapter.

The interpretation of balance sheet ratios

The calculation of ratios is a straightforward affair. The problems arise in interpreting the 'cold' facts in a sensible and logical manner.

If a series of balance sheets show a clear trend (be it good or bad) the student of balance sheets is in a position to make a superficial comment on the situation. If, however, he is to get to the nub of the matter he must interpret the figures that were used in the initial ratio calculation. This may mean, for example, investigating the creditors shown in the balance sheet. Are any of them overdue for payment? Are any being paid small sums to keep the account turning over? Is there a particularly large creditor who may create a financial crisis if his debt is 'called'?

It may be that some of the debtors of the business are overdue to pay, some debts may indeed be bad or one debtor may be so dominant that by failing to pay on time he adversely affects the liquidity of the business.

Table 5.2 illustrates the type of questions the analyst should be asking when trying to interpret balance sheet figures.

So far in this chapter balance sheets have been looked at in historical terms but in the final analysis management must look to the future.

Table 5.2 Looking behind balance sheet figures

Liabilities	Assets
Sundry creditors	*Debtors*
What are the terms of trade?	What are the terms of trade?
Is there a dominant creditor?	Are debtors well spread?
Are creditors' businesses sound?	Is there a dominant debtor?
Are tax and VAT up to date?	Are the debtors considered good?
Financial institutions	*Stock*
What are servicing costs?	Is stock undervalued?
What are the repayment terms?	Is stock value vulnerable to short-term market fluctuations?
What is the bank overdraft limit?	Would a forced sale result in a significant reduction in value?
	What stage of maturity have the animals reached which are valued in the stock figure?
	Is there a demand for the stock?
	Machinery
	What is market value?
	What is the life of machinery?
	What are the tax implications?
	Land and buildings
	What is market value?
	Is there likely to be compulsory purchase?
	Are the buildings in good condition or obsolete?

The use of balance sheets for interpreting the consequences of management decisions on future business structure

Apart from using balance sheets to review past performance, it is important to make use of the knowledge of balance sheet structure to establish the effect management decisions will have on the future structural health of a business. What will happen to the liquidity of a business if a new parlour is purchased and financed on bank overdraft? Will the purchase of 100 weaned calves on overdraft improve or reduce the liquidity of the business? These questions can be answered by taking the existing balance sheet and adjusting for the proposed changes as illustrated in Example 5.6.

Example 5.6 How to structure borrowing

The farmer decided to update his old herringbone parlour with a new herringbone at a cost of £30 000 after grant. There was also a beef unit on the farm which was only half utilised and so it was planned to purchase an additional 100 calves at £100 each to bring it to capacity.

The balance sheet for Cirencester Farms as at 31 March 1997

Liabilities	£	Assets	£
Sundry creditors	10 000	Cash	—
Bank overdraft	25 000	Debtors	18 000
		Stock	25 000
Current liabilities	35 000	Current stock	43 000
		Breeding stock	32 000
Net worth	100 000	Plant and machinery	60 000
	135 000		135 000

To fund these changes the farmer needed to borrow money and asked his advisor which finance method would be the best one to use.

What would be the effect of financing by bank overdraft?

(1) *Parlour at £30 000*. Financing the parlour by way of overdraft would drain the liquidity of the business quite dramatically and the current ratio would change from 1.23:1 to 0.66:1. This is because a fixed asset (parlour) is being purchased with a current liability (bank overdraft). It is important to note that the net worth remains unchanged because the additional liability (i.e. the increased overdraft) is balanced by an asset (the parlour) of equal value

Liabilities	£	Assets	£
Sundry creditors	10 000	Cash	—
Bank overdraft	55 000	Debtors	18 000
		Stock	25 000
Current liabilities	65 000	Current assets	43 000
		Breeding stock	32 000
Net worth	100 000	Plant and machinery	90 000
	165 000		165 000

(2) *Purchase of calves*. The purchase of calves on overdraft only marginally affects the liquidity ratio and the business remains relatively liquid (1.18:1). This is because the current asset (stock) is being purchased with a current liability (bank overdraft)

Liabilities	£	Assets	£
Sundry creditors	10 000	Cash	—
Bank overdraft	35 000	Debtors	18 000
		Stock	35 000
Current liabilities	45 000	Current assets	53 000
		Breeding stock	32 000
Net worth	100 000	Plant and machinery	60 000
	145 000		145 000

What if the parlour were financed by a bank loan?

(1) The parlour is an asset which will have a useful working life of probably ten years or more and therefore it would seem reasonable to arrange payment over a period of years. Thus a ten year loan for purchasing the parlour would affect the balance sheet as shown below and the liquidity of the business is unaltered if a bank loan is used.

Liabilities	£	Assets	£
Sundry creditors	10 000	Cash	—
Bank overdraft	25 000	Debtors	18 000
		Stock	25 000
Current liabilities	35 000	Current assets	43 000
Bank loan	30 000	Breeding stock	32 000
Net worth	100 000	Plant and machinery	90 000
	165 000		165 000

(2) The calves are part of the working assets and will be reared and sold probably within 18 months. It is therefore inappropriate to borrow money over a long period to finance the purchases of calves and it has been seen that they only marginally alter the liquidity of the business.

If the parlour is financed by bank loan and the calves by way of overdraft the revised balance sheet would be:

Liabilities	£	Assets	£
Sundry creditors	10 000	Cash	—
Bank overdraft	35 000	Debtors	18 000
		Stock	35 000
Current liabilities	45 000	Current assets	53 000
Bank loan	30 000	Breeding stock	32 000
Net worth	100 000	Plant and machinery	90 000
	175 000		175 000

The balance sheet is still liquid and the borrowing is reasonable in relation to the equity of the farmer.

This balance sheet projection can be a very useful aid to preventing financial decisions being taken which might lead to cash flow problems at a future date.

Sources and applications of funds by balance sheet analysis

The lifeblood of any business is cash and it is therefore very important that optimum use is made of the cash available.

In analysing past performance it is essential to investigate what happened to the cash available to the business, where did it come from and where did it go? This information is determined by preparing a statement of the sources of funds and their application.

One method of discovering cash movement would be to analyse the cash book but this can be a tedious exercise and it is simpler to compare the opening and closing balance sheets for a particular year. This method enables the analyst to establish the net cash movements to and from the individual items in the balance sheet. If the stock figure decreased from £50 000 to £40 000 then clearly £10 000 cash would have been released and this would be a source of cash for the business. The fact that the farmer bought £200 000 worth of stock and sold £210 000 is not relevant; the important figure in this type of analysis is the net movement of funds.

All sources and application of funds statements must balance as all cash which comes from a source must be applied somewhere even if it is only to reduce the bank overdraft or increase a credit balance at the bank.

Fig. 5.7 Sources and application of funds statement prepared from an analysis of balance sheets

Balance sheets of Ben Clarke, farmer, as at 1 April 1996 and 31 March 1997

Liabilities	1/4/96	31/3/97	Assets	1/4/96	31/3/97
Sundry creditors	5 000	8 000	Cash	1 000	1 000
Bank overdraft	6 000	7 000	Debtors	8 000	9 000
			Livestock	4 000	6 000
	11 000	15 000		13 000	16 000
AMC	20 000	18 000	Breeding		
			Livestock	12 000	13 000
			Plant and machinery	31 000	34 000
Proprietor's			Land	40 000	40 000
capital or equity	65 000	70 000			
	96 000	103 000		96 000	103 000

Figure 5.7 demonstrates a simple sources and application of funds statement derived from analysing the changes between opening and closing balance sheets for a particular year. (There were no revaluations between the two balance sheets.)

Figure 5.8 summarises the changes in sources and applications shown in Fig. 5.7.

Fig. 5.8 Summary of sources and applications

Sources	£	Applications	£
Increase in creditors	3000	AMC loan repayments	2000
Increase in overdraft	1000	Increase in debtors	1000
Increase in proprietor's		Increase in livestock	2000
capital	5000	Increase in breeding	
		livestock	1000
		Increase in plant and	
		machinery	3000
	9000		9000

Summary of sources and applications

(1) Creditors increased by £3000 thus providing the business with more funds (source)
(2) The bank overdraft increased, providing an additional £1000 cash (source)
(3) AMC were repaid £2000 which meant an outflow of cash (application)
(4) The capital account increased by £5000, probably from retained profits (source)

(5) Debtors increased and these required funding (application)
(6) Livestock increased and required funding (application)
(7) Breeding stock increased and utilised cash to fund (application)
(8) Machinery was purchased and used up cash (application)

Adjustment of balance sheet figures for non-cash items

The figures appearing in balance sheets frequently mask a number of cash and non-cash items which must be allowed for if a true picture of cash movement within the business is to be revealed. Depreciation and revaluation are two classic examples of non-cash items. Profit and taxation are two cash items. As these are not balance sheet items it is helpful to have available the relevant profit and loss account to ensure that the appropriate adjustments can be made to the balance sheet figures. Figure 5.9 shows how the balance sheet figures might be adjusted for depreciation. This information enables a fuller analysis to be made of the sources and application statement in Fig. 5.7.

(1) The assumption was made in Fig. 5.7 that the increase of £5000 in the proprietor's capital account was a source of funds attributable probably to retained profits. In fact the profit before depreciation and last year's tax charge was £12 000 (source)
(2) It was assumed that the investment in machinery was £3000 but it is now known that a 'non-cash' depreciation charge of £6000 was made in the year (application)
(3) £1000 was utilised in paying tax (application)

Fig. 5.9 Adjustments to the sources and application of funds statement from information in the profit and loss account. The profit and loss account for Ben Clarke for the year ending 31 March 1997.

Profit after tax and depreciation	5000
Depreciation	6000
Taxation	1000

If we now combine all the information then the sources and application of funds statement would be as shown in Fig. 5.10.

From Figs 5.7–5.9 it can be seen that the business generated a total of £16 000 cash of which three-quarters (£12 000) was from profit, the remainder coming from creditors and the bank; £9000 of these funds were injected into machinery purchase, £3000 were put to increased livestock, £2000 repaid long-term loans with £1000 being used to fund debtors and £1000 to the taxman.

It is therefore clear that although there is a reasonable profit (£12 000) this has not been reflected in a dramatic reduction in overdraft but has instead been invested in machinery and livestock, hopefully to help generate future profits.

Fig. 5.10 Summary of adjusted sources and applications

Sources	£	Applications	£
Profit for year (before) depreciation and tax	12 000	AMC loan repayment	2 000
Increase in creditors	3 000	Increase in debtors	1 000
Increase in overdraft	1 000	Increase in livestock	2 000
		Increase in breeding livestock	1 000
		Increase in plant and machinery (34 000 – 31 000 + 6 000)	9 000
		Taxation paid	1 000
	16 000		16 000

By taking a forward budget and balance sheet it is possible to use this technique to look at the likely sources and application of funds in the future.

Landlord and tenant's capital

Capital, along with land and labour, is a key resource in any business. Without capital a business cannot survive but in practice it is always a limited resource.

Maximising the return on the capital employed in a business should be a key task. The capital invested in a farm business can be split into two types which identify the capital associated with the generation of short to medium-term profits and the longer-term capital associated with long-term and investment profits.

These divisions in the capital structure of the farm business are applied whether the farm is tenanted or owner-occupied and in this way it is possible to carry out comparative analysis on all farms.

Tenant's capital

This represents the capital which would normally be invested in a farm business by a tenant and includes the more liquid assets of cash, debtors, livestock and crops, cultivations, plant and machinery and in some cases the tenant's fixtures, such as buildings, where they are not provided by the landlord.

An average value of the tenant's capital in the farm business is assessed by taking the value at the beginning of a trading year and again at the end of the year and averaging these figures.

With the knowledge of this average figure there are a number of calculations and comparisons which can be made to help assess the performance of the business. The investment per hectare may be compared with other farms and also the return on the tenant's capital calculated and compared with other farms of like structure. Return on the tenant's capital varies tremendously but averages around 12% and in some cases will reach 30%+.

Landlord's capital

The landlord's capital is made up of land, buildings and improvements attributable to the landlord. The return on the landlord's capital is significantly lower than on the tenant's capital, basically because the rents associated with land are very low in relation to the capital cost of land. For example a let cereal farm is likely to have a rent of around £150 to £225 per hectare with the capital value around £4750 to £6800 per hectare, giving a return on capital in the region of 2.5–4%.

6 Enterprise Studies

The agricultural industry in the UK exhibits a great diversity in enterprises both in size and in type. Table 6.1 illustrates the size of the major enterprises, while Table 13.2 gives an indication of the level of self-sufficiency in the UK.

Enterprise studies

In Chapter 3 the gross margin concept of analysing farm enterprises was discussed where the variable costs of production are deducted from gross output to give the gross margin for an enterprise. The gross margin for a particular enterprise can vary significantly from farm to farm, so that it is quite feasible within the UK agricultural industry to find, for example, two dairy herds with the same breed of cow and the same management system producing gross margins as diverse as £650 per cow and £1100 per cow. Cereal gross margins on neighbouring farms can show a similar range, with wheat gross margins ranging from £424 per hectare to £827 per hectare. Table 3.2 indicates the likely range in production standards and the associated gross margins for 1997.

With the wide divergence in gross margin returns evident within a particular sector, there must be key factors, both physical and financial, which will determine the ultimate level of profitability. Some of these factors will relate to output, others to the efficiency of utilisation of inputs. Each enterprise will have its own key factors and it is important that the farmer is aware of the factors which affect the profitability of the enterprises on his own farm.

Factors affecting the profitability of a dairy herd

From the statistics presented in Table 6.1, the importance of the dairy industry to British agriculture is evident and it is therefore appropriate to look in some detail at the factors affecting its profitability.

Figure 6.1 shows the way that a tree diagram can be developed to highlight the major factors affecting the gross output from a dairy cow.

The analysis of gross output can be taken further if required by looking at the components of calving index, yield, milk quality and so on.

The emphasis to be placed on any particular component of milk sales or net replacement cost will depend on the economic situation at the time and the physical constraints on a business. The introduction of milk quotas is an economic and physical constraint with which dairy farmers have, in recent times, had to cope. This

Table 6.1 UK land use and livestock numbers 1966 to 1996 ('000 ha or '000 hd)

Year	1966	1976	1986	1988	1989	1990	1991	1992	1993	1994	1995	1996
Crop area (× 1000 ha)												
Wheat	955	1 240	1 997	2 083	2 083	2 013	1 981	2 067	1 759	1 811	1 859	1 976
Barley	2 368	2 172	1 916	1 878	1 652	1 517	1 393	1 297	1 164	1 106	1 192	1 267
Oats	396	233	97	120	119	107	104	101	92	109	112	96
Others	36	31	14	12	20	21	23	22	16	16	18	18
Total cereals	3 755	3 676	4 024	3 896	3 874	3 658	3 501	3 487	3 031	3 042	3 180	3 357
Potatoes	286	223	178	180	175	178	177	180	170	164	171	177
Sugarbeet	183	207	200	198	197	194	196	197	197	195	196	199
Oilseed rape	0	48	299	348	321	390	440	422	377	404	354	356
Vegetables (field)	152	203	146	141	141	142	139	135	126	127	129	132
Orchard fruit	79	52	38	37	36	34	34	33	32	32	28	28
Soft fruit	19	16	15	15	15	15	15	14	13	13	12	12
Others (inc. fallow)	451	375	387	492	444	466	517	565	620	536	472	495
Total tillage	4 925	4 800	5 287	5 311	5 203	5 077	5 019	5 033	4 566	4 513	4 583	4 756
Grass (< 5 years)	2 539	2 156	1 723	1 613	1 535	1 580	1 583	1 562	1 561	1 436	1 387	1 376
Total arable	7 464	6 956	7 010	6 924	6 738	6 657	6 603	6 595	6 127	5 949	5 970	6 133
Grass (> 5 years)	4 946	5 064	5 077	5 161	5 249	5 263	5 252	5 213	5 209	5 322	5 309	5 289
Rough grazing	7 175	6 511	6 045	5 948	5 972	5 949	5 912	5 910	5 840	5 775	5 764	5 725
Other land (inc. woods)	n/a	415	543	562	623	680	616	632	678	708	728	745
Set-aside	—	—	—	—	—	—	97	160	677	728	633	509
Total area	19 585	18 956	18 676	18 595	18 581	18 549	18 479	18 511	18 530	18 482	18 404	18 401

Livestock numbers (×1000 head)												
Dairy cows	3 188	3 233	3 138	2 909	2 865	2 847	2 770	2 682	2 667	2 715	2 602	2 587
Beef cows	1 088	1 765	1 308	1 373	1 495	1 599	1 666	1 699	1 751	1 775	1 805	1 829
Heifers in calf	776	927	879	834	793	756	733	762	798	771	771	813
Total cattle and calves	12 164	14 035	12 533	11 872	11 975	12 079	11 885	11 804	11 729	11 834	11 733	11 913
Ewes †	11 908	11 297	14 252	19 017	16 205	20 424	20 334	20 385	20 563	20 544	20 507	20 277
Lambs < 1 year						22 036	21 950	22 341	22 132	21 510	21 093	20 168
Total sheep and lambs	29 584	28 231	37 016	40 942	42 988	43 799	43 621	43 998	43 901	43 205	42 771	41 530
Breeding sows	733	743	716	703	660	660	678	672	687	680	644	639
Gilts in pig	131	139	108	101	97	109	107	108	115	104	100	105
Total pigs	7 473	7 908	7 937	7 980	7 509	7 449	7 596	7 609	7 756	7 797	7 534	7 496
Laying fowls	51 324	49 500	38 096	37 389	33 957	33 468	33 286	33 206	32 695	32 543	31 692	n/a
Table fowls	33 921	58 979	63 807	75 305	70 176	73 588	75 701	73 298	79 451	75 205	76 621	n/a
Growing pullets	23 143	18 214	12 502	11 236	9 411	10 452	11 016	10 764	10 653	10 293	10 098	n/a
Total poultry (millions)	121	140	121	131	120	124	127	124	130	126	126	n/a

* Including ponds, derelict, farm roads, yards and buildings.
† Including shearlings from 1990.
Source: MAFF.

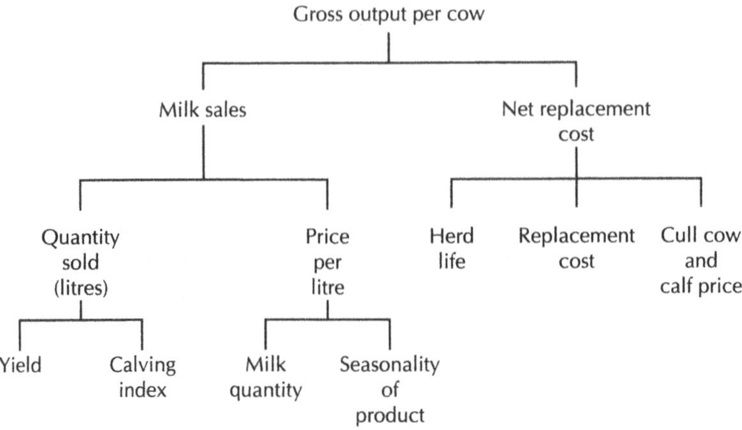

Fig. 6.1 Factors affecting the gross output from a cow.

has meant careful consideration of yield requirements per cow in relation to input factors such as concentrate usage.

Farmers must try to assess the factors over which they believe they have most influence and optimise performance in those areas.

The impact of improving performance in a particular area can be demonstrated by looking at the improvement in profitability if the calving index improves from 405 days to the optimum of 365 days, as illustrated in Example 6.1.

Example 6.1 The impact of calving index on the value of milk produced per annum per cow

Lactation yield (litres)	Price per litre (pence)	Calving index (days)		
		365	385	405
4000	20	800	758	721
4500	20	900	853	811
5000	20	1000	948	900
5500	20	1100	1043	991
6000	20	1200	1138	1081

The improvement in financial performance as a result of this improvement in husbandry is quite dramatic. If the lactation yield of a cow was 5000 l and the calving index was 365 days, then at 20p per litre, the milk sales would be £1000. If the calving index slipped to 405 days, then the value of milk would only be £900, a reduction of £100. The impact on a 100 cow herd would be dramatic – £10 000! Example 6.1 also illustrates that the milk sales from a cow with a lactation yield of 5000 l and a calving index of 405 days is the same as the milk sales from a cow with a lactation yield of 4500 l and a calving index of 365 days.

Gross output per cow is only one facet of profitability for milk production from a dairy herd. Profitability will also be influenced by the level of inputs both variable and fixed.

The variable costs, of which concentrates is the largest, are by definition variable and rather more under the short-term control of the good manager. The measurement and comparison of the efficiency of the use of variable inputs is an important tool in the fight for maximum profit. The impact of differing concentrate usage on milk yield in association with nitrogen usage and stocking density is illustrated in Table 6.2. The Table shows that output per cow can be increased by raising the concentrate input. This will improve output per cow and therefore output per hectare but initially a more profitable method of raising output per hectare is by increasing nitrogen input, and/or grassland management, so that stocking density can be improved. Where quotas restrict ultimate production, the exercise will be one of input cost reduction rather than output expansion, and again the better utilisation of forage will prove the most profitable path to follow.

Table 6.2 Interrelationship between milk yield, concentrates, nitrogen and stocking density

Fertiliser (N kg/ha)	125	250	375
Milk yield (l/cow)		Concentrates (t/cow)	
4500	1.35	1.15	1.0
5000	1.95	1.60	1.4
5455	2.80	2.25	2.0
Stocking density (cows/ha)	1.82	2.08	2.31

The overhead cost equation is in practice the most difficult to change in the short term. It is, however, important that the impact of decisions on size and mix of enterprises on overhead costs is assessed.

Figure 6.2 shows the relationship between gross output, variable and overhead costs resulting in the 'net margin' from a dairy enterprise which represents the enterprise's contribution to profit.

Similar analysis can be made of other livestock enterprises as shown in Figures 6.3 and 6.4.

Arable enterprises can be analysed in the same way as livestock enterprises. Table 6.1 shows the importance of wheat and barley production in the UK. Many farmers believe that growing cereals is relatively simple, compared to production of livestock and livestock products, but this is not necessarily the case; to perform at the highest level requires high skills and attention to detail.

Factors affecting the profitability of a cereal enterprise

As with livestock enterprises, profitability is affected by the level of output, which is dependent on yield and price, the input levels of variable and overhead costs associated with the enterprise. Figure 6.5 outlines the major factors affecting the profitability of a cereal enterprise, while Figs 6.6 and 6.7 highlight the factors affecting the profitability of potatoes and sugarbeet.

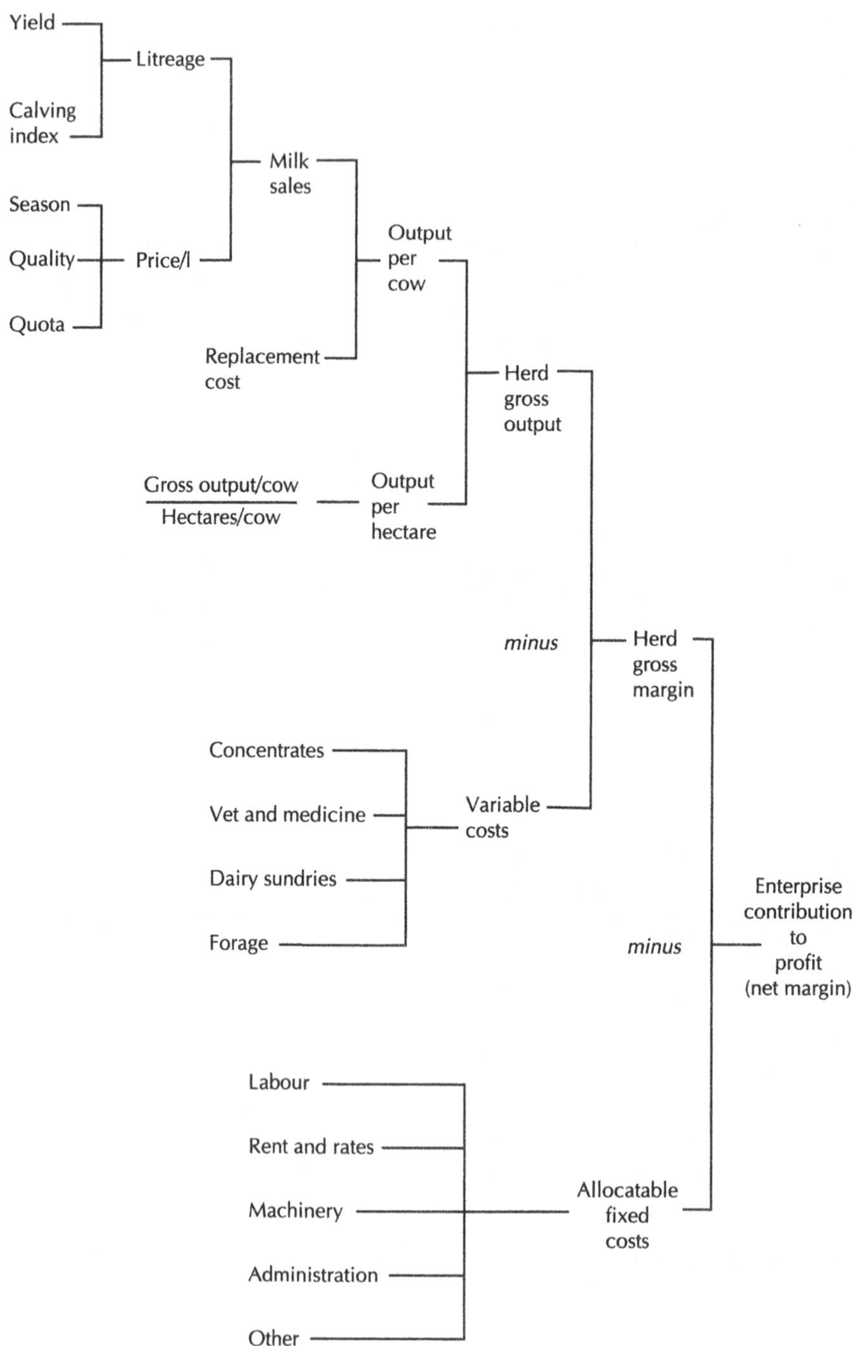

Fig. 6.2 Factors influencing the profitability of milk production.

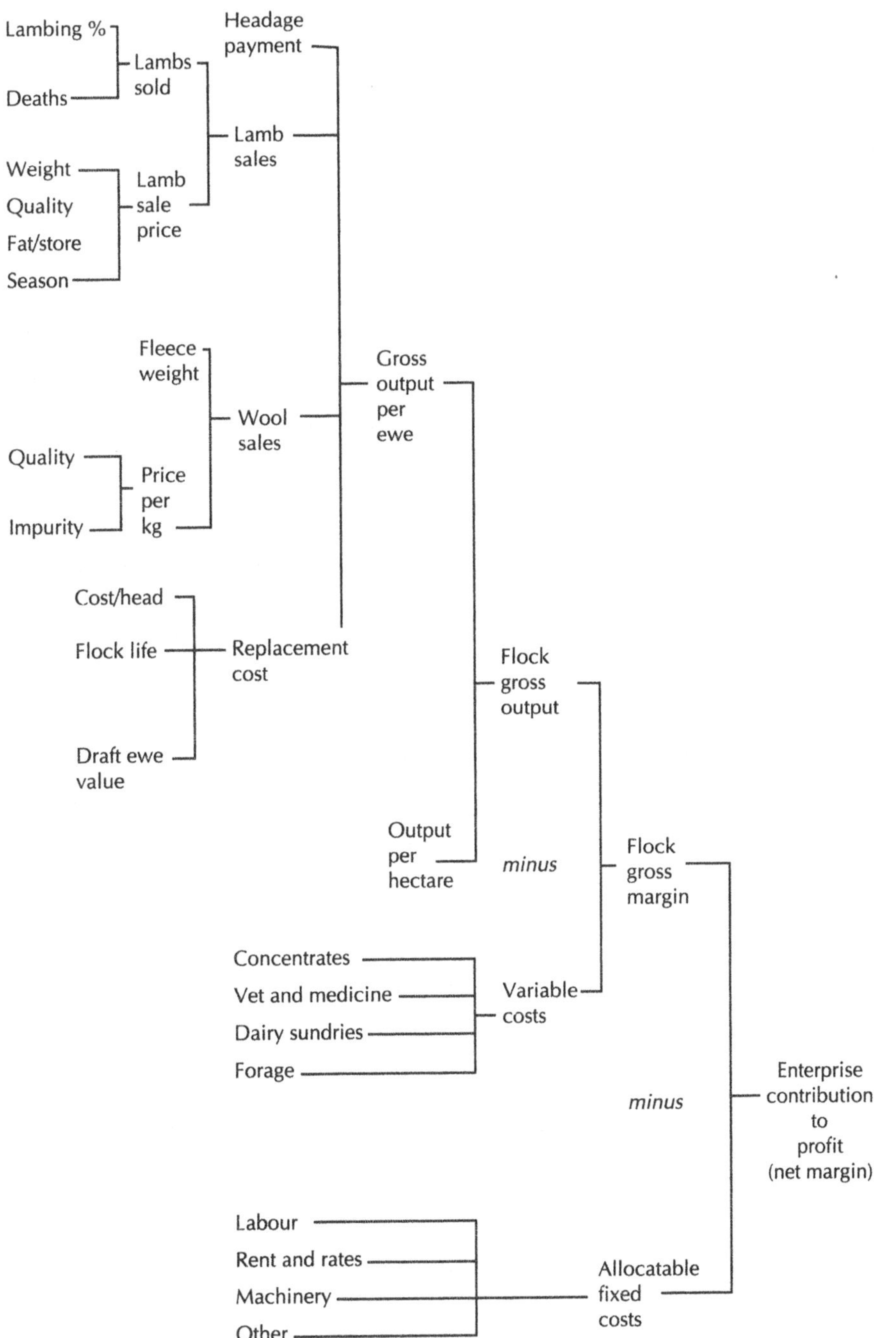

Fig. 6.3 Factors influencing the profitability of a sheep enterprise.

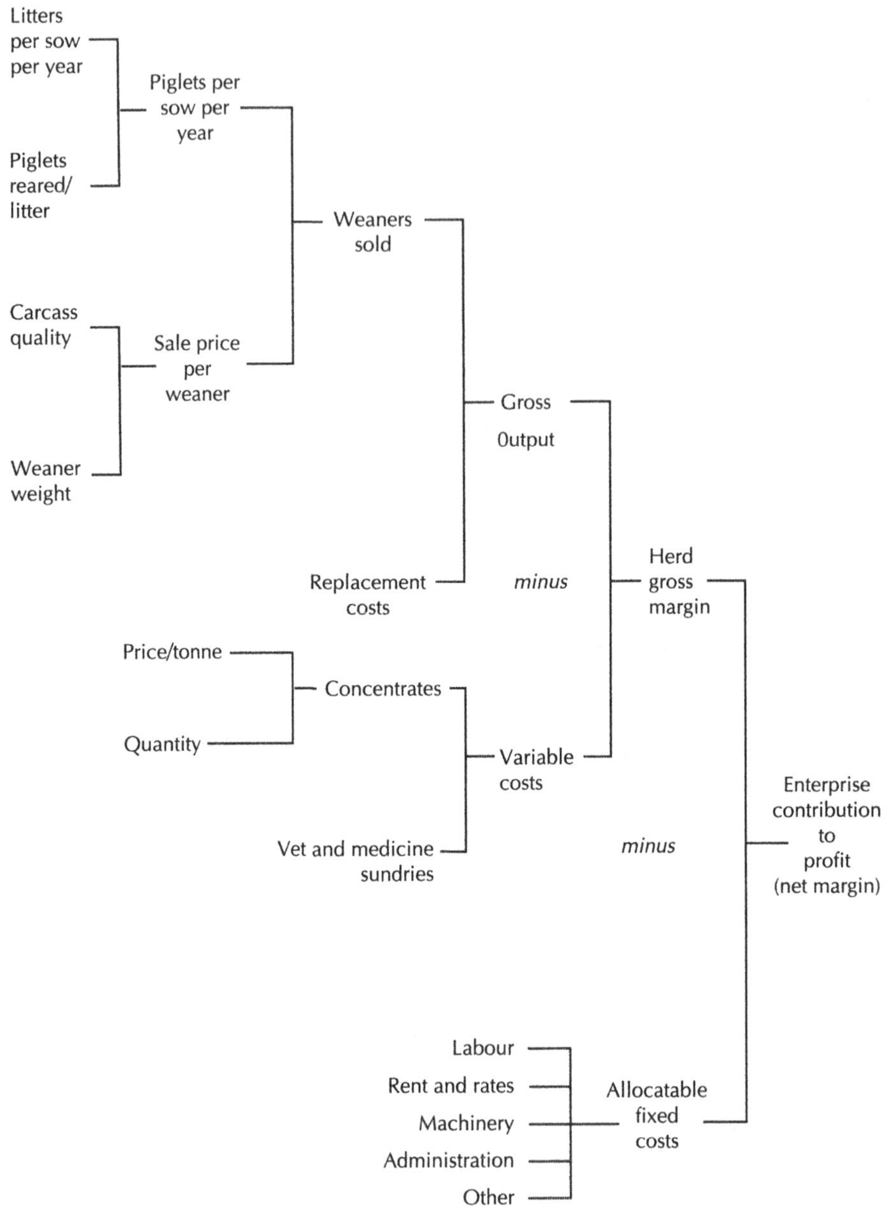

Fig. 6.4 Factors influencing the profitability of a sow and weaner unit.

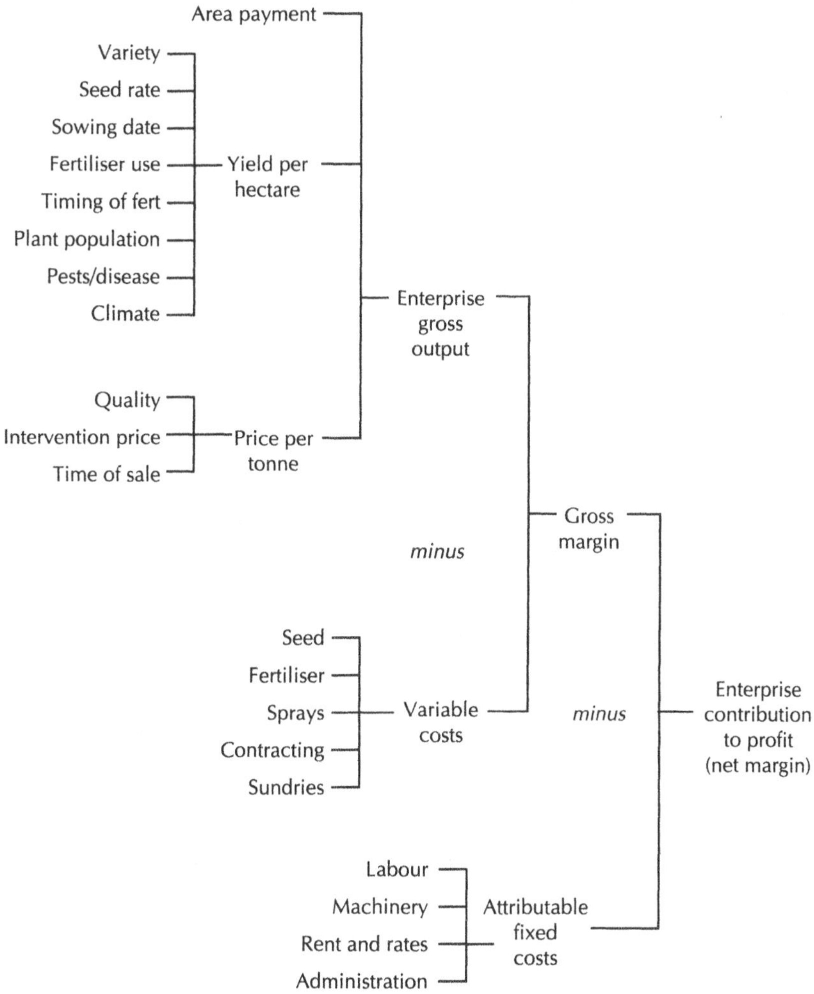

Fig. 6.5 Factors influencing the profitability of cereal production.

Enterprise efficiency within the context of the total farm business

Technical and financial enterprise efficiency are of paramount importance but it must be borne in mind, when looking at enterprise efficiency, that there are other considerations which may influence the ultimate enterprise mix on a farm and indeed the production criteria for a particular enterprise. For example, where land is a limiting resource, it may be economically viable to 'buy' land for a livestock enterprise, in the form of concentrates or roughage, to allow increased unit size. The performance per livestock unit might be lower but the enterprise return per hectare could be the optimum solution to a particular farm business problem.

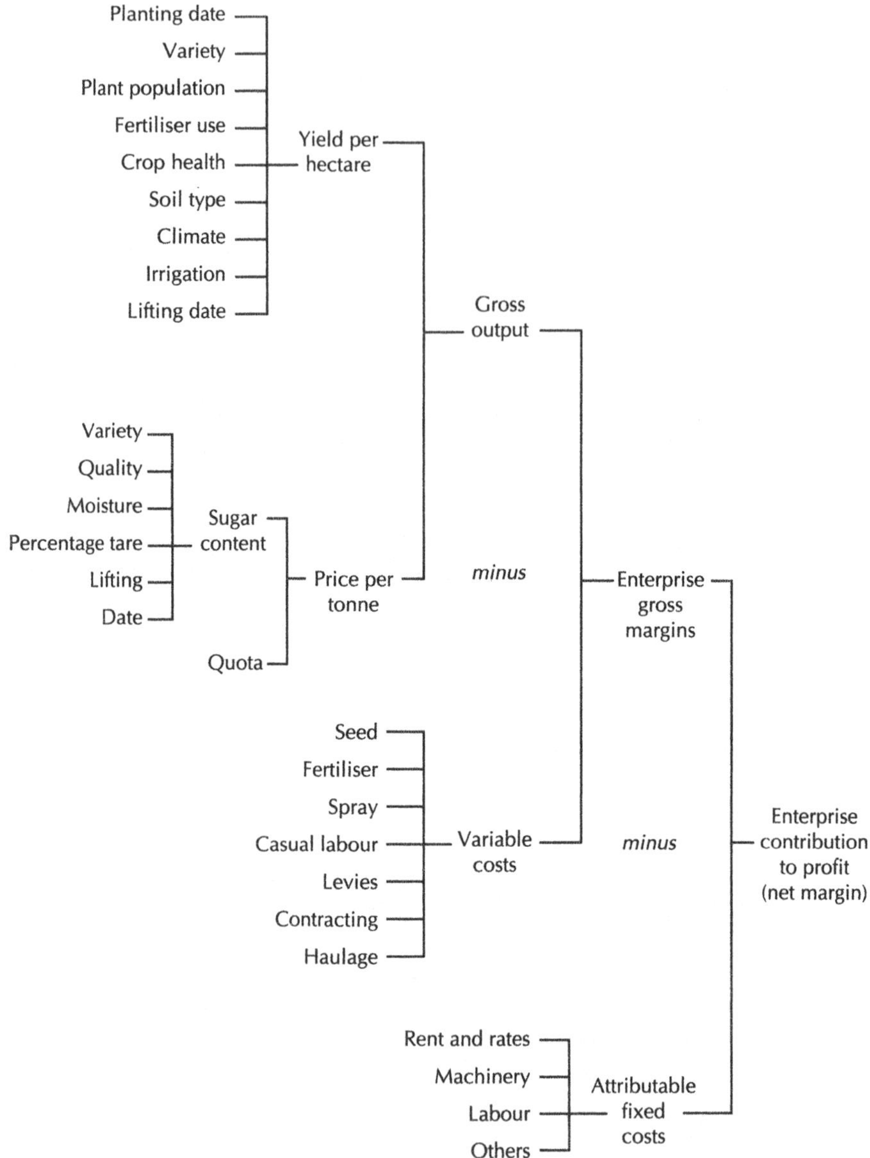

Fig. 6.6 Factors influencing the profitability of sugarbeet production.

Enterprise efficiency has to be related to the total resources available to the farm business. Indeed, it is important to consider individual enterprise efficiency against total farm enterprise mix. Where there are limiting resources (which will usually be the case), such as labour, land or capital, it will be necessary to optimise the return to that limiting resource. In some cases an individual enterprise will not in itself be

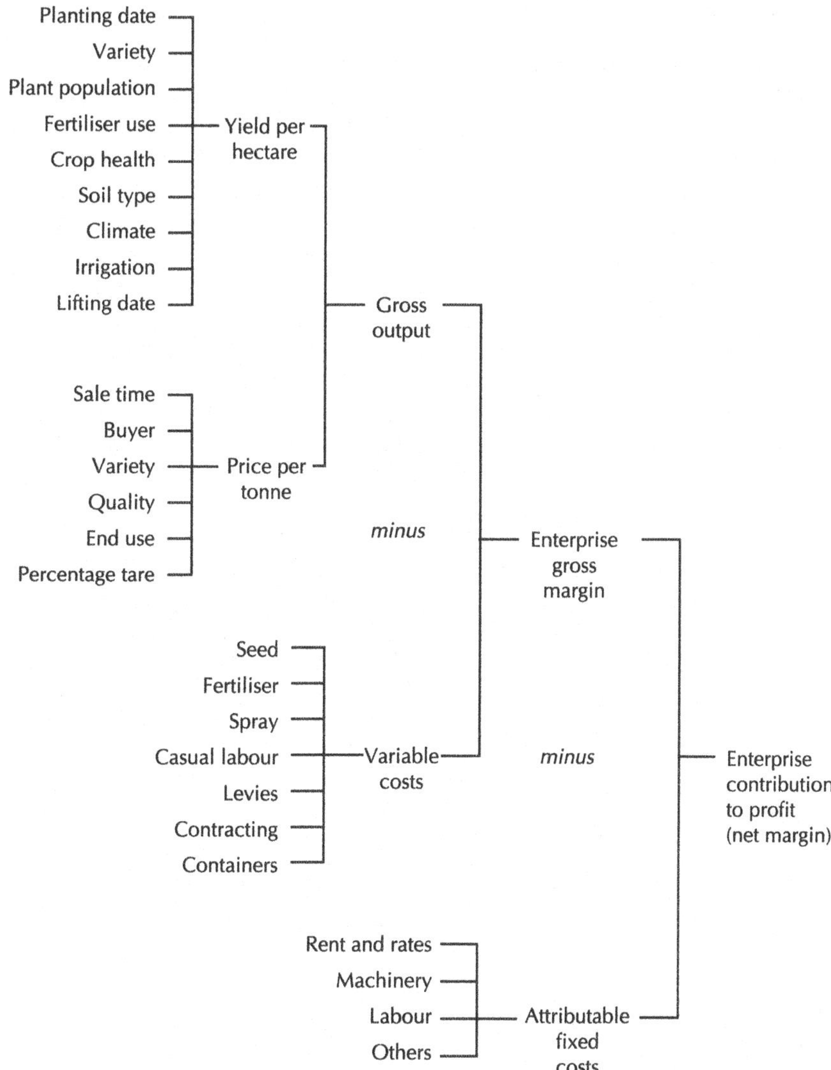

Fig. 6.7 Factors influencing the profitability of potato production.

giving its maximum return but in the context of the overall farm business profit will be optimised.

This does not diminish the importance of looking for the highest technical and financial efficiency from an enterprise within the parameters imposed on the overall farm business.

The impact of overhead costs on enterprise size and mix

The concept of overhead and variable costs was explained in Chapter 3. In looking at the technical and financial efficiency of an enterprise earlier in this chapter, overhead costs were ignored. However, in the final analysis they impinge on the degree to which a business can clinically look at maximising the technical performance of an individual enterprise and therefore maximising returns. Overheads are a limiting factor in determining the ultimate size of an enterprise; for example, the dairy parlour may limit the size of a dairy unit, although the financial returns per hectare from the dairy herd may be the best return from all enterprises on the farm.

It may be that the dairy enterprise performance is so good that it is capable of covering the increased costs of a bigger parlour and still producing the best return per hectare, although in practice this is unlikely.

The limitations on enterprise size and mix are frequently forced on a farm business by the overhead cost structures existing within that business.

The law of diminishing returns

The response of crops and livestock to the various inputs is explained by the law of diminishing returns (variable proportions). At first the yield response is very good (stage 1), then it decreases but remains positive (stage 2), and finally the yield falls (stage 3). It is possible to analyse the effect of any input but for crops nitrogen is usually measured, being the most important. The producer wishes to know how much nitrogen to use on the crop to obtain:

(1) The best physical relationship between nitrogen and yield: technical efficiency
(2) Maximum profit: economic efficiency

The principle of diminishing returns is illustrated in Example 6.2, with nitrogen being added to 1 ha of wheat. All other inputs are assumed to be at the optimum and are held constant.

Example 6.2 The response of wheat to different levels of nitrogen

Nitrogen/ha (40 kg)	Yield (kg)	Average yield (kg)	Marginal yield (kg)
1	900	900	900
2	2 900	1 450	2 000
3	5 000	1 666	2 100
4	7 000	1 750	2 000
5	8 500	1 700	1 500
6	9 500	1 583	1 000
7	10 250	1 464	750
8	10 750	1 343	500
9	10 750	1 194	0
10	10 000	1 000	–750

Average yield

The average yield is calculated by dividing the total yield by the lots of nitrogen applied. Thus at the yield of 2900 kg in Example 6.2 the average yield is

$$\frac{2900}{2} = 1450\,\text{kg}$$

Average yield is used as a measure of the physical (technical) response of wheat to nitrogen. Technical efficiency occurs where the average response is at its highest. This point can be determined from the data in Example 6.2 as being at 4 lots of nitrogen when total yield is at 7000 kg. The results of plotting the data are shown in Fig. 6.8.

Technical efficiency is achieved at yield OY using ON of nitrogen. This is at the point where the average yield is at its highest and denotes the end of stage 1 in the input:output relationship.

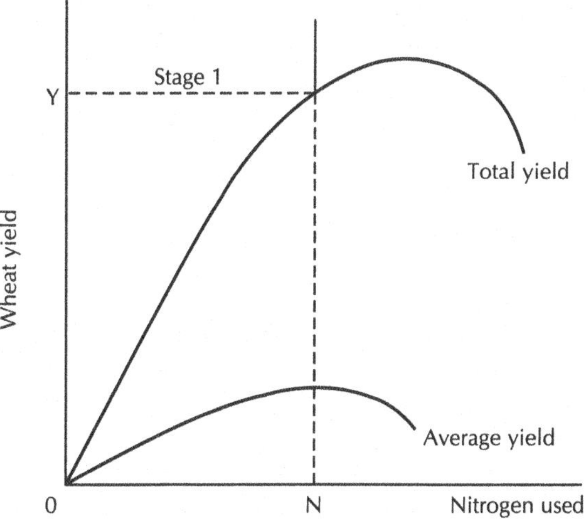

Fig. 6.8 Technical efficiency of nitrogen application.

Marginal yield

The marginal yield is the extra yield generated by using more nitrogen. Thus as the amount of nitrogen is raised from 3 to 4 lots, in Example 6.2, the marginal yield is

$$7000\,\text{kg} - 5000\,\text{kg} = 2000\,\text{kg}$$

When plotted it is important to note that marginal figures are drawn in between 0 and 1, 2 and 3, 3 and 4 lots of nitrogen, etc.

Maximum profit (economic efficiency)

Maximum profit is calculated by introducing the costs of inputs and value of output to the analysis. When this is done, with nitrogen at 35 p per kg, and wheat at £100 per tonne, it is possible to see where there is the greatest difference between the total value of wheat and the total cost of producing it. This can be measured and shown graphically, as in Fig. 6.9.

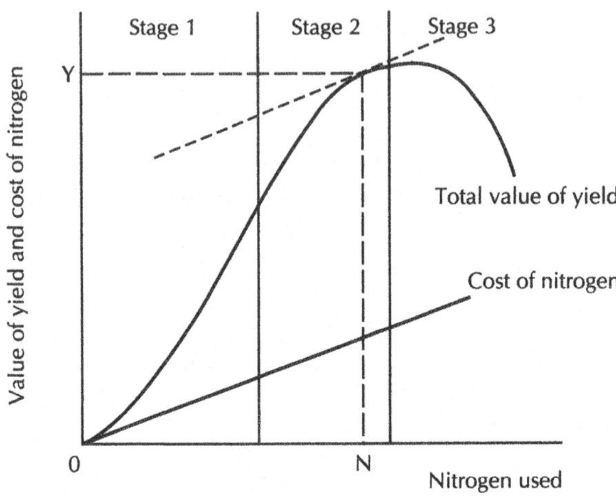

Fig. 6.9 Maximum profit (economic efficiency).

Maximum profit is achieved at a yield of OY using ON of nitrogen. This position is determined by drawing a line parallel to the cost line, and where it becomes tangential to the total value line, maximum profit will be made. This will occur in stage 2, and will normally not be at the same point as technical efficiency.

The farmer, however, is most likely to be interested in the effect of small changes in inputs to the levels he is currently using. Marginal analysis is therefore more relevant for farmers determining the profit maximising output than is total analysis.

Maximum profit (using marginal analysis) is measured by equating the cost of extra nitrogen (the marginal cost) to the value of the extra yield (marginal revenue) which it creates. When the extra nitrogen costs less than the value of extra yield, then it is worth using more nitrogen. If the extra nitrogen costs more than the value of the increased yield then it is not worth using it. Maximum profit occurs when the extra value of yield (marginal revenue) is equal to (or just greater than) the extra cost (marginal cost) of nitrogen used to create it. This can be illustrated in graphical form, as in Fig. 6.10.

Profit is maximised at OY using ON of nitrogen. It is important to note that determining the profit-maximising position using marginal analysis gives exactly the same answer as using total analysis.

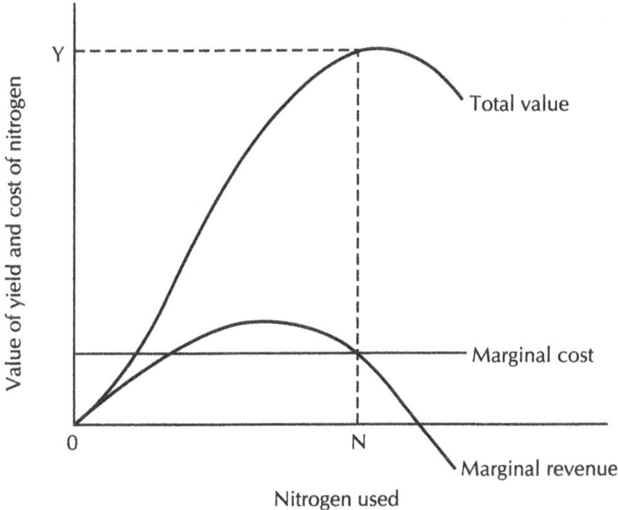

Fig. 6.10 Marginal analysis to determine the profit-maximising output.

Summary of diminishing returns

Stage 1 occurs between the origin and the highest point of the average yield curve. Here returns (yield increases) to the variable input (nitrogen) are increasing at an increasing rate. It is therefore worth going beyond stage 1 in terms of inputs used because the gains (marginal revenue) are greater than the costs (marginal cost).

Stage 2 occurs between stage 1 and stage 3 (surprise, surprise!). Here returns (yield increases) to the variable input (nitrogen) are decreasing but remain positive. The profit maximising output will occur somewhere in stage 2; exactly where will depend upon the relative cost of nitrogen and the value of wheat. The more expensive nitrogen is, the closer the profit-maximising output will be to stage 1. At present the relationship between nitrogen cost and the yield it creates is very good for farmers and it would take a large fall in wheat prices or an enormous increase in nitrogen cost to cause a reduction in nitrogen usage on farms.

Stage 3 occurs to the right of the highest point of the total value curve and since returns (and profit) are negative in this area it is obviously not a desirable place to produce.

Conclusion

Maximum profit will occur in any enterprise where

Marginal revenue is $\geqslant$ marginal cost

Sensitivity analysis

We live in an imperfect world and unfortunately the best-laid plans and budgets are only as good as the assumptions behind them.

In Chapter 8, the concept of variance analysis is discussed as a method of analysing variance from budget. Sensitivity analysis is a method of looking, in advance, at the impact of variance from budget on ultimate profitability. The impact of changes in performance level of key factors on enterprise efficiency and profitability should be considered. For example, with a dairy herd it would be sensible to look at the effect of an increase or decrease in concentrate usage on returns or an increase or decrease in stocking rate, as illustrated in Table 6.2.

A similar exercise can be done on a cereal enterprise where the impact of an increase or decrease in yield on budget performance can be assessed. It is also possible to look at the yield:price relationship or the input:output relationship to maintain performance, as shown in Figs 6.11 and 6.12.

In Fig. 6.11 the price of milling wheat achieved is £125 per tonne. The following year the farmer, budgeting forward, plots a histogram correlating the percentage yield increase necessary to maintain his previous returns, if the price per tonne falls. If the price falls to £120 per tonne, then a 4% increase in yield will be necessary to maintain income. Should the price fall to £80 per tonne, then a 56% increase in yield will be required to maintain income.

In Fig. 6.12 the relationship between costs and price is illustrated. Assuming that the yield remains at last year's level (7.3 tonnes per hectare) as the price falls from £125 per tonne to £100 per tonne, then to maintain income (as last year) costs will have to be reduced by £182 per hectare. Clearly this is unlikely and income is going to fall.

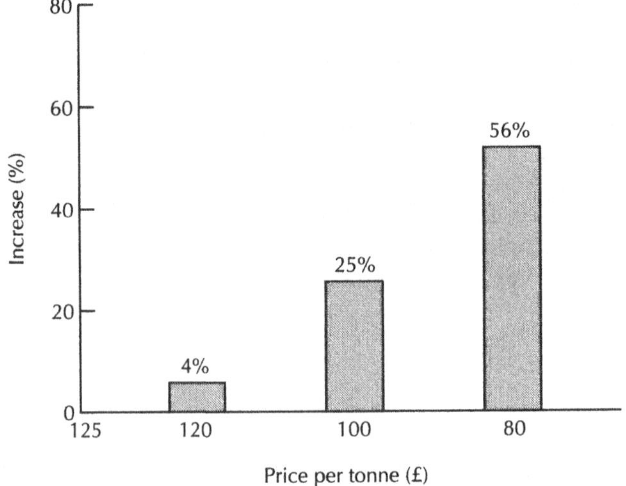

Fig. 6.11 Yield increase to maintain income as price falls (excluding area payments).

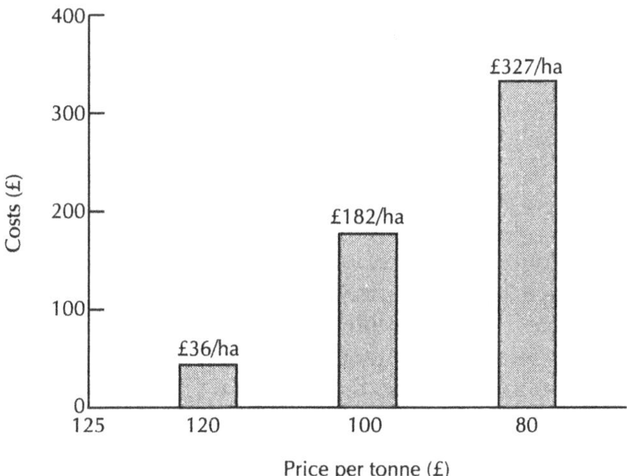

Fig. 6.12 Cost reduction to maintain income as price falls (excluding area payments).

Summary

The importance of understanding the factors which will affect the ultimate profitability of an enterprise cannot be overemphasised. It is the physical efficiency levels which will determine the gross margins and ultimately the level of profitability of the business.

Whilst an understanding of the factors influencing profitability is fundamental to the success of the production process, it has to be remembered that even the best-organised businesses will be subject to variations in performance, both in terms of physical output (e.g. yield of wheat or milk), market returns and input prices. It is therefore important to look at the sensitivity of the business to possible changes in these areas.

This chapter has looked only briefly at the subject of enterprise studies. Further details can be investigated in the other texts which are available.

7 Planning: Profitability

In this book the viability of a business is constantly under review – from performance in the market place to performance in the field, farm office and world of high finance. The techniques for establishing profitability (Chapter 7), feasibility (Chapter 8) and worthwhileness (Chapter 9) are explained in detail. The modern businessman needs to address each of these separate yet interdependent facets of the farm business if he is to progress in today's economic and political environment.

What constitutes a viable farm business?

The *Concise Oxford Dictionary* defines viability in terms of living things. A plant or animal is viable if it is able to live or exist in a particular climate or environment. The viability of a business can be defined in a similar way. A business has to be able to survive in the political, social and economic environment in which it operates (see Chapter 1).

There are three key indicators of the viability of a business:

(1) Profitability
(2) Feasibility (cash flow)
(3) Worthwhileness (return on capital)

It is important to test the past, present and potential future performance of a business in these areas.

Profitability

For any business to survive in the medium to long term it has to make and retain profits on an annual basis. Income must exceed expenditure (including depreciation, tax, personal drawings and any expansion).

Feasibility (cash flow)

No business can survive in the short, medium or long term unless it has sufficient cash to fund its trading operation. In the short term cash flow is more important than profitability or worthwhileness although, in the final analysis, all three ingredients are essential to the survival of a business.

Worthwhileness (return on capital)

In order to survive and grow, a business must show an acceptable return on the money (capital) invested in it. A business which over the medium and long term is not making a sensible return on its capital (between 8 and 20%) will be restricted in its development and will probably be unable to fund the expansion which is necessary simply to stand still in an inflationary world.

So far (Chapter 4 and 5) we have only looked into the past to analyse the business. Now it is time to look towards the future, which is far more exciting!

We have seen in Chapters 4 and 5 how accounts can be used to analyse the past performance of a farm business. Enterprise input and output information can be compared with local and regional standards; physical and financial inefficiencies can be identified.

In Chapter 6 the main factors which influence the profitability of individual enterprises have been indicated and realistic levels of achievement that could be expected are illustrated in Chapters 3 and 4 (Tables 3.2 and 4.2).

This chapter shows how reasonable husbandry improvements can be incorporated into a farm budget. It also examines other opportunities for change and introduces the concept of the 'partial budget'.

Consider Example 7.1 which is a gross margin account. It is a 'normalised' (the average performance over the last three years) account based on the present cropping and stocking regime and the present levels of husbandry performance of a 269 ha farm, having £1750 capital invested per hectare.

Weaknesses already identified, by using the techniques described in Chapter 4, can now be adjusted to realistic performance levels. The effect is staggering, as shown in Example 7.2.

The new accounts are based on the same area and combination of cropping and stocking and the same input/output, cost and price structure. The only difference is that these adjusted accounts incorporate improvements to enterprises' physical performance and (perhaps overhead costs) that could reasonably be expected. Profit, in this case, can be expected to increase by about £48 000. Of course the new standards of performance have still to be achieved. That is what management is all about. If the decision is taken to improve levels of husbandry to a higher degree of excellence, then subsequent performance must be carefully monitored to ensure that the targets set are actually being met.

Adjusting the normalised accounts is an essential step in farm planning for two reasons:

(1) It measures, as shown, the effect that improved husbandry can have on farm profit
(2) It establishes a 'bench-mark' profit level for subsequent farm reorganisation. Unless this is done, the full financial implications of replanning the farm business cannot be identified

In some cases, this stage marks the end point of a farm planning exercise. What must be done is to take action, make the proposed husbandry changes and monitor them closely to check that the desired effects are achieved.

Example 7.1 Bleak Farm gross margin account (year ending 30 September 1997)

Cropping	*ha*	*Livestock*	*Numbers*	*GLU*
Wheat: winter	30	Dairy cows	150	150 (150 @ 1.0)
Barley: winter	45			
Barley: spring	35	Sheep: ewes	300	39 (300 @ 0.13)
Grass: silage	50	lambs	435	21 (435 @ 0.05)
Grass: grazing	80			
Set-aside	19			
Roads and waste	10	Pigs: sows	45	
		bacon	810	
Total	269			210
				Stocking density: 1.28 GLU/ha

Enterprise	*Wheat*		*Barley*	*(winter)*	*Barley*	*(spring)*
	per ha	total	per ha	total	per ha	total
Hectares		30		45		35
Yield	5.2	156	5.0	225	4.5	158
Price £/tonne	96.0		92.0		100.0	
Sales	499.2	14 976	460.0	20 700	450.0	15 750
Area payment	145.4	4 362	145.4	6 543	145.4	5 089
Gross output	644.6	19 338	605.4	27 243	595.4	20 839
Seed	30	900	30	1 350	28	980
Fertiliser	50	1 500	35	1 575	34	1 190
Sprays	25	750	40	1 800	28	980
Other	16	480	6	270	0	0
Total variables	121	3 630	111	4 995	90	3 150
Gross margin	524	15 708	494	22 248	505	17 689

	Grass silage				Grass grazing	
	per ha	total			per ha	total
Hectares		50				80
Sales		0				0
Gross output		0				0
Seed	12	600			18	1 440
Fertiliser	90	4 500			80	6 400
Sprays	20	1 000			12	960
Total variables	–122	6 100			110	8 800
Gross margin	–122	–6 100			–110	–8 800

Enterprise	Cows		Sheep		Sows	
	per cow	total	per ewe	Total	per sow	total
Average number		150		300		45
Yield: litres	5 500	825 000				
lamb %			145	435		
piglets					18	810
Price/unit	20					
Sales: (milk)	1 100	165 000				
(wool)			3.30	990		
Headage payment			18.00	5 400		
Stock sold		145		365		0
Price	120.8	18 125	48.66	14 600		0
Transfers out		0		70		810
Value			9.33	2 800	450	20 250
Culls sold		41		66		16
Price	136.6	20 500	4.62	1 386	14	640
Sub-total	1 357	203 625	83.91	25 176	464	20 890
Livestock in		41		69		18
Price	219	32 800	11.50	3 450	24	1 080
Gross output	1 138	170 825	72.41	21 726	440	19 810

Example 7.1 *(continued)*

Enterprise	*Cows*		*Sheep*		*Sows*	
	per cow	total	per ewe	total	per sow	total
Concentrates	0.28 kg/l		0.05 t/ewe		1.8 t/sow	
Price/tonne (£)		133		140		150
Cost (£)	205	30 723	7	2 100	270	12 150
Bulk feed (£)	10	1 500				
Vet and AI (£)	40	6 000	1.5	450	25	1 125
Other (£)	15	2 250	2.3	690	10	450
Total variables excluding forage	270	40 473	10.8	3 240	305	13 725
Gross margin excluding forage	868	130 352	61.61	18 486	135	6 085
Forage costs (£) (see below for calculation)	71	10 642	14.2	4 258		
Gross margin after forage	797	119 710	47.41	14 228		

Forage cost allocation

	Silage: cost £6 100	Grazing: costs £8 800	Total £
Cows:	$\frac{150}{210}$ GLU × £6 100 = £4 357	$\frac{150}{210}$ × £8 800 = £6 285	10 642
Sheep:	$\frac{60}{210}$ GLU × £6 100 = £1 743	$\frac{60}{210}$ × £8 800 = £2 515	4 258

Enterprise	*Bacon* £/head	*Pigs* Total	
Sold or transferred: 753	60	45 180	
Sub-total	60	45 180	
Stock transferred in: 810	25	20 250	
Gross output	35	24 930	
Feed	29.65	22 326	
Vet and medicine	1.50	1 130	
Other	1.30	979	
Total variables	32.45	24 435	
Gross margin	0.66	495	(based on an average transfer number of 753)
Farm gross margins	£/ha/head	Total £	
Wheat	524.00	15 708	
Barley winter	494.00	22 248	
Barley spring	505.00	17 689	
Dairy cows	797.00	119 710	
Sheep	47.41	14 228	
Pigs: sows	135.22	6 085	
bacon	0.66	495	
Set-aside	261.60	5 072	
Total gross margin	748.09	201 235	

Example 7.1 *(continued)*

	£/ha	Total £
Overhead costs		
Labour	137.54	37 000
Machinery depreciation	18.17	4 889
repairs	31.92	8 587
fuel and power	32.77	8 816
contractor	11.90	3 200
miscellaneous	2.38	638
Building repairs/depreciation	13.54	3 642
General insurance	5.79	1 557
Miscellaneous	19.72	5 306
Rent (notional)	56.97	15 326
Total	330.71	88 961
Net farm income (total gross margin – overhead costs)	417.38	112 274
Less physical labour	22.30	6 000
Management and investment income (MII)	395.08	106 274
Add		
Notional rent	56.97	15 326
Deduct:		
Loan/mortgage interest	95.72	25 750
Farm profit	356.32	95 850

Variable costs	£
Seeds	5 270
Fertilisers	15 165
Sprays	5 490
Crop misc.	750
Concentrates	67 299
Bulk feed	1 500
Vet and AI	8 705
Total	104 179

Tenant's capital £1750/ha

MII as % return on capital 22.6%

Note: there are marginal differences between farm total figures and £/ha due to rounding of the figures.

Example 7.2 Bleak Farm gross margin account (year ending 30 September 1997)

Cropping	*ha*	*Livestock*	*Numbers*	*GLU*
Wheat: winter	30	Dairy cows	150	150 (150 @ 1.0)
Barley: winter	45			
Barley: spring	35	Sheep: ewes	300	39 (300 @ 0.13)
Set-aside	19	lambs	435	21 (435 @ 0.05)
Grass: silage	50			
Grass: grazing	80			
Roads and waste	10	Pigs: sows	45	
		bacon	875	
Total	269			210
				Stocking density: 1.62 GLU/ha

Enterprise	*Wheat*		*Barley*	*(winter)*	*Barley*	*(spring)*
	per ha	total	per ha	total	per ha	total
Hectares		30		45		35
Yield	7.28	218	6.99	315	5.50	193
Price £/tonne	96.00		92.00		100.00	
Sales	699.88	20 966	643.08	28 939	550.00	19 250
Area payment	145.40	4 362	145.40	6 543	145.40	5 089
Gross output	844.28	25 328	788.48	35 482	695.40	24 339
Seed	43	1 290	30	1 350	32	1 120
Fertiliser	75	2 250	67	3 015	55	1 925
Sprays	55	1 650	53	2 385	45	1 575
Other	6	180	6	270	0	0
Total variables	179	5 370	156	7 020	132	4 620
Gross margin	665	19 958	632	28 462	563	19 719

Example 7.2 *(continued)*

Enterprise	*Grass silage*		**Grass grazing**	
	per ha	total	per ha	total
Hectares		50		80
Sales		0		0
Gross output		0		0
Seed	12	600	18	1 440
Fertiliser	90	4 500	80	6 400
Sprays	22	1 000	12	960
Total variables	122	6 100	110	8 800
Gross margin	–122	–6 100	–110	–8 800

Enterprise	*Cows*		*Sheep*		*Sows*	
	per cow	total	per ewe	total	per sow	total
Average number		150		300		45
Yield: litres	6 909	1 036 350*				
lamb (%)			174	522		
piglets					20	900
Price/unit	20 p					
Sales (milk)	1 381.80	207 270				
(wool)			3.30	990		
Headage payment			18.00	5 400		
Stock sold		145		452		0
Price	120.8	18 125	60.29	18 080		0
Transfers out		0		70		900
Value			9.33	2 800	500	22 500
Culls sold		41		66		16
Price	136.6	20 500	4.62	1 386	14	640
Sub-total	1 639.2	245 895	95.54	28 656	514	23 140

* It is assumed that the farm had its own quota available to allow an increase in production without leasing in or buying. Leasing at 10p per litre would reduce dairy returns and profit by £21 135.

Livestock in		41		69		18
Price	219	32 800	11.5	3 450	24	1 080
Gross output	1 420	213 095	84.04	25 206	490	22 060
Concentrates:	0.32 kg/1		0.05 t/ewe		1.8 t/sow	
Price/tonne(£)		133		140		150
Cost (£)	294	44 107	7	2 100	270	12 150
Margin/concs (£)	742					
Bulk feed (£)	10	1 500				
Vet and AI (£)	40	6 000	1.5	450	25	1 125
Other (£)	15	2 250	2.3	690	10	450
Total variables excluding forage	359	53 857	10.8	3 240	305	13 725
Gross margin excluding forage	1 061	159 238	73.24	21 966	185	8 335
Forage costs (see below for calculation)	(£) 71	10 642	14.2	4 258		
Gross margin after forage	990	148 596	59.04	17 708		

Forage cost allocation

	Silage: cost £6 100	Grazing: costs £8 800	Total £
Cows:	$\frac{150}{210}$ GLU × £6 100 = £4 357	$\frac{150}{210}$ × £8 800 = £6 285	10 642
Sheep:	$\frac{60}{210}$ GLU × £6 100 = £1 743	$\frac{60}{210}$ × £8 800 = £2 515	4 258

Example 7.2 *(continued)*

Enterprise	*Bacon* £/head	*Pigs* Total	
Sold or transferred: 875	60	52 500	
Sub-total	60	52 500	
Stock transferred in: 900	25	22 500	
Gross output	35	30 000	
Feed	29.65	25 944	
Vet and medicine	1.50	1 313	
Other	1.30	1 138	
Total variables	32.45	28 395	
Gross margin	1.83	1 605	(based on an average transfer number of 875)

Farm gross margins	£/ha/head	Total £
Wheat	665	19 958
Barley winter	632	28 462
Barley spring	563	19 719
Dairy cows	990	148 596
Sheep	59.04	17 708
Pigs: sows	185	8 335
bacon	1.83	1 605
Set-aside	261.7	5 075
Total gross margin	927.34/ha	249 455

Overhead costs			*Variable costs*	
Labour	137.54	37 000		£
			Seeds	5 800
Machinery depreciation	18.17	4 889	Fertilisers	18 090
repairs	31.92	8 587	Sprays	7 570
fuel and power	32.77	8 816	Crop misc.	450
contractor	11.90	3 200	Concentrates	84 038
miscellaneous	2.38	638	Bulk feed	1 500
Building repairs/depreciation	13.54	3 642	Vet and AI	8 888
General insurance	5.80	1 557	Total	126 336
Miscellaneous	19.72	5 306		
Rent (notional)	56.97	15 326		
Total	330.71	88 961		
Net farm income (total gross margin – overhead costs)	596.63	160 494	Tenant's capital £1 750/ha	
			MII as % return on capital 32.82%	
less physical labour	22.30	6 000		
Management and investment income	574.33	154 494		
Add:				
Notional rent	56.95	15 326		
Deduct:				
Loan/mortgage interest	95.72	25 750		
Farm profit	535.58	144 070		

Note: there are marginal differences between farm total figures and £/ha due to rounding of the figures.

It is more likely, however, that the critical analysis of the business (Chapter 4) will have revealed indications that even greater profits could be made by making further changes to the business.

The next step in planning is to look at the opportunities that could exist and the restrictions which might stand in their way.

Opportunities and constraints

Before discussing the possible courses of action which are open to a farmer, three main questions must be asked:

(1) Are there any further alternative methods of production at which it could be worth looking?
(2) What present enterprises might be increased or reduced in scale, or even discarded altogether?
(3) What additional enterprises could be adopted and what performance levels might be achieved?

In each case it is essential to investigate the effect on gross margins, overhead costs, the demands on the scarce resources of the farm business and the inter-relationship between each enterprise.

Similarly, there will be restrictions on some of the proposed changes which will help determine the final plan. These can best be classified under five broad headings. The five constraints are:

(1) Land and rotational limitations
(2) Institutional limits in the form of contracts and quotas
(3) Restrictions imposed by the availability of labour, machinery and buildings
(4) Cash or credit limits
(5) Personal ability, preferences and dislikes

Profitability

The objective of farm planning is to deploy the available resources of land, labour and capital to produce the best profit, subject to the opportunities and constraints that apply to the business. Many techniques are available to this end, some of which are described in this chapter and others in Chapter 10.

The simplest technique and, because of it, perhaps the most widely used in farm management, is that of partial budgeting.

Partial budgets

Partial budgeting is a quick and useful method of evaluating the average annual effect on farm profit of (usually but not necessarily) minor changes to the business:

changes which do not have a major effect on the overhead resources or the overall structure of the farm, but which have shown in farm business analysis the need for further study.

The procedure is based simply on the balance of credits and debits likely to result from the proposed change, just as they would appear in the farm trading and profit and loss account (so capital amounts invested are not included, but the annual costs of making that capital investment are included). The main difference is that in partial budgeting we are endeavouring to calculate the 'average' expected annual profit and not a particular year's profit. For this reason, average capital invested is used when calculating interest charges attributable to the project, and not year 1 charges, when the outstanding overdraft balance would attract the highest interest charges. Neither can we disregard interest, as might happen in the final year of the project, where principal has been fully repaid. Other than this acceptable principle, the technique of partial budgeting follows the basic rules of a trading and profit and loss account, except that it isolates and includes only those items which are involved in the proposed change. That is why it is called a partial budget.

Structure and layout of a partial budget

The basic information of the 'losses' and 'gains' to the business can be listed under the following headings:

Losses		Gains	
	£		£
Extra costs		Costs saved	
Revenue foregone		Extra revenue	
Gain		Loss	
Total		Total	

The next step is to inspect the items that must be included within each heading.

Extra costs

Depreciation

Investment in buildings, drainage or machinery which might be made to accomplish the change will attract interest and depreciation charges. For management purposes, depreciation is relatively simple.

Buildings and other long-term items, such as drainage, are written off over ten years in farm planning, which implies a flat rate charge of 10% per annum. Machinery is depreciated over five years, i.e. 20% per annum (as described in Chapter 3).

In some ways, depreciation is really an unusual item, for it does not appear in the cash flow, yet is a definite charge against profit. In its simplest form, it can be looked on as the amount of money a farmer must set aside each year to replace the machine or building when it has reached the end of its useful life. However, the money is normally put into the farm bank account, which effectively reduces the overdraft on the project in hand. For management purposes, therefore, depreciation can be 'likened' to replacement of principal, which means annual interest charges are decreased annually.

Interest charges

Interest charges in a partial budget are accordingly based on the average capital involved for buildings and machinery. It is quite easy to calculate this charge, for all that is needed is to halve the value of the investment, and then charge the product at the going interest rate, thus arriving at an average interest charge. For livestock, however, since the investment in the herd or flock remains constant (because cull animals are replaced), interest is charged on the full capital.

Other extra costs

Extra costs may include such items as extra labour or overtime or contractors' charges. Additional overhead costs of fuel, repairs, insurance, rent and others may also occur.

Exclusions from a partial budget

Variable costs, like fertilisers, seeds, sprays and feeding stuffs, can be excluded from a partial budget if gross margins are used, rather than using all the inputs and outputs that would normally appear in the trading account (this is further explained under 'revenue foregone').

Revenue foregone

If enterprises are reduced in scale or eliminated, output is lost and revenue is sacrificed, but costs of production are saved.

We can better use the gross margins, which have already been calculated, and thus avoid using all this input/output information in the budget. One of the joys of using the gross margin system is that calculations of this nature become much easier.

Costs saved and extra revenue

Each item involved is identical to the 'losses' side of the partial budget and is treated in the same way. The result is an overall gain or loss to the profit figure.

Example 7.3 shows how a partial budget can be applied to a real farm situation, to help the farmer arrive at a decision.

A farmer currently grows 13 ha of sugarbeet, 12 ha of potatoes, 50 ha of winter wheat and 25 ha of spring barley. The gross margins are:

sugarbeet £600/ha
potatoes £1000/ha
w. wheat £500/ha
s. barley £400/ha

He has an opportunity to expand his potato contract to 20 ha but labour shortage would mean that he must discard the sugarbeet enterprise. He can get £2000 for the equipment and £25 ha per year for renting the quota. He can buy an extra 8 ha of potato quota costing £1600 in total. Disposing of the sugarbeet enterprise releases enough autumn labour to grow the extra potatoes and increase the winter wheat by 10 ha, allowing him to reduce the spring barley by 5 ha. There are no set-aside problems.

He will have to buy another potato harvester; net cost £6000. The storage buildings will cost £10 000 to accommodate the extra potatoes; there is adequate grain storage.

Interest rate is 10% per annum. Machinery is depreciated at 20% and buildings at 10% per annum. Should he make the change?

Example 7.3 A partial budget to determine the profitability of a change

Losses	£	Gains	£
Extra overhead costs		*Overhead costs saved*	
Machinery bought £6000		Machinery sold £2000:	
Depreciation 6000 × 20%	1 200	Depreciation 2000 × 20%	400
Interest on average		Interest on average	
Capital $\frac{6000}{2}$ × 10%	300	capital $\frac{2000}{2}$ × 10%	100
Buildings cost £10 000			
Depreciation £10 000 × 10%	1 000		
Interest on average			
capital $\frac{10\,000}{2}$ ×10%	500		
Sub-total	3 000		500
Income foregone		*Extra income*	
13 ha sugar beet at gross margin £600/ha	7 800	8 ha potatoes at gross margin £1 000 ha	8 000
5 ha spring barley at gross margin £400/ha	2 000	10 ha winter wheat at gross margin £500/ha	5 000
Sub-total	9 800		13 000
Gain	700		
Total	13 500		13 500

The partial budget above shows that he will lose revenue in the form of gross margins from 13 ha of sugarbeet and 5 ha of spring barley. On the other hand, he gains 8 ha of potatoes and 10 ha of winter wheat.

Extra costs occur in additional interest and depreciation associated with the new investment in buildings and machinery, but he saves interest and depreciation on the machinery he is selling. The resulting partial budget shows a gain in favour of the change of £700.

The principles shown and used in this example follow the accepted procedure in partial budgeting.

Of course there are short cuts, as there are in most accounting methods, one of which could be used in the partial budget in Example 7.3. For instance, the items of machinery bought and sold appear on both sides of the equation under interest and depreciation charges lost and gained. It would be simpler to use the net cost of machinery, i.e. £4000, and charge the net extra interest and depreciation as a single entry on the losses side of the budget. The result is shown in Example 7.4.

Example 7.4

Losses	£	Gains	£
Extra overhead costs		*Overhead costs saved*	
Machinery net cost £4000			
Depreciation 4000×20%	800		
Interest on average			
capital $\frac{4000}{2} \times 10\%$	200		
Building cost £10 000			
Depreciation 10 000×10%	1 000		
Interest on average			
capital $\frac{10\,000}{2} \times 10\%$	500		
Sub-total	2 500		—
Income foregone		*Extra income*	
13 ha sugar beet at gross margin £600/ha	7 800	10 ha winter wheat at gross margin £500/ha	5 000
5 ha spring barley at gross margin £400/ha	2 000	8 ha potatoes at gross margin £1000/ha	8 000
Sub-total	9 800		13 000
Gain	700		
Total	13 000		13 000

'Grey' areas in partial budgets

Small quirks exist in budgeting which may confuse beginners, but are not difficult to understand and apply. An example of this occurs in the interest charge calculation.

Interest charges included in partial budgets are calculated on average capital for items which depreciate, such as machinery, buildings and other investments. These are written off within a specified period for the reasons outlined earlier.

There are, however, some investments which do not attract depreciation charges, such as a dairy cow or a breeding sow, although clearly 'she' could be written off over her life in the herd and discarded like a piece of machinery. We choose to calculate herd or flock depreciation in a different way from that which is used for machinery.

In Chapter 3 herd replacement cost was incorporated into the gross margin calculation, which means that a breeding livestock enterprise maintains a constant age (and therefore constant value) during its lifetime. In turn this means that in a partial budget interest charges should be based on the total initial investment, which in this case also happens to be the average investment in a breeding unit.

A similar problem arises with working capital. In any single trading year the working capital required for growing the crops, feeding the stock and other items, such as paying wages, is slowly built up to a peak. It can be argued that in a partial budget the average working capital employed should be used for the calculation of interest charges. In most farm businesses, however, especially those involving cereals and beef, working capital will extend well into a second farming year, which is why we choose to use total working capital when calculating interest charges in a partial budget.

We also acknowledge that in this there is a debatable 'grey area'.

Uses and limitations of partial budgets

Partial budgeting is essentially a method of evaluating the expected average annual profitability of a predetermined course of action. As a result of this evaluation, the farm plan may be changed, but the partial budget is not the means whereby the change is conceived. Mostly the change will be derived from an intuitive thought by the farmer or by a new cropping or stocking opportunity which appears. The deficiency of partial budgeting is that it does not focus attention on the demands on the overhead resources of land, labour and capital for which other enterprises may be competing. More formal planning procedures which consider these demands are described in Chapter 10.

The biggest weakness of all is that partial budgeting can only evaluate the profitability of a project and does not take into account the cash flow (feasibility) or the return on capital (worthwhileness).

8 Planning: Cash Flow

Chapter 7 demonstrated the first step in establishing the 'viability' of a project: The partial budget example showed how profits could be increased through an opportunity for a minor change to the business.

A proposal can thus be shown to be profitable; but it may not be feasible.

It is possible that a plan which has profit potential cannot be brought to fruition because of a shortage of cash to fund the individual stages involved.

This chapter is concerned with the methods available to test feasibility, in terms of cash.

Cash flow

The preparation of cash flow projections has two main functions:

(1) To establish that sufficient cash is available to the business to make a profitable farm plan feasible
(2) To assist in monitoring and controlling the performance of a business to ensure, as far as possible, that the targets set in forward budgets are achieved

The various methods of assessing the past performance and the likely future profitability of changes within a farm business were discussed earlier. Profitability itself does not guarantee that a business will thrive. To establish the viability of a farm plan it must also be shown that the proposals are feasible in terms of cash and a worthwhile return on capital is made. A farm budget could show a farm plan to be highly profitable, but in the final analysis there might be insufficient cash available to fund the plan whilst the profit is being made. History shows that many businesses have failed to survive, not because they were necessarily unprofitable, but because they had insufficient cash to pay creditors at the appropriate time.

The preparation of cash flows

There are four fundamental principles to remember when preparing a cash flow:

(1) *Cash flows are only concerned with the movement of cash.* When preparing cash flows from budgets, always ignore valuation changes, depreciation charges, notional receipts and payments in kind as they are not cash items, and whilst they will affect the profitability of a plan they will not influence its feasibility
(2) *Timing of cash receipts and cash expenditure is critical.* To be able to identify when cash will be received or paid out demands that a breakdown is made of

the timing of sales and purchases. Having established the timing of the sales and purchases it is then important to know:

(a) The terms of trade of debtors (e.g. the Milk Marque or grain merchants). This will indicate how long after selling goods the cash will be received.
(b) The terms of trade of creditors (e.g. feed merchants or garage). This will identify how long after purchasing goods cash will have to be paid out.

The importance of timing is illustrated in Example 8.1 where failure to identify the terms of trade of debtors and creditors correctly resulted in a failure to reveal the peak cash deficit of £4000 which must be financed to make the plan feasible.

Example 8.1 The significance of the timing of receipts and payments in cash flow preparation

A farmer anticipates that over a six month period, August to January, he will sell 100 tonnes of wheat at £100 a tonne: 10 tonnes will be sold in August, 20 tonnes each September, November and December and 30 tonnes in October.

On the expenses side a local contractor is to be used to resurface the farm drive at a cost of £7200 which will be paid in instalments. The work completed in August and September will be £2500 a month, £1000 of work will be completed in October and £600 a month in November and December.

If the farmer assumed that he would receive cash from his grain merchant in the month the grain was collected and that he could take a month's credit from the contractor, then the cash flow forecast would be:

	Aug	Sep	Oct	Nov	Dec	Jan	Total
Forecast receipts	1 000	2 000	3 000	2 000	2 000	—	10 000
Forecast expenses	—	2 500	2 500	1 000	600	600	7 200
Monthly cash surplus (deficit)	1 000	(500)	500	1 000	1 400	(600)	2 800
Cumulative cash surplus (deficit)	1 000	500	1 000	2 000	3 400	2 800	

This cash flow shows no cumulative cash deficit and therefore no cash funding will be necessary.

If the actual terms of trade, however, were such that the grain merchant paid one month after delivery of the grain and the contractor required a cash settlement for his work as soon as he invoiced the farmer then the actual cash flow would be:

	Aug	Sep	Oct	Nov	Dec	Jan	Total
Actual receipts	—	1 000	2 000	3 000	2 000	2 000	10 000
Actual expenses	2 500	2 500	1 000	600	600	—	7 200
Monthly (cash) surplus/(deficit)	(2 500)	(1 500)	1 000	2 400	1 400	2 000	2 800
Cumulative cash surplus/deficit	(2 500)	(4 000)	(3 000)	(600)	800	2 800	

This cash flow indicates that the plan would show a cash deficit for the first four months peaking at £4000 in September.

(3) *Cash flows should be constructed on a monthly basis.* It is essential that cash flows are prepared on a monthly rather than a quarterly basis for the first year, to avoid the masking of significant peaks in cash requirements, as shown in Example 8.2. If a two-year budget is being prepared, then the cash flow for the second year should be prepared on a quarterly basis because of the inaccuracies that would be inherent in trying to predict the monthly cash position of the business so far into the future.

Example 8.2 The significance of preparing monthly cash flows

	Month 1	Month 2	Month 3	Quarter 1
Receipts:				
milk	7 000	9 000	19 000	35 000
Total receipts (a)	7 000	9 000	19 000	35 000
Expenses:				
concentrates	2 500	5 500	8 000	16 000
fertilisers	8 000	4 000	—	12 000
vet and medicine	300	300	400	1 000
Total expenses (b)	10 800	9 800	8 400	29 000
Monthly cash surplus/(deficit) (a) – (b)	(3 800)	(800)	10 600	6 000
Opening balance	—	(3 800)	(4 600)	
Cumulative balance	(3 800)	(4 600)	6 000	

A quarterly cash flow would mask the fact that the business has a cash deficit in two months during the quarter. The peak overdraft requirement of £4600 arises in Month 2.

(4) *The cash flow from an annual budget can take up to two years to unfold.* Where an annual budget is prepared for enterprises that have a production cycle which is longer than 12 months (e.g. 18 months for beef) the cash receipts and expenses shown in the 12-month cash flow, for the budget period, can vary significantly from the returns shown in the budget. The same principle applies where there is a cash flow for the budget year; detailed information will also be required for livestock, crops and stores on hand at the end of the previous year, plus details of the debtors and creditors outstanding at that time.

How to approach cash flow preparation

In the context of farm management there are limitations in looking at cash flow in isolation. It is better to record on the same document the physical flow of resources and outputs relating to the flow of cash. This is particularly important when using cash flow as an aid to monitoring and controlling budgets.

The basis of a good cash flow starts at a very early stage in the farm planning process. The secret lies in designing budget data collection forms which allow sufficient detailed physical and financial information to be recorded for the construction of a monthly cash flow.

The forms need not be over complicated and will vary in design according to the requirements of the individual farm business. Figs 8.1–8.3 illustrate some of the data collection forms used by the authors. An integral part of all the forms is the recording of the assumptions on which the figures are based.

Cash flow preparation should be approached in a methodical way to minimise the chance of cash items being overlooked. This is facilitated by dividing the cash flow of a business into three categories:

(1) *The trading net cash flow*. The trading net cash flow includes all the cash items that are found in the forward budget or profit and loss account. This includes receipts as well as expenses.

    Collection of information. Data for the trading net cash flow is recorded on the budget data collection forms as illustrated in Figs 8.1–8.3. Similar forms can be designed for other enterprises and for the recording of fixed cost data. When all the information has been gathered, it is transferred to the cash flow form.

(2) *The capital net cash flow*. The capital net cash flow shows the net flow of cash relating to farm business transactions occurring in the balance sheet of a business, such as land and machinery.

    When planning the capital expenditure and receipts on a farm, a capital expenditure and receipts budget should be drawn up as shown in Fig. 8.4. This budget gathers together information for the annual budget and for the cash flow. While most of the cash flow information relates to the capital cash flow, it is logical to record the running costs of the equipment, which are part of the trading cash flow. All the information affecting the cash position of the business is transferred to the cash flow document.

(3) *The personal net cash flow*. The personal net cash flow records the movement of all personal cash that goes into or out of the business. If private funds are kept separate from the business and will not be utilised in the business, then they should not be recorded in the cash flow.

    The design of a personal budget data form is usually a very simple affair, frequently taking the form of details on drawings from the business and tax payments. All the information is transferred from the budget data form to the cash flow.

The recording of the cash flow

All the information relating to the movement of cash is taken from the various budget data collection forms and recorded on the cash flow form in the appropriate

Fig. 8.1 An annual stock flow chart for a dairy herd

Month	Cows	Heifers	Purchases		Cull cows		Calvings		Calf sales				Calf deaths	
	Numbers transferred in	Numbers transferred in	No.	Price £	No.	Price £	Heifers	Bulls	Heifer No	Price £	Bull No	Price £	Heifers	Bulls

Fig. 8.2 Monthly milk production and concentrate use for a dairy herd

Month	Litres per month	Price per litre	Monthly milk cheque	Tonnes per month	Price per tonne	Monthly payment
Opening debtor				Opening creditor		
Totals						
Assumptions						

Fig. 8.3 Arable production chart for budget and cash flow preparation

Crop	Area ha	Yield ha	Price £	Total t	Output £	Value of crops in store at start	Actual sales in period	Value of crop in store at end	Seed kg	Seed £	Fertilisers kg	Fertilisers £	Sprays l	Sprays £
Winter wheat Winter barley Winter oil seed rape														
Totals														
Opening debtor														
Assumptions, e.g. pattern of sales terms of trade of merchants basis of output and input data														

Fig. 8.4 Capital expenditure and receipts

Date	Item purchased	Gross cost £	Grant %	Grant £	Depreciation %	Depreciation £	Running costs £	Source and cost of capital	Items sold	Sale price £

Fig. 8.5 Cash flow with quarterly review and variance analysis

		Month 1				Month 2			
		Budget		Actual		Budget		Actual	
		No.	£	No.	£	No.	£	No.	£
Trading receipts	Cattle								
	Sheep								
	Crops								
	Milk								
Total receipts (a)									
Trading payments	Livestock								
	Cattle								
	Sheep								
	Pigs								
	Sub-total								
	Variable costs								
	Feeds								
	Seeds								
	Concentrates								
	Sub-total								
	Fixed costs								
	Power and machinery								
	Labour								
	Insurance								
	Rent and rates								
	Bank charges								
Sub-total									
Total payments (b)									
Trading net cash flow (a) – (b) (I)									
Capital and personal	Capital introduced								
	Sale of property								
	Mortgages and loans								
	Grants								
	Cash introduced								
	Total capital introduced (c)								
	Capital payments								
	Cost of farm								
	drawings								
	Total capital and personal payments (d)								
Capital and personal net cash flow (c) – (d) (II)									
Net cash flow (I) + (II)									
Opening bank balance									
Cumulative bank balance									

Month 3				Review of year to date				Variance from budget	
Budget		Actual		Budget		Actual			
No.	£	No.	£	No.	£	No.	£	No.	£

month when the payment will be made or the income received. Whether each category of cash flow is recorded on a separate document and then a summary transferred to a master cash flow for the whole business or, alternatively, the details of each category recorded on a single document with a summary at the bottom, will depend on the individual business requirements. Where there is a large and complicated business, the three categories are frequently recorded separately and a summary cash flow prepared. In contrast, smaller businesses tend to prepare a composite cash flow from the start.

However the cash flow may be compiled, it is most important that the farmer is able to:

(1) Identify his capital requirements in cash terms
(2) Have a format which will aid him in monitoring and controlling the farm budget

An acceptable format for presentation of cash flow data is shown in Fig. 8.5. This records physical and financial budget data on a monthly basis, and also actual performance, plus a three-monthly review of performance for the year to date. The variances from budget are also recorded.

The calculation of interest charges on overdrawn balances

One of the practical problems in preparing cash flows is deciding how to assess the interest charge which should be inserted. Deciding the interest rate is a matter of discussion and making an assessment of likely trends in rates; it is the calculation of the interest to be paid once the rate has been decided which can cause problems.

Loan interest is easily worked on the basis of the capital outstanding for the period in question, the resulting interest figure being recorded in the appropriate month of the cash flow. Bank interest is charged on a quarterly basis in March, June, September and December.

In practice the interest charged on overdrafts is calculated on a daily overdrawn balance outstanding. Computers help the banks to calculate the interest they are owed by customers to a specific day of the month which they charge to the account a few days later. If, for example, the date for charging interest to customers' accounts is the 20th day of the month, the banks will have worked the charge up to, say, the 17th of the month. The few days between recording and charging the interest are used to carry out checks and make any amendments necessary. Clearly when preparing a cash flow manually it is impossible to achieve this degree of accuracy and it is also impossible to allow for the effect of uncleared effects, i.e. cheques which the farmer has paid into his account but for which his banker has not received value, which normally takes three days.

The simplest method of calculating the interest is to base the interest charges on any overdrawn balances for the three months preceding the month in which interest is charged, as shown in Example 8.3.

Example 8.3 The calculation of interest charges on an overdraft for cash flow purposes

Consider interest charges for December.

Procedure

(1) Calculate the cumulative bank balance for September. If the balance is positive, there will be no interest charge for the month. Where the balance is negative, an estimate of the interest to be charged is made as follows:

$$\text{Interest charge for month} = \frac{\text{overdrawn balance} \times \text{annual interest rate as \%}}{12 \text{ months}}$$

(2) Repeat the calculation for October and November
(3) Add together the interest charges for the three months
(4) Record the total interest charge in the December month under bank charges and interest

Sensitivity tests – how accurate is a cash flow?

The point was made earlier that cash flows and budgets are based on assumptions which in practice may or may not turn out to be accurate. In an effort to assess the implications of variance from the major assumptions, sensitivity tests are carried out. This is achieved by looking at the physical and financial data used in the budget and testing what effect specific increases or decreases in returns or performance would have on both the budget and cash flow.

For example, in the case of a dairy enterprise it would be logical to test the effect of a reduction (or increase) in milk output (say 10%) or the effect of a 10% increase (or decrease) in concentrate usage. Similarly, the result of an increase (or decrease) in concentrate or fertiliser price could be investigated.

How to use a cash flow to monitor and control a business

A great deal of time and effort is expended in planning, budgeting and preparing cash flows to try to assess the viability of differing business management decisions. This effort can be largely wasted unless, having acted on the information, control is exercised to ensure that targets are achieved and the plans brought to fruition. The monitoring and control of the forward budget through the medium of the cash flow of the business is a basic aid to the manager in establishing whether his targets, and therefore objectives, are being achieved. On many occasions he will record significant variances from plan and hopefully, by knowing about these deviations when they occur, the good manager will be able to analyse the reason for failure and take action to correct the position, or at least try to ensure the same problem does not arise again.

With the monitoring and control function in mind, the cash flow form shown in Fig. 8.5 records physical and financial information for the budget and actual performance. It is not always easy to rationalise a monthly variance from budget, so it is expedient to carry out a quarterly review of the performance to date, and any variance should be analysed.

Variance analysis – what is it and how does it work?

There are three areas where variance from budget can occur:

(1) A variance in the level of activity, i.e. the scale of operation is different from forecast
(2) A variance in the unit price of a commodity or input
(3) A variance in the unit quantity of a commodity produced or an input used

These variances may be positive or negative.

(1) *Variance in the level of activity (plan differences).* At the planning stage significant errors may have been made about the scale of a particular enterprise. For example, 50 ha of potatoes were grown instead of the planned 75 ha. This sort of variance will affect the overall profitability of the plan (unless the 25 ha are utilised by an enterprise giving a similar return). No measure is gained of the enterprise management from this information.

    Calculation of activity variance. This variance is calculated by multiplying the difference in the level of activity by the budgeted yield and price as illustrated in Example 8.4.

(2) *Variance in the price of a commodity or input (price differences).* The sale price of a commodity or the purchase price of an input frequently vary from the budget figure and it is helpful from a management viewpoint to analyse how much of the overall variance, from plan, is due to price differences alone. Once this difference has been established the reasons for the divergence can be investigated.

    There are many causes of price differences, including straightforward supply and demand effects, poor marketing or buying and, of course, bad budgeting.

    Calculation of price variance: Price variance is the difference between the actual price received (or paid) and the budgeted price multiplied by the actual quantity produced (or used) as illustrated in Example 8.4.

(3) *Variance in the quantity of a commodity or input (volume variance).* There are many reasons why output or input volume can vary from budget. Management performance may, in practice, be below the standard assumed when budgeting or it may be due to circumstances beyond the control of the farmer (e.g. poor silage made in adverse weather conditions, resulting in increased concentrate usage).

    Calculation of volume variance. Volume variance is the difference between the actual volume and the budgeted volume multiplied by the standard price as shown in Example 8.4.

Example 8.4 Demonstration of the use of variance analysis techniques

A farmer planned to plant 75 h of potatoes and budgeted for a yield of 30 tonnes per hectare at £50 per tonne. The actual hectarage planted was 55 and the yield 25 tonnes per hectare. The sale price of the potatoes exceeded expectations and averaged £70 per tonne. The farmer wants to analyse the deviation from his budget.

	Budget £	Actual £	Positive variance £	Negative variance £	Total variance £
Total output from potatoes	112 500	96 250			(16 250)
Plan difference –20 ha @ 30 t @ £50				(30 000)	
Price difference + 55 ha @ 25 t @ £20			27 500		
Volume difference –55 ha @ 5 t @ £50				(13 700)	
			27 500	(43 750)	(16 250)

In this example the farmer failed to plant the planned hectarage of potatoes and so there was a reduction in income of £30 000 for this reason. The lower yield achieved, which may have been due to bad management, further reduced farm income compared to budget. Fortunately price was higher than planned, and this resulted in a positive cash income which compensated, to some extent, for the negative variances. The net effect was that actual returns were £16 250 below budget.

Clearly where an analysis of cash flow variance is being made the timing variable will play a significant part in any analysis.

Summary

Cash flows are an essential feature of good business management and need careful preparation. Furthermore it is critical that a clear record is made of all the assumptions (including physical data) on which the cash flow is based.

As well as satisfying the main functions of determining the feasibility of a plan and acting as a means of monitoring and controlling performance against budget, cash flows have additional uses:

(1) Bank managers are always very keen to see projections of the monthly cash position of a business because they are able to gauge the likelihood of a business being able to service and repay any credit they provide and also establish the peak overdraft required by the farmer

(2) Farmers who have budgeted and monitored the budget through cash flows tend to have increased knowledge of the financial and marketing aspects of their business

The necessity for a farmer to prepare detailed cash flows will, in general, be governed by the capital and cash position of his business. A highly geared business will have little alternative to producing cash flows. The farmer who has little or no borrowing is making good profits and with no plans for capital expenditure financed by creditor funds may well feel cash flows are unnecessary. He may, however, find them useful for predicting when surplus funds will be available and the truth is that all businesses will benefit from the discipline imposed by forward budgeting and cash flow preparation.

Six aids to convincing a bank manager or other financier that your cash flow is meaningful

(1) Ensure you do not underestimate the amount of interest you will have to pay!
(2) Remember to allow for capital repayments on loans
(3) Do not give the impression that you are taking more than a reasonable amount of cash out of the business for living expenses (only one Rolls Royce per annum!) in relation to the annual retained profits in the business
(4) Do not give the impression that you can live on fresh air by skimping on living expenses
(5) Check that the fixed costs in your budget and cash flow are not lower than those shown in the audited accounts you gave the bank, which are now two years old with inflation having run at 10%! There will hopefully be instances when fixed costs are lower, but this needs justifying
(6) Check the cross-casts and additions in your cash flow

The relationship between cash flow, trading profit and capital

A positive cash flow over a 12-month period does not necessarily mean a business is making a trading profit. Equally a negative cash flow over a 12-month period does not necessarily mean a business is losing money.

A cash flow accounts purely for cash movements within a year and whether the cash relates to a cost or receipt associated with the inputs or output for that year is not considered. It is therefore necessary, to determine the trading net profit, to adjust the cash flow to take account of these facts.

The major items which are likely to be recorded in a cash flow which do not appear in a TPL account are:

(1) Receipts associated with a previous trading period (opening debtors)
(2) Expenses associated with a previous trading period (opening creditors)
(3) Capital receipts (including grants) or capital expenditure
(4) Loan repayments
(5) Proprietors' drawings
(6) VAT payments and receipts
(7) Tax payments
(8) Opening bank balance
(9) Capital introduced, e.g. aunt's bequest

There are also some items in a TPL which do not appear in a cash flow; these are:

(1) Receipts and expenses for the year in question which have not been received or paid (closing debtors and creditors)
(2) Changes in valuation between the beginning and end of the year
(3) Depreciation of fixed assets
(4) Use of homegrown produce by the farmer
(5) The benefit of the farmer's house as a private dwelling
(6) Transfers between enterprises on the farm

Example 8.5 shows how to convert an annual cashflow, surplus or deficit, into a trading profit or loss for the same period by accounting for the items:

(1) Which appear in the cash flow when they do not appear in the TPL account
(2) Which are omitted from a cash flow but need to be included in a TPL account

Example 8.5 The relationship between cash flow and net profit or loss

A farmer has monitored his quarterly cash flow for the year from 1 April 1996 to 31 March 1997. At the year end his figures were:

Quarterly cash flow for year ending 31 March 1997

	Apr/Jun (Quarter one £	Jul/Sep (Quarter two) £	Oct/Dec (Quarter three) £	Jan/Mar (Quarter four) £	Total £
Income					
Milk sales	20 000	16 000	25 000	35 000	96 000
Culls	5 000	3 000	1 500	2 000	11 500
Wheat	3 500	5 000	15 200	15 800	39 500
Aunt's bequest			18 000		18 000
Sub-total	28 500	24 000	59 700	52 800	165 000
Less:					
Expenditure					
Concentrates	2 500	1 000	8 000	13 500	25 000
Fertilisers	10 750		8 000		18 750
Sprays	6 000	2 000	2 000		10 000
Vet and medicine	500	600	800	400	2 300
Machinery repairs	2 250	4 350	2 100	300	9 000
Other expenses	2 300	2 450	2 220	1 530	8 500
Bank interest	4 000	4 500	5 500	5 000	19 000
Loan repayment	3 000	3 000	3 000	3 000	12 000
Drawings	2 500	2 500	2 500	2 500	10 000
Sub-total	33 800	20 400	34 120	26 230	114 550
Net cash flow	(5 300)	3 600	25 580	26 570	50 450
Opening balance	(107 140)	(112 440)	(108 840)	(83 260)	
Closing balance	(112 440)	(108 840)	(83 260)	(56 690)	

This cash flow shows a cash surplus of £50 450 in the year.

At the start of the year the business was owed £12 120 (opening debtors) and at the end of the year was owed £1220 (closing debtors). At the start of the year the farmer owed £4650 (opening creditors) and at the end of the year he owed £5870 (closing creditors). During the year the valuation reduced from £170 000 to £160 000. The breakdown of debtors and creditors was as follows:

	Opening debtor £	Closing debtor £	Opening creditor £	Closing creditor £	Change £
Milk	8 000				(8 000)
Culls	620	1 220			600
Wheat	3 500				(3 500)
Concentrates			2 000	4 000	2 000
Fertilisers			750		(750)
Vet and medicine			900	150	(750)
Machinery repairs			1 000	1 720	720
Total	12 120	1 220	4 650	5 870	

When the farmer's agricultural consultant called to review the performance for the year ending 31 March 1997 he produced a TPL as follows:

	£		£
Opening valuation	170 000		
Expenditure:		Sales:	
Concentrates	27 000	milk	88 000
Fertilisers	18 000	culls	12 100
Sprays	10 000	wheat	36 000
Vet and medicine	1 550		
Machinery repairs	9 720	Closing valuation	160 000
Others	8 500		
Bank interest	19 000		
Depreciation	6 000		
Net profit	26 330		
Total	296 100		296 100

The farmer immediately challenged the profit figure because he could not see how a cash flow surplus of £50 450 could possibly equate with a net profit of only £26 330.

The consultant took the cash flow surplus and adjusted it, to arrive at the net profit, in the following way:

		£
Cash surplus		50 450
Less:		
Opening debtors	12 120	
Closing creditors	5 870	17 990
		32 460
Add:		
Opening creditors	4 650	
Closing debtors	1 220	5 870
		38 330
Less:		
Non-trading income	18 000	
Depreciation	6 000	
Valuation change	10 000	34 000
		4 330
Plus:		
Loan repayments	12 000	
Drawings	10 000	22 000
Net profit		26 330

The TPL account sales figures and some of the expenditure items did not appear, at first sight, to equate with the figures on the cash flow. The differences are attributable to the adjustments made for the debtor and creditor position, as analysed below:

	Cash flow figure £	Opening debtor £	Opening creditor £	Closing debtor £	Closing creditor £	TPL £
Milk	96 000	(8 000)				88 000
Culls	11 500	(620)		1 220		12 100
Wheat	39 500	(3 500)				36 000
Concentrates	25 000		(2 000)		4 000	27 000
Fertiliser	18 750		(750)			18 000
Vet and medicine	2 300		(900)		150	1 550
Machinery repairs	9 000		(1 000)		1 720	9 720
Total	202 050	(12 120)	(4 650)	1 220	5 870	192 370

This example while relatively simple illustrates the importance of understanding the relationship between cash flow and profit. In this example the cash surplus of £50 450 translates to a profit of £26 330 – just over half the cash surplus. Out of this profit the farmer must pay his own living expenses (£10 000 per annum), fund any capital expenditure in the business and make a return on his investment.

The relationship between cash flow, TPL account and the capital account

Using the figures shown in Example 8.5, let us consider the impact of cash flow and TPL on the capital account, or surplus, of a business. Example 8.6 illustrates the

Example 8.6 The effect of the annual TPL account and cash flow on the capital in a business

Opening balance sheet as at 1 April 1996

Liabilities	£	Assets	£
Creditors	4 650	Debtors	12 120
Bank overdraft	107 140	Valuation	170 000
Bank loan	250 000	Machinery	60 000
Capital account		Land	500 000
(proprietor's equity)	380 330		
Total	742 120		742 120

Closing balance sheet as at 31 March 1997

Liabilities	£		Assets	£
Creditors	5 870		Debtors	1 220
Bank overdraft	56 690		Valuation	160 000
Bank loan	238 000		Machinery	54 000
				(opening value 60 000)
				(less depreciation 6 000)
Capital account				
(proprietors equity)			Land	500 000
Opening balance	380 330			
Plus:				
Capital introduced	18 000			
Trading profit	26 330			
	424 660			
Less:				
Drawings	10 000			
		414 660		
Total		715 220		715 220

balance sheet at 1 April 1996 (the first day of the trading year, and cash flow) and the balance sheet on 31 March 1997 (the last day of the trading year).

Example 8.5 shows the trading profit at 31 March 1997 to be £26 330. This, however, ignores the deduction of the farmer's personal drawings, loan repayments and capital injections into the business.

These three items are recorded in the cash flow but the cash flow does not reflect retained profit because it will not have accounted for the opening and closing debtors or creditors, or for depreciation.

Thus the closing balance sheet draws on information from both the cash flow and TPL account to arrive at the proprietor's capital at the end of the year.

In Examples 8.5 and 8.6 there is a positive cash flow of £50 450 and the TPL shows a profit of £26 330. The farmer's capital account, however, increased by £34 330 which, after allowing for the reduction in bank loan and the depreciation charge against the machinery, represented the difference between the capital introduced and the drawings (net £8000) plus the trading profit.

In more complicated situations (for example, where tax payments or bad debt adjustments are necessary) the calculation includes more items but the broad principles remain the same.

9 Planning: Capital Investment Appraisal

Chapters 7 and 8 looked at two aspects of establishing if a plan is viable, namely profitability and feasibility. This chapter continues this theme and looks at the methods of determining if a capital project is firstly worthwhile and secondly feasible.

Capital investment appraisal

It is important to establish the viability (i.e. feasibility, profitability and worthwhileness) of an investment before committing funds to it. This will facilitate the selection of good projects and the avoidance of poor ones which may result in liquidity problems and possible insolvency.

The calculation of the profitability of a farm plan is discussed in Chapter 7, and the same methods are used for determining the profitability of a capital investment.

What are the aims in calculating worthwhileness and feasibility?

(1) *To establish the return (as a percentage) on the capital that has been invested and compare this return with:*
   (a) The cost of any capital which has to be borrowed, i.e. the interest rate charged by a financial institution
   (b) The opportunity cost associated with a business's own capital invested in a different project
   (c) The returns which an alternative investment project might offer
(2) *To see if the anticipated return is likely to be sufficient to justify the risk involved with an investment*
(3) *To ensure that the capital requirements of a project can be satisfied over the term of the project*

How is capital investment calculated?

Capital investment falls into two major categories:

(1) *Investment in fixed assets*

Fixed assets include land, buildings, plant and machinery. When considering a new investment it is relatively simple to calculate the value of the investment. The cost of the assets to be purchased can be established with the appropriate vendor or merchant. To this cost must be added any additional costs arising from the purchase of particular assets, such as legal costs with land purchase or installation costs with a dairy parlour.

Where an existing asset is to be used in a new project, the value that should be attributed to the asset is its current cash value if it were to be sold, not the original purchase price or its replacement cost or indeed its written down value in the books of the business.

(2) *Investment in working capital.* This represents the additional money that is necessary to fund the production cycle, i.e. increased debtors, increased stock and additional cash. The working capital will always be available as cash at the end of a project to be utilised again.

It is important to remember that it is the cash value of the investment which should be compared to the net cash return.

How is the life of a project determined?

The importance of a realistic assessment of the life of a project cannot be over-emphasised. The longer the life attributed to a project, the greater potential there is for cash income to be generated and therefore the higher will be the return achieved.

Realism is the key; over-optimistic forecasts of the effective life of machinery, plant or buildings can only lead to an inaccurate estimate of the viability of an investment.

Calculation of worthwhileness

Worthwhileness is determined by calculating the percentage return that is forthcoming from an investment. Several factors are instrumental in determining whether a particular level of return is acceptable. These are:

(1) If the farmer is using his own capital then he will expect to achieve a return which is at least as good as (or better than) alternative investments available to him, i.e. building society, gilt-edged securities
(2) If the farmer is borrowing money, then the return will need to cover the cost and provide a margin for the farmer himself
(3) Whether the money is borrowed or not, the decision regarding what is an acceptable return will be influenced by the risks associated with a particular investment. The greater the risks then, generally speaking, the higher should be the return to the farmer for undertaking the investment

There are three methods of assessing the worthwhileness of an investment: rate of return, net present value, and internal rate of return (discounted yield).

Rate of return

Rate of return is based on the formula

$$\frac{\text{Rate of return}}{\text{(as a percentage)}} = \frac{\text{Average additional annual profit} \times 100}{\text{Additional capital required}}$$

where the additional profit is calculated after deducting depreciation but before deducting interest and taxation [Example 9.1, solution (a)]. This calculation is now rarely used.

Depreciation is frequently worked on a 'reducing straight-line' basis assuming a nil residual value. The writers prefer, however, to calculate depreciation by assessing the life of the project, estimating the residual value of the asset at the end of its useful life and then using the following formula:

$$\text{Depreciation} = \frac{\text{Cash value of investment} - \text{Estimated residual value of asset}}{\text{Effective life of project}}$$

The rate of return is calculated before interest because one of the objects of the exercise is to establish the cost of money (i.e. interest rate) which the project will support.

Some authorities use the average additional capital as the capital employed, on the basis that repayment of principal will occur during the lifetime of the project and the profit calculation in this method has already included depreciation which represents this capital repayment. This approach, however, tends to overstate the return on capital [Example 9.1, solution (b)].

Example 9.1 The calculation of rate of return

Cash flows from the projects would be as follows:

Year £	Drainage £	Pig building £
1	3 000	1 500
2	3 000	1 500
3	3 000	2 500
4	3 000	2 500
5	2 000	2 500
6	2 000	2 500
7	2 000	2 500
8	2 000	2 500
9	1 000	2 000
10	1 000	2 000
	22 000	22 000

A farmer has the choice of investing £10 000 in draining 30 ha of wet, heavy clay land or putting up a new fattening house for pigs. It is estimated that both investments would have an effective life of ten years. The residual value of the investment would be nil.

$$\text{Depreciation in both cases will be} = \frac{£10\,000 - \text{nil}}{10}$$
$$= £1000 \text{ per annum}$$

$$\text{Average additional annual profit} = \frac{£22\,000 - £10\,000}{10}$$
$$= £1200 \text{ per annum}$$

Solution (a)

$$\text{Rate of return as a percentage of initial capital} = \frac{£1200 \times 100}{£10\,000}$$
$$= 12\%$$

Solution (b)

$$\text{Rate of return as a percentage of average capital} = \frac{£1200 \times 100}{£5\,000}$$
$$= 24\%$$

Both formulae used for the calculation of rate of return are unable to distinguish between the investment in land drainage or pigs though we know that the pattern of receipts from the two projects will differ on an annual basis.

The limitations of the rate of return method of calculating worthwhileness

The rate of return method is simple and quick to calculate and is widely used, but there are distinct disadvantages.

(1) No account is taken of the timing of receipts and their real value. Cash received today is worth more than the same amount of cash to be received in the future. If the money was available now it would be possible to invest it and so earn more money.
(2) By dealing in average returns, the rate of return calculation cannot distinguish between two investment opportunities which show exactly the same total profit over the life of the investment. It will show the same rate of return although the pattern of the cash flows may be completely different.

Net present value and internal rate of return methods

These methods are more refined than the simple rate of return method but they are more time-consuming to calculate. They are based on the concept of discounting and it is therefore possible to take account of the timing of receipts and their true value at the time they are received.

The concept of discounting is related to that of compounding.

The concept of compounding

One pound received today is worth more than £1 to be received in one year's time because the £1 received today can be invested to earn interest. This is compounding and is illustrated in Example 9.2.

If £100 was available for investment today at 10% interest, then in five years' time it would be worth £161.

Example 9.2 To show the effective return on an initial investment of £100 at a compound interest rate of 10% pa over a five-year period

Year	Capital at start of year	Interest rate	Annual interest earned	Capital at end of year
1	100	10	10	110
2	110	10	11	121
3	121	10	12.1	133.1
4	133.1	10	13.3	146.4
5	146.4	10	14.6	161

This means that £100 invested today at 10% for five years is just as valuable as receiving £161 in five years' time. The formula for calculating this compounding effect is

$$\pounds\ (1+i)^n$$

where n = number of years and i = interest rate expressed as a decimal

It is important to remember that compounding has nothing to do with inflation and should not be confused with the inflationary effect which is discussed later in this chapter.

The concept of discounting

It follows from the statement on compounding that £1 received in one year's time is worth less than £1 received today because there will have been no investment opportunity for the £1 due in one year's time. This is illustrated in Example 9.3.

This concept of reducing future receipts to their present value is known as discounting and can also be rationalised in an equation which is the reciprocal of compounding:

$$\text{Present value of future £} = \frac{1}{\text{Future value of the present £1}}$$

$$= \frac{1}{(1+i)^n}$$

where n = number of years and i = interest rate expressed as a decimal

Example 9.3 To show the initial investment (£62) required to yield £100 in five years' time given a compound interest rate of 10% pa.

Year	Capital at start of year	Interest rate	Annual interest earned	Capital at end of year
1	62.0	10	6.2	68.2
2	68.2	10	6.8	75.0
3	75.0	10	7.5	82.5
4	82.5	10	8.3	90.8
5	90.8	10	9.2	100.0

If £100 were to be received in five years' time and in the intervening years money could have been invested at 10%, then the £100 due in five years would only be worth the same as £62 received today.

The greater the investment opportunity for the present £1, then the lower will be today's value of £1 when received at some future date.

Example 9.4 Example to show that the greater the investment opportunity for the present investment of capital, the greater will be the rate of growth of that capital and the lower will be the value of a £1 received in the future

	Interest rate 10%			Interest rate 20%		
Year	Capital at start of year	Annual interest earned	Capital at end of year	Capital at start of year	Annual interest earned	Capital at end of year
1	100.0	10.0	110.0	64.7	12.9	77.6
2	110.0	11.0	121.0	77.6	15.6	93.2
3	121.0	12.1	133.1	93.2	18.6	111.8
4	133.1	13.3	146.4	111.8	22.4	134.2
5	146.4	14.6	161.0	134.2	26.8	161.0

A sum of £161 to be received in five years' time would be equivalent to £100 today if the £100 could be invested at 10% per annum, or £64.70 if the investment rate was 20% per annum.

Tables of discount factors at varying discount rates are available and these facilitate the use of discounting techniques. The tables generally work on the assumption that cash is received at the end of each year though tables are available where the discount factors are based on money being received at six-monthly intervals. The more frequently the interest payment is received, the greater will be the discounting effect.

The principle of discounting allows any payment, whatever its size or date or receipt, to be expressed at its present day value. This is particularly useful where a farmer wants to compare returns from a series of alternative investment opportunities and these returns vary on an annual basis.

Which discount rate (interest rate) should be used in a particular situation?

The selection of the discount rate is crucial. In a borrowing situation it is normal practice to discount initially at the rate corresponding to the cost of the borrowed money (interest rate) over the period of the project. This will mean making an assumption about the rate of interest that is likely to prevail over the life of the investment. The reason for selecting the cost of money as the initial rate at which to discount is to ensure that the investment will at worst have recovered the capital invested and any interest paid.

Where no borrowing is involved then a discount rate should be used which reflects the minimum return on the investment which the proprietor of the business will accept. This is likely to equate with the rate of interest he can earn on his money by investing in gilt-edged securities, county council bonds or even a building society.

Net present value (NPV) method

Definition

Net present value is the net result, in terms of present values, of discounting all expected cash flows (including the initial outlay) at some approporiate rate of discount (interest rate). The cash flow is calculated and interest and depreciation are excluded.

If, as a result of discounting, the NPV is positive, then there is a higher return on the investment than the rate of discount (interest) used. If the NPV is negative, then the rate of return is less than the rate of discount used.

It is important to remember that the NPV does not give the rate of return on an investment (unless NPV equals zero); it simply shows whether the return is more or less than the selected discount rate. NPV will, however, measure the profit or loss promised by an investment at a given cost of capital.

Consider a capital investment of £100 000 which generates the net cash flow over four years shown in Example 9.5 and would need financing with borrowed money attracting interest at 10%. The residual value of the investment is nil.

Example 9.5 Example of an NPV calculation for a given cash flow using an annual discount factor of 10% and comparing this value with the money 'value' return for the same cash flow

Year	Net cash flow before interest and depreciation (a) £	Discount factor 10% (b) £	Product (a) × (b) £
0	(10 000)	1.000	(10 000)
1	2 000	0.909	1 818
2	3 000	0.827	2 481
3	5 000	0.751	3 755
4	5 000	0.683	3 415

Money value = + 5 000 Net present value = +1 469

NB:

(1) The initial capital investment is discounted in year 0 and the discount factor is 1. This is logical as the object is to relate the value of future income to the original value of the investment. Whatever the discount rate used, the discount factor in year 0 will be 1 as money received today will always equal its present day value.
(2) The depreciation factor for the investment was accounted for by assessing the residual value of the asset. In this example it was nil and so no adjustment to the cash flow was necessary. If the residual value of the asset had been, say, £2000, then this sum would have been shown as a positive cash flow in year 5 and discounted accordingly.
(3) The NPV is calculated by adding the positive products in years 1 to 4 and then deducting the negative product for year 0, i.e. NPV is the value of the future discounted cash flows less the value of the initial investment,

In this example it is now known that the cash flow generated by the investment of £10 000 will cover, at present values, repayment of capital and interest charged at 10% and yield a profit of £1469.

This can be demonstrated in the following manner. If £11 469 (£10 000 initial investment plus £1469 profit) is borrowed at 10% interest, £10 000 of this money is invested in the project to give the forecast cash flow and £1469 spent on a holiday to the Seychelles, then repayment of the loan would be achieved as illustrated in Example 9.6.

The fact that the cash flow is sufficient to service and repay the loan demonstrates the validity of the discounting technique.

If it is only necessary to determine that the return from an investment is higher than the likely interest costs or minimum investment return that can be achieved on the money market, then the calculation of NPV will suffice. If the precise return from a project is required, then the internal rate of return (discounted yield) must be calculated.

Example 9.6 Example to show the validity of the NPV calculation

An initial loan of £11 469 on which interest is charged at 10% per annum compound on the outstanding balance is repaid by an identical cash flow to Example 9.5.

Year	Loan outstanding	Annual interest charged at 10% (a)	Cash flow before interest and depreciation (b)	Surplus for loan repayment (b) – (a)
1	11 469	1 147	2 000	853
2	10 616	1 062	3 000	1 938
3	8 678	868	5 000	4 132
4	4 546	454	5 000	4 546
5	0			

Internal rate of return (discounted yield) method

Definition

The internal rate of return (IRR) is defined as the rate of discount at which net present value is nil. It represents the rate of interest, charged on borrowed capital, which a project could just repay on bank overdraft terms (i.e. repayments can be made at any time).

The IRR does not measure the profit from an investment, only the worthwhileness.

Consider the same capital investment, net cash flow and interest rate used in Example 9.5 for calculating NPV.

It was established that NPV was + £1469. By definition the IRR is found when NPV is nil, therefore it is known that IRR is greater than 10% (because NPV is + £1469).

To find the break-even point, the net cash flow is discounted at a higher rate of discount to try to determine when NPV is nil. Deciding which discount rate to use is a matter of guesswork, but it should always result in negative NPV from which the break-even point can be pinpointed.

In this example a 20% discount rate is appropriate.

Example 9.7 To illustrate the method for establishing the IRR for an investment project

By discounting at 20% it can be seen that the IRR is greater than 10% but less than 20% (because at 20% discount NPV is negative).

	Discount rate 10%			Discount rate 20%		
Year	Net cash flow	Discount factor	Product	Net cash flow	Discount factor	Product
0	(10 000)	1.000	(10 000)	(10 000)	1.000	(10 000)
1	2 000	0.909	1 818	2 000	0.833	1 666
2	3 000	0.827	2 481	3 000	0.694	2 082
3	5 000	0.751	3 755	5 000	0.579	2 895
4	5 000	0.683	3 415	5 000	0.482	2 410
	NPV at 10%	1 469		NPV at 20%	(947)	

There are three methods of calculating the IRR from the information now available:

(1) *Arithmetic method*
NPV at 10% discount rate = + £1469
NPV at 20% discount rate = – £947
Difference in NPV between 10% and 20% = + £1469 + (–947)
A difference of 10% in discount rates gives a difference in NPV = 2416.

If it is assumed that the relationship between discount rate and NPV is linear where the discount rates chosen are 10% or less apart, then: 1% movement in discount rate represents a movement in NPV of:

$$\frac{£2416}{10} = £241.6$$

The IRR is established where NPV is zero. This is determined by calculating the percentage discount rate represented by the NPV of +£1469.
Since a 1% movement in discount rate equates to a movement of £241.6 in NPV this is an easy calculation.

$$\text{Discount rate \%} = \frac{1469}{241.6}$$
$$= 6.08\%$$

However, the cash flow has already been discounted at 10%.
Therefore the IRR $= 10\% + 6.08\%$
$= 16.08\%$

(2) *Formula method*

$$\text{IRR} = \frac{\begin{matrix}\text{Lower} & & \text{Difference} & & \text{Net present} \\ \text{discount} & + & \text{between} & \times & \text{value at lower} \\ \text{rates} & & \text{rates} & & \text{rates}\end{matrix}}{\text{Difference between NPVs}}$$

$$= 10 + \frac{(20 - 10) \times 1469}{1469 - (-947)}$$

$$= 16.08\%$$

(3) *Graphical method*

This method entails plotting the NPV against the discount rate used, and reading the IRR from the graph (Fig. 9.1)

Where the line joining the points cuts the X axis gives the IRR.

It is now known that the project will give a return sufficient to repay the capital and cover interest charged at 16.08%.

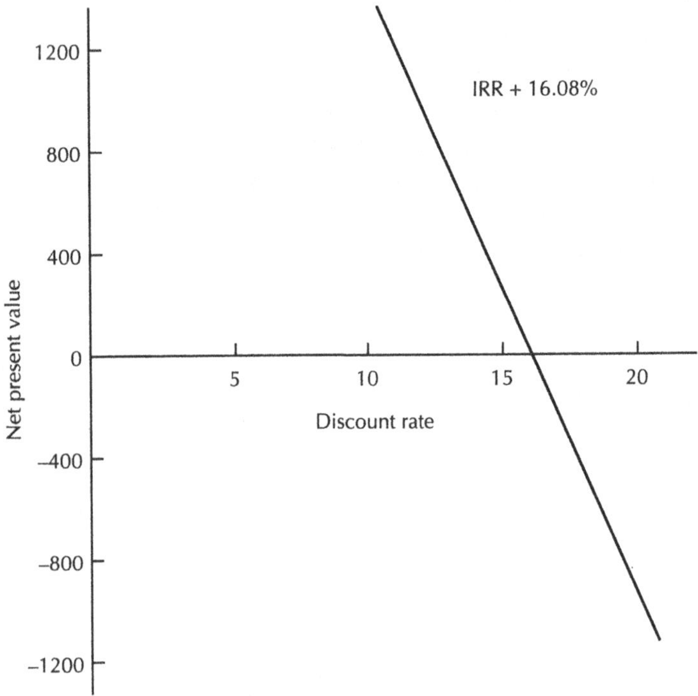

Fig. 9.1 Graphical method of calculating the IRR.

NB:

(1) When calculating IRR an assumption is made that the relationship between the net present values at differing discount rates is linear. This is not true but is sufficiently accurate where the difference between discount rates is 10% or less.
(2) No account is taken of the risk element associated with a particular investment project; this must be assessed separately.
(3) The net cash flow is calculated before deducting interest and depreciation. Depreciation is accounted for by assessing the residual value (which may be nil) of the investment at the end of its life and treating this as a cash credit in the year it would be received and then discounting it at the appropriate discount factor for the year in question.
(4) Where working capital is tied up in a project, then at the end of the project it will be released. The present value of this working capital should be assessed by multiplying its money value by the appropriate discount factor for the year it is released.
(5) By using this method of calculating worthwhileness in Example 9.1 the land drainage project shows an IRR of 22.6% and the pig project an IRR of 17%. Thus it is possible to distinguish between the two projects (see Table 9.1).

Table 9.1 Calculation of the internal rate of return

Year	Drainage cash flow	Discounted cash flow		Pig building cash flow	Discounted cash flow	
		15%	25%		15%	25%
0	(10 000)	(10 000)	(10 000)	(10 000)	(10 000)	(10 000)
1	3 000	2 610	2 400	1 500	1 305	1 200
2	3 000	2 268	1 920	1 500	1 134	960
3	3 000	1 974	1 536	2 500	1 645	1 280
4	3 000	1 716	1 230	2 500	1 430	1 025
5	2 000	994	656	2 500	1 243	820
6	2 000	864	524	2 500	1 080	655
7	2 000	752	420	2 500	940	525
8	2 000	654	336	2 500	817	420
9	1 000	284	134	2 000	568	268
10	1 000	247	107	2 000	494	214
	+12 000	+2 363	−737	+12 000	+656	−2 633

Internal rate of return $15 + \frac{(25 - 15) \times 2363}{3100}$ $= 22.6\%$ $\qquad 15 + \frac{(25 - 15) \times 656}{3289}$ $= 17\%$

Calculation of feasibility

An investment may appear to be very profitable and show a good return on the capital invested (worthwhileness) but to be viable it must be established that there is sufficient capital, either the proprietor's own or creditor capital, to finance the project throughout its life. The consequences of proceeding with an investment where there is insufficient capital available have obvious implications.

To determine whether an investment project is feasible, the payback period is calculated.

Payback period

The feasibility of a project is assessed by calculating the time taken to recover the capital invested, plus any interest payable on borrowed funds. The cash income is taken before deduction of interest and depreciation because:

(1) The interest chargeable is based on the annual capital outstanding and is deducted from the cash income, giving the amount of cash available to repay the capital.
(2) Depreciation is a non-cash item and is therefore excluded from the cash flow of an investment.

This calculation is easily worked by using the format shown in Example 9.8.

Example 9.8 The calculation of the payback period for an initial investment of £10 000 given interest at 10% per annum on the outstanding capital

Year	Cash flow before interest and depreciation	Annual interest (10%)	Annual payback	Capital investment outstanding at end of year
0				10 000
1	2 000	1 000	1 000	9 000
2	3 000	9 000	2 100	6 900
3	5 000	690	4 310	2 590
4	5 000	259	4 741	(2 151)

From this example it can be seen that the initial capital invested plus interest will be repaid some time between years 3 and 4. If it is assumed that the cash available for capital repayment comes in equal monthly amounts, then the monthly instalment will be as follows:

$$\frac{2590 - (-2151)}{12} = \frac{4741}{12} = 395$$

At the end of year three, £2590 of capital still has to be repaid. Therefore at £395 a month it will take:

$$\frac{£2590}{£395} \text{ months to repay} = 6.5 \text{ months}$$

The payback period is, therefore, three years and 6.5 months.

If a loan had been agreed at 10% interest, repayable over five years, the project would have been feasible because repayment of interest and capital would have been achieved within the term of the loan. If the funds had only been available for three years, then alternative finance would have been required.

A measure of the period for which the investment is at risk can be gauged from establishing the payback period.

Limitations of the payback method

In the payback method, no account is taken of the cash flow after the capital and interest has been recovered, and therefore no measure is gained of the sensitivity of a project, in terms of overall income potential, to changes in economic conditions which might adversely affect profit levels over the period of the investment project.

The effect of taxation on calculation of worthwhileness and feasibility

The tax effect on the cash flow from an investment can be important, resulting in a significant change in the return on capital employed and the payback period.

Tax allowances, however, can help to reduce and delay the implications of taxation. Ignoring personal allowances, there are two areas where tax allowances, which directly relate to an investment, can alter the cash flow:

(1) *Capital allowances.* These are available where fixed assets (machinery, plant, buildings) have been acquired, and they can be set off against potential taxable income. Capital allowances are not available on any working capital invested in a project. The impact of these allowances has reduced in the past few years because of changes in tax law.

(2) *Interest payments.* In most instances interest can be charged against profits before the taxable income is assessed.

It should be remembered, however, that where a business is not making profits and therefore has no tax liability it may not be able to gain any tax advantage until some time in the future, or never if future profits are not made!

To refine the techniques for assessing worthwhileness and feasibility of an investment project the tax implications can be built into the calculation as shown in Example 9.9.

Example 9.9 The impact of taxation on the calculation of feasibility and worthwhileness

A farmer is considering improving his farm by draining 20 hectare of his wetter land. Before reaching a decision on whether to proceed with the investment he wants to establish if the project will be (1) profitable; (2) worthwhile; and (3) feasible.

The capital cost, after grant, will be £600 per hectare. The improved land will allow a change in the four-year rotation from:

5 ha winter wheat	*to*	5 ha potatoes
5 ha winter wheat		5 ha winter wheat
5 ha winter barley		5 ha winter wheat
5 ha winter barley		5 ha winter barley

There will be no need for additional machinery, as potatoes are already grown on the farm and sufficient labour is available to cope with the increased potato hectarage. There will, however, be an additional working capital requirement of £2000.

The gross margins per hectare achieved, before and after drainage, are expected to be:

Crop	Gross margin Before drainage £	 After drainage £
Winter wheat	400	500
Winter barley	300	400
Potatoes	—	700

The project, if undertaken, will have to be financed entirely on borrowed money which, it is estimated, will attract interest at 15% over the period of the project. An eight year bank loan can be arranged. After careful discussion with the local ADAS advisor it is determined that the drainage scheme should have an effective life of ten years. It was assumed that the grant was received immediately (in practice this is optimistic).

The farmer's marginal rate of tax is 29%.

To assist in assessing the viability of the investment project the farmer has decided to calculate:

(1) The profitability using a partial budget
(2) The worthwhileness using rate of return, not present value and internal rate of return
(3) Feasibility using the payback period before tax
(4) Feasibility using the payback period after tax

Calculating profitability using a partial budget

The investment is profitable, as demonstrated by the annual gain in income of £1100, shown in the partial budget below which determines average profit:

Losses		*Gains*	
Extra costs	£	*Costs saved*	£
Interest:			
on drainage $\frac{12\,000}{2} \times 15\%$	900		
on working capital	300		
Depreciation:			
on drainage (over 10 years)	1 200		
Income foregone		*Extra income*	
5 ha barley @ £300/ha	1 500	5 ha potatoes @ £700/ha	3 500
Gain	1 100	10 ha wheat @ £100/ha	1 000
		5 ha barley @ £100/ha	500
	5 000		5 000

NB:

(1) The interest on the drainage is based on half the total capital invested in drainage
(2) The interest on the working capital is based on the total working capital
(3) Depreciation of the drainage is based on a ten year life

Worthwhileness calculated by the rate of return method formula

$$\frac{\text{Average additional annual profit before interest but after depreciation} \times 100}{\text{Additional capital}}$$

$$= \frac{1100 + 900 + 300)}{(12\,000 + 2000)} \times 100$$

Rate of return = 16.4%

Worthwhileness calculated by the net present value and internal rate of return methods

Excluding tax implications shown in Table 9.2, a 10% difference in discount rates gives a movement in NPV as follows:

$$+£3998 + (-£1329) = £5327$$
$$\text{therefore IRR} = 15\% + 7.5\%$$
$$= 22.5\%$$

NB: The working capital tied up in the project is released in year 11.

Table 9.2 The effect on NPV at different discount rates

Year	Capital investment	Discount rate 15%			Discount rate 25%		
		Gross annual cash flow	Discount factor	Product	Gross annual cash flow	Discount factor	Product
0	14 000		1.000	–14 000		1.000	–14 000
1		3 500	0.870	3 045	3 500	0.800	2 800
2		3 500	0.756	2 646	3 500	0.640	2 240
3		3 500	0.658	2 303	3 500	0.512	1 792
4		3 500	0.572	2 002	3 500	0.410	1 435
5		3 500	0.497	1 740	3 500	0.328	1 148
6		3 500	0.432	1 512	3 500	0.262	917
7		3 500	0.376	1 316	3 500	0.210	735
8		3 500	0.327	1 145	3 500	0.168	588
9		3 500	0.284	994	3 500	0.134	469
10		3 500	0.247	865	3 500	0.107	375
11		2 000	0.215	430	2 000	0.086	172
		NPV = + 3 998			NPV = – 1 329		

Worthwhileness can also be calculated by the NPV and internal rate of return methods including tax implications. In Table 9.3 the NPV at 15% is £1827 and at 25% is – £2629. The IRR is derived as follows:

$$\text{Internal rate of return} = 15 + \frac{(25 - 15) \times 1827}{1827 - (-2629)}$$

$$= 15 + \frac{10 \times 1827}{4456}$$

$$= 19.1\%$$

NB:

(1) In this example, capital allowances were taken at 4% per annum on the drainage investment. The capital allowances delayed the timing of tax payments
(2) When taxable income arose, the payment of tax on this taxable income did not occur until the following year, hence a tax liability in year 11
(3) At the end of the project the working capital was recovered but its present value was greatly reduced. No tax liability was attached to the repayment of working capital
(4) If there had been no working capital released in year 11 the tax payment of £876 would have given rise to a negative product. This product would have been deducted, along with the original capital, from the sum of the products (years 1 – 10) to give NPV

Table 9.3 Calculation of NPV (at 15% and 25%) including tax

Year	Capital invested	Gross annual cash	Tax allowances: Capital allowances machines buildings drainage £	Tax allowances: Interest at 15%	Tax allowances: Total allowances £	Taxable income £	Tax at 29% £
0	14 000						
1		3 500	480	2 100	2 580	920	
2		3 500	480	1 890	2 370	1 130	267
3		3 500	480	1 687	2 167	1 333	328
4		3 500	480	1 466	1 946	1 554	387
5		3 500	480	1 219	1 699	1 801	451
6		3 500	480	944	1 424	2 076	522
7		3 500	480	639	1 119	2 381	602
8		3 500	480	300	780	2 720	690
9		3 500	480		480	3 020	789
10		3 500	480		480	3 020	876
11		2 000					876

Net annual cash flow £	15% Discount factor	15% Product £	25% Discount factor	25% Product £	Feasibility: Interest £	Feasibility: Repayment of principal £	Feasibility: Balance £
	1.000	(14 000)	1.000	(14 000)			(14 000)
3 500	0.870	3 045	0.800	2 800	2 100	1 400	(12 600)
3 233	0.756	2 444	0.640	2 069	1 890	1 343	(11 257)
3 172	0.658	2 087	0.512	1 624	1 687	1 485	(9 772)
3 113	0.572	1 781	0.410	1 276	1 466	1 647	(8 125)
3 049	0.497	1 515	0.328	1 000	1 219	1 830	(6 295)
2 978	0.432	1 286	0.262	780	944	2 034	(4 261)
2 898	0.376	1 090	0.210	609	639	2 259	(2 002)
2 810	0.327	919	0.168	472	300	2 510	(+508)
2 711	0.284	770	0.134	363	—	—	
2 624	0.247	648	0.107	281	—	—	
1 124	0.215	242	0.086	97	—	—	
Net present value	= + 1 827		= −2 629				

Feasibility calculated by determining the payback period before tax

The calculation to establish the payback period for an initial investment of £14 000 given interest at 15% per annum on the outstanding balance is as follows:

Year	Cash flow before interest and depreciation £	Annual interest at 15% £	Annual payback capital £	Capital investment outstanding £
0				14 000
1	3 500	2 100	1 400	12 600
2	3 500	1 890	1 610	10 990
3	3 500	1 649	1 851	9 139
4	3 500	1 371	2 129	7 010
5	3 500	1 051	2 449	4 561
6	3 500	684	2 816	1 745
7	3 500	262	3 238	(1 493)

Payback per month in year 6 = £1745 – (–1493) = £270

$$\text{Therefore £1745 can be paid back in } \frac{1745}{270} \text{ months} = 6.5$$

Therefore payback period = 6 years 6.5 months

NB: The interest chargeable on an annual basis differs from the average figure shown in the partial budget. The interest is assessed on the outstanding balance.

Feasibility calculated by payback method after tax

Columns 14–16 in Table 9.3 show how payback after tax and interest is measured. To arrive at the precise time of payback the following calculation can be made:

$$\text{Payback per month in year } 7 = \frac{\text{£}2002 + \text{£}508}{12} = 209$$

$$\text{Therefore £2002 can be paid back in } \frac{2002}{209} = 9.6 \text{ months}$$

The payback period is therefore 7 years 9.6 months which is feasible as an eight-year loan is available.

Inflation and capital appraisal

When assessing new projects, it is important to distinguish between real and money returns, and real and money costs of capital.

Real returns and real costs of capital

'Real' means that inflation is excluded from any of the figures used in the capital appraisal calculation, that is, no inflation is assumed in the forecast income and expenses associated with the annual cash flow, meaning that constant prices are used.

Money returns and money costs of capital

'Money' means that inflation is included in the figures used in the forecasts of income and expenses, that is, there is a stated level of inflation assumed in any forecasts, although inflation of prices will not necessarily be constant.

Clearly when discussing the returns from projects it is important that all parties are talking in the same terms, as there is a great danger that misinterpretation could occur, as shown in Example 9.10. It is important not to mix the two concepts when budgeting.

Example 9.10

Consider the case of an accountant telling his farmer client that he estimates the project to put up additional pig housing for 50 sows will show a return of 8%. The farmer could well say 'But I can get 6.5%, tax paid, from a building society without any risk.'

In the majority of cases the accountant will have calculated the return from the new development in real terms, i.e. inflation excluded. The return from the building society, however, will only be available whilst inflation runs at a high level and interest rates remain high. The return is quoted in money terms, and in real terms it could be negligible.

What approach should be adopted concerning the problem of inflation?

(1) Ensure that when discussing returns from capital expenditure all parties can distinguish between real and money returns
(2) Present a consistent approach when forecasting future cash flow. Adopt a policy of working in real terms or money terms, not a mixture of both
(3) Where it is decided to operate in money terms, ensure that the level of inflation assumed (it may vary for different elements of the calculation) is clearly recorded
(4) If the assessment is made in real terms (constant prices) then test the sensitivity of the calculation to differential inflation, i.e. where inputs (fertilisers, concentrates) are increasing in price at a faster rate than returns (grain prices, milk)
(5) If constant prices are assumed, then it is reasonable to accept lower returns from a project. If, however, inflation is built into returns, the percentage return on capital at which a project becomes acceptable will be higher. The target level of return, therefore, should be related to the level of inflation used in the calculation

(6) It should be realised that the real cost of borrowing money is generally lower than the apparent money cost, though this is not always true, as evidenced in 1985 and 1986 (see Examples 9.11 and 9.12)

Example 9.11 The cost of borrowing money at 20% APR

The cost of borrowing money from a bank for 12 months is estimated to be 20% annual percentage rate. Inflation over the period in question runs at 15% per annum. A farmer borrows £100 for 12 months at 20% interest.

	£	
Capital borrowed	100	
Interest for 12 months	20	(the money cost of borrowing)
Total capital and interest	120	

As inflation is running at 15% the *real* cost of the loan is:

$$£120 \times \frac{100}{115} = £104.34$$

Example 9.12 The cost of borrowing money at 15% APR

The cost of borrowing money from a bank for 12 months is estimated to be 15% annual percentage rate. Inflation over the period in question runs at 6% per annum
A farmer borrows £100 for 12 months at 15% interest.

	£	
Capital borrowed	100	
Interest for 12 months	15	(the money cost of borrowing)
Total capital and interest	115	

As inflation is running at 6% the *real* cost of the loan is:

$$£115 \times \frac{100}{106} = £108.49$$

In this example the 'real cost' of borrowing is higher than in Example 9.11 although the interest rate charged is lower!

Summary of capital appraisal techniques

To enter into major capital investment projects without testing the viability of the investment is foolhardy. However, any management technique is only as good as the information used. Results should be interpreted in relation to the initial information from which they are derived. All the appraisal methods have their place, but the more sophisticated techniques have, to some extent, overcome the weaknesses in the simple calculations traditionally used.

It should be remembered that appraising the viability of capital investment by calculating profitability, worthwhileness and feasibility is only one facet of the overall decision-making process. The farmer will also be assessing other factors, such as risk, inflationary effects, social considerations and the general political and economic outlook before reaching a final decision on investment.

10 Planning: Labour and Machinery

Labour and machinery costs usually form a large part of the overhead costs of a farm business. It is therefore necessary to effectively measure and control the use of these two resources. A labour efficiency index calculation allows some conclusions to be drawn and perhaps even suggests some action to be taken; a labour profile is even more revealing. Peaks and troughs can be observed, and changes can be made to the enterprise mix to make allowances for them.

A labour profile is constructed by calculating the hours required by each enterprise per month. These figures can be obtained from standard published data or from records kept on the farm. The hours available for work on each enterprise per month are then calculated. By comparing the two sets of figures together (or by drawing a graph or histogram) it is possible to determine the peak labour demand and whether there are periods when labour is in surplus. The next step is then to take action to minimise peaks (by better mechanisation, use of contractors, or altering the enterprise mix), utilise surplus labour better, or to reduce the workforce if large surpluses of labour are apparent.

It is at this point that some formality can be applied to farm business planning.

H. Burr and D.B. Wallace (Cambridge University) introduced the concept of 'planning by intuitive but ordered sequence', which became the starting point of many other planning techniques relating to labour and machinery usage.

Planning by intuitive but ordered sequence

The first step is the same, as in all planning exercises, and that is to establish the resources available and then develop the opportunities that exist within the imposed limitations to the business.

Clearly, an adjusted gross margin account is essential as the initial benchmark for further plans (see Chapter 4). The procedure is to examine ways in which farm profit may be increased under one of the following headings:

(1) By increasing total gross margins at the same level of overhead costs
(2) By increasing total gross margins while accepting a smaller increase in the level of overhead costs
(3) By maintaining present gross margins but reducing overhead costs
(4) By reducing total gross margins and decreasing overhead costs by a greater amount

It is important to note that these proposed changes are not being created through improvements to husbandry practices. These have already been suggested in converting a normalised account to an adjusted account to reach this stage. The changes suggested in this method involve reorganisation of the structure of the business. By working in the order suggested above it is possible, first of all, to seek a combination of enterprises that yields a higher total gross margin than is currently being achieved within existing resources, and without any increase in overhead costs. Step by step, other enterprises and extra gross margins can be investigated until ultimately extra overhead resources are needed to allow the change to take place. This may mean that extra labour or an additional piece of machinery must be acquired, the effect of which can be measured (using a partial budget) and incorporated in the ultimate profit calculation.

Similarly, decreases in overhead costs can be investigated while total gross margins are maintained or even reduced. This is not just a case of 'pulling in the belt' as far as overhead costs are concerned; it involves a detailed study of areas where overhead costs can consciously be reduced.

The problem occurs when some real measure is needed of how much overhead costs are likely to fall or rise under a new regime of cropping and stocking. This is especially so in respect of labour and machinery costs which are always the largest overhead cost items.

At this point we can apply the labour charts mentioned earlier (Table 10.1).

Example 10.1 Use of labour profiles in farm planning

It is assumed that the enterprises shown in the labour profile in Table 10.2 combine to produce a total gross margin (as taken from the adjusted accounts) of £131 743.

Table 10.2 also reveals a huge labour surplus in nearly every month except March. It is possible to re-organise the enterprise mix and also reduce the work force by one man (Table 10.3).

By growing 20 ha less spring barley and 10 ha extra winter wheat and winter barley, the total gross margins increase to £133 453, a gain of £1710. This enterprise rearrangement releases enough labour at peak times to allow the labour force to be reduced by one man. Overhead costs would be reduced by £6000, creating a total gain for the change of £7710 over the original plan. There is now a small deficit of labour in March which can be overcome using casual labour or overtime.

The March labour deficit is not unsurmountable. The labour profile for the remainder of the year is much better balanced than before, but still shows surpluses of labour available, especially in June and July. Clearly other changes could be investigated in a similar manner.

Thus one of the main advantages of this procedure is demonstrated in that it does focus attention on the seasonal labour and machinery requirements of different enterprises and different crop and stock combinations. Indeed the labour profile itself may indicate some of the possible changes that could be considered. It is one means of testing the feasibility as well as the profitability of a new farm plan; only

Table 10.1 Standard figures for a labour profile

	ha	Labour requirements: man hours per ha or per head												
		Oct	Nov	Dec	Jan	Feb	Mar	Apr	May	Jun	Jul	Aug	Sep	Total
W. wheat	30	4.8	0.8	0	0	0	0.6	1.1	0.5	0	0	3.1	2.9	13.8
W. barley	45	2.3	0.6	0	0	0	0.7	0.7	0.3	0	1.5	4.5	2.9	13.5
S. barley	35	0.8	1.3	0.8	0	0	3.8	0.6	0.3	0	0	3.5	1.8	12.9
Potatoes	0	0	28.4	8.2	0.7	0	6.2	12.5	1.6	3.3	3.4	0.8	7.4	72.5
Dried peas	0	1.2	0.7	0.2	0	0.2	3.2	1.2	2.5	0.2	2.2	2.5	0.7	14.8
Oilseed rape	0	0.5	0	0	0	0	1.0	0	0	0	2.0	5.5	2.0	11.0
F. roots	0	1.0	1.0	0	0	0	1.5	1.5	1.0	0	0.5	0.5	0	7.0
Silage	50	0.3	0.3	0.2	0	0.6	0.8	0.9	5.7	3.9	2.0	0.2	0.6	15.5
Grazing	80	0.3	0.3	0.2	0	0.6	0.8	0.9	0.6	0.6	0.6	0.2	0.6	5.7
Waste	10													
Total	250													
	No.													
Dairy cows	150	3.0	3.3	3.3	3.3	3.3	3.3	3.0	2.7	2.7	2.7	2.7	2.7	36.0
Heifers	0	3.5	3.5	3.5	3.5	3.5	3.5	1.6	1.6	1.6	1.6	1.6	1.6	30.6
Beef	0	2.0	2.0	2.0	2.0	2.0	2.0	2.0	2.0	2.0	2.0	2.0	2.0	24.0
Ewes	300	0.3	0.2	0.2	0.3	0.3	1.0	0.4	0.3	0.4	0.2	0.2	0.2	4.0
Sows	45	2.5	2.5	2.5	2.5	2.5	2.5	2.5	2.5	2.5	2.5	2.5	2.5	30.0
Piglets	875	0.05	0.05	0.05	0.05	0.05	0.05	0.05	0.05	0.05	0.05	0.05	0.05	0.6
Man hours available														
General farmworker (GFW)		162	98	92	102	100	158	176	216	224	231	216	188	1963
Stockman		200	200	200	200	200	200	200	200	200	200	200	200	2400

Table 10.2 Labour profile for standard farm (units are hectares for crops and numbers for livestock)

		Hours needed per enterprise (rounded up)												
	Units	Oct	Nov	Dec	Jan	Feb	Mar	Apr	May	Jun	Jul	Aug	Sep	Total
W. wheat	30	144	24	0	0	0	18	33	15	0	0	93	87	414
W. barley	45	104	27	0	0	0	32	32	14	0	68	203	131	611
S. barley	35	28	46	28	0	0	131	21	12	0	0	123	63	452
Silage	50	15	15	10	0	30	40	45	285	195	100	10	30	775
Grazing	80	24	24	16	0	48	64	72	48	48	48	16	48	456
Dairy cows	150	450	495	495	495	495	495	450	405	405	405	405	405	5 400
Ewes	300	60	60	60	90	90	300	120	90	120	60	60	75	1 200
Sows	45	113	103	113	113	113	113	113	113	113	113	113	113	1 356
Piglets	875	44	44	44	44	44	44	44	44	44	44	44	44	528
Total/month		997	848	766	742	820	1 237	930	1 026	925	838	1 067	996	11 192
Add 15% for maintenance		150	127	115	111	123	186	140	154	139	126	160	149	1 679
Total required		1 147	975	881	853	943	1 423	1 070	1 180	1 079	964	1 227	1 145	12 871
Man hours available														
GFW 3		486	294	276	306	300	474	528	648	672	693	648	564	5 889
Stockmen 5		1 000	1 000	1 000	1 000	1 000	1 000	1 000	1 000	1 000	1 000	1 000	1 000	12 000
Total available		1 486	1 294	1 276	1 306	1 300	1 474	1 528	1 648	1 672	1 693	1 648	1 564	17 889
Surplus or deficit		339	319	395	453	357	51	457	468	593	729	421	419	5 018
Total gross margin														131 743

Table 10.3 Change 1 to standard farm enterprises (units are hectares for crops or numbers for livestock)

		Hours needed per enterprise (rounded up)												
	Units	Oct	Nov	Dec	Jan	Feb	Mar	Apr	May	Jun	Jul	Aug	Sep	Total
W. wheat	40	192	32	0	0	0	24	44	20	0	0	124	116	552
W. barley	55	127	33	0	0	0	39	39	17	0	83	248	160	746
S. barley	15	12	20	12	0	0	56	9	5	0	0	53	27	194
Silage	50	15	15	10	0	30	40	45	285	195	100	10	30	775
Grazing	80	24	24	16	0	48	64	72	48	48	48	16	48	456
Dairy cows	150	450	495	495	495	495	495	450	405	405	405	405	405	5 400
Ewes	300	75	60	60	90	90	300	120	90	120	60	60	75	1 200
Sows	45	113	113	113	113	113	113	113	113	113	113	113	113	1 356
Piglets	875	44	44	44	44	44	44	44	44	44	44	44	44	528
Total/month		1 052	835	750	742	820	1 175	936	1 027	925	853	1 073	1 018	11 199
Add 15% for maintenance		158	125	113	111	123	176	140	154	139	128	161	153	1 681
Total required		1 210	960	863	835	943	1 351	1 076	1 181	1 064	981	1 234	1 171	12 880
Man hours available														
GFW 3		486	294	276	306	300	474	528	648	672	693	648	564	5 889
Stockmen 4		800	800	800	800	800	800	800	800	800	800	800	800	9 600
Total available		1 286	1 094	1 076	1 106	1 100	1 274	1 328	1 448	1 472	1 493	1 448	1 364	15 489
Surplus or deficit		76	134	213	271	157	–77	252	267	408	512	214	193	2 609
Total gross margin														133 453

this time we are testing the flow of labour through a farming year and not the cash flow as previous feasibility tests have attempted to show.

Unlike cash, however, labour cannot be put aside for some use in the future, when it may be needed. If a surplus of regular labour exists as indicated in the labour profile, then it is wasted and can not be transferred to other periods of the year when labour may be scarce.

Similarly, an increase or reduction in labour requirement cannot be split into convenient portions. Regular labour comes as a 'chunky' input, and whilst use of overtime, casual labour and contractors allows some flexibility within the business, it is not possible to regularly employ part of a man.

The technique of 'planning by intuitive but ordered sequence' is enhanced by testing the feasibility of new plans with labour profiles. The weakness occurs in the use of standard labour performance data which may not apply to many farm situations where specialist gangs of men and machines accomplish major seasonal tasks such as silage making or potato, sugar beet or cereal harvesting. A labour feasibility test would be much more credible if actual farm performance data for men and machinery combinations were available.

Gang-work day charts

Nearly all farmers know how much work can be achieved by a given combination of men and machines in an average day. It is the application of that knowledge that has led to the use of gang-work day charts in testing whether a given labour force can accomplish a given set of tasks during main peak labour periods. The following example examines an isolated autumn labour peak, which is nearly always the time on farms when labour is the most limiting resource. If, therefore, the objective of planning is to maximise returns to the most limiting resource and the plan evolved creates an autumn peak, it is essential that the feasibility is thoroughly tested as in Example 10.2(a).

Example 10.2(a) Use of gang-work day charts in planning

The first step, as always, is to collect sufficient accurate data about the resources to allow a meaningful gang-work day chart to be drawn up.

The cropping regime of the 120 ha farm in question is established as:

		Gross margin/ha £
Winter wheat	40 ha	700
Spring barley	40 ha	600
Potatoes	15 ha	1 400
Sugar beet	15 ha	1 200
Dried peas	10 ha	650

(Livestock enterprises are excluded in this example.)

Other information required

(1) The number of days available for field work during the peak period involved is needed. This may seem an unreasonable demand when it can be argued that no two seasons ever produce identical weather conditions for field work to take place. Observations over many years in different regions on different soil types have, however, been made, and a fund of knowledge is now available on this subject. It is possible now for a farmer, in any area, to refer to books and obtain a useful estimation of the number of days he could expect to have, in an average year, to accomplish his field work. This data makes allowance for such things as overtime, use of headlights, illness, machinery maintenance and other contingencies. Clearly some years will be less favourable than others, but until long-range weather forecasts become reliable, planning must revolve around these figures!

In this example the days available for field work are: August, 24; September, 22; October, 17; November, 14; December, 10

(2) The remaining necessary information includes (Table 10.4):
   (a) A list of activities relating to each enterprise
   (b) The number of men required for each job
   (c) The rate at which each job can be accomplished (ha per day)
   (d) Any restrictions that apply to the timing of specific tasks

Table 10.4

Task	Time needed (days)	Gang size	Notes
Harvest (Ww/Sb)	6.5	3	Start not earlier than 10 August
Baling	8	1	All straw baled
Bale carting	3	1 (+ casual labour)	
Dung carting	1	1	Potato land (two men and spreaders could be used)
Subsoiling	3	1	All root crops
Potato harvest:			
(a) Sold off field	0.4	3 (+ casual labour)	Earliest start mid-September
(b) To store	0.5	3 (+ casual labour)	Half potatoes must be sold off
or (c) To store	0.6	4 (+ casual labour)	the field. Must end by mid-November
Sugar beet harvest	0.8	3	
Plough	2.4	1	Can be doubled up if required
Drill w. wheat	6.5	2	Includes cultivations and combine drilling. Must end by mid-November

Tasks and work rates

Total men available: 5

The procedure now is to combine this information and convert it for the gang-work day chart. For example, 80 ha of corn must be harvested at a rate of 6.5 ha per day, i.e. 12.3 days needed. In practice, this would probably be rounded to 13 days. Thus three men are employed fully for 13 days on the gang-work day chart, and whilst these are rarely consecutive days, it is convenient to plot them as one block on the chart.

Similarly, the time taken for other tasks can be calculated and plotted accordingly (Table 10.5).

Table 10.5

Task	Time needed (days)	Gang size
Cereal harvest	13	3
Bale	10	1
Bale cart	27	1 (+ casuals)
Dung cart	15	1
or	8	2
Subsoil	10	1
Pot harvest (1/2)	19	3 (+ casuals)
Pot harvest (2nd 1/2)	15	3 (+ casuals)
or	13	4 (+ casuals)
Sugar beet harvest	19	3
Plough	24	1
or	12	2
Drill w. wheat	7	2

When all these are plotted on the gang-work day chart, it can be seen that few problems exist during this 'so-called' autumn peak within the limits set. There are in fact gaps in the gang-work day chart created by surplus labour which, unless used constructively, are wasting money. It is not difficult to see that other changes to the plan should be investigated to increase profits. Such changes could involve:

(1) Increasing the area of high gross margin crops at the same level of overhead costs
(2) Buying sophisticated root harvesting equipment to allow even more potatoes or sugar beet to be grown
(3) Maintaining the same enterprise mix but reducing the labour force by one man
(4) Reducing intensive enterprises to a three-man regime, even if practices such as minimal cultivations have to be considered

Many alternatives could be considered and tested by this method, but it is a long and tedious process; there is a short cut.

Simplify then rebuild

The total gross margin for the enterprise combination in Example 10.2(a) is shown in Example 10.2(b).

Example 10.2(b) Total gross margin

		£
40 ha w. wheat	@ GM £700/ha	28 000
40 ha s. barley	@ GM £600/ha	24 000
15 ha potatoes	@ GM £1 400/ha	21 000
15 ha sugar beet	@ GM £1 200/ha	18 000
10 ha peas	@ GM £650/ha	6 500
120 ha	Total	97 500

A labour force of five men at a cost of £12 000 each (total £60 000) leaves a residue of £37 500 from which the other overhead costs of power and machinery, rent, interest charges and sundry overheads must be deducted. It is worthwhile investigating if a larger residue could be achieved within the known constraints.

The procedure for the technique of 'simplify then rebuild' requires the advisor to do precisely what the title suggests; that is, to postulate the simplest system that the farm could follow. From there the farm business can be gradually intensified and the marginal return to labour and machinery calculated as the number of men and the amount of capital invested in mechanisation increase. As each stage of the progressively more intensive rotation is reached, it is necessary to construct a gang-work day chart to test feasibility and discover how many men are required; just as a partial budget is needed when investment is made in a new item of machinery.

Example 10.3 Simplifying the plan

Simplification of the crop combination used in Example 10.2 would revolve around three men which is the minimum number needed to harvest corn. (More sophisticated harvesting tackle could reduce this still further and a contractor would eliminate regular labour altogether, but this example now sets out to investigate a three-man gang.)

It must be assumed that only 80 ha of straw is required as before and that peas can only be prudently grown one year in six from a rotational point of view.

Cropping could be 20 ha of peas and 100 ha of spring barley; a total gross margin of £73 000. Overhead costs are reduced by two men (i.e. £24 000), giving a residue for other overhead costs of £37 000, which is a drop in income of £500, albeit at this stage no account has been taken of the costs saved through discarding sugar beet and potatoes and the associated equipment. A gang-work day chart of this regime reveals a huge surplus of labour with no formal tasks to accomplish after mid-October. If instead that time was spent harvesting potatoes and sugar beet within a three-man labour force and the time periods specified, the cropping system could become:

Gross margins after enterprise changes		£
20 ha peas	@ GM £650/ha	13 000
78 ha s. barley	@ GM £600/ha	46 800
8 ha potatoes	@ GM £1 400/ha	11 200
14 ha sugar beet	@ GM £1 200/ha	16 800
	Total	87 800

This would leave a residue (after the labour costs of three men only) of £51 800; a gain of £14 300 over the original plan. Clearly a change of this nature is worth considering, but even this is not the end point. A whole host of new adjustments can be considered before the most profitable (but feasible) plan is derived. Each marginal substitution can be measured as the plan becomes more complex. However, when many enterprises are involved the procedure becomes very complex indeed, which is why other, more mathematical techniques have evolved.

Planning by returns to a limiting resource

In some ways this is the converse of the 'simplify then rebuild' method, in that after listing the opportunities and constraints, it commences by selecting enterprises which show the highest return to land. These are nearly always the most intensive enterprises, as shown in Example 10.4, a planning procedure based on 'creep budgeting'.

Example 10.4 A combination of gang-work day charts and marginal substitution in 'creep budgeting'

An autumn peak is again considered, but in this example the problem area occurs after the normal cereal harvest has been completed and the period involved is from late September through to Christmas. The enterprises available for combination are:

Enterprise performance and restrictions	GM per hectare £
Main crop potatoes	800
Sugar beet	600
Winter wheat	450
Spring barley	350
Spring beans	300
Restrictions to planning are:	
Total arable area	200 ha
Maximum cereals	150 ha
Maximum winter wheat	100 ha

- If sugar beet is grown, at least half must be harvested by mid-November (see periods shown in Table 10.6 below)
- All potatoes must be lifted by the end of October
- Two corn drills are available but all wheat must be drilled by mid-November
- Ploughing cannot be doubled-up but can continue into Period IV.
- Arable labour force is six men and it may be assumed that labour is limiting only at this time in the year

This data has been collected from the farm in question, although if it was not available, a reasonable estimate of work rates for any known combination of men and machinery could be obtained from the standard data available.

Table 10.6 Periods and rates of work

Period	I (late Sept and Oct)	II (Early Nov)	III (Late Nov and early Dec)	IV (Mid-Dec)
Days available	21	10	16	5
Task or gang-size		Rate of work (ha/day)		
Harvest potatoes:	0.9	n/p	n/p	n/p
Harvest beet:				
(a) three-man gang	1.2	1.2	0.8	0.8
(b) two-man gang	0.2	0.8	0.6	0.6
Plough:				
one-man gang	4.5	4.0	3.5	3.0
Cultivate and drill w. wheat				
(a) three-man gang	8.0	6.0	n/p	n/p
(b) two-man gang	6.0	4.0	n/p	n/p

n/p: not possible.

Planning procedure

Step 1

The first step is to construct an initial farm plan which is feasible within the given constraints. This may well be the existing plan if it can be seen to work. It is more likely that the planner will want to follow an original path, in which case a simple selection process can be applied. Selection by returns to land (i.e. in order of gross margins per hectare) testing feasibility with a gang-work day chart, will derive a useful starting point.

Plan 1

(a) *Potatoes* to limit of October harvest
21 days available at 0.9 ha/day allows 18.9 ha to be grown (five men)
(*NB:* ploughing must also start in Period I if it is to be completed within the limits set)

(b) *Sugar beet* to limit, half by mid-November
Ten days at (maximum) 1.2 ha/day allows 12 ha to be harvested in this Period II, i.e. 24 ha altogether

(c) *Winter wheat* to limit of two men available in Period II.
Ten days at 4 ha/day allows 40 ha

(d) *Spring barley* to limit of cereals, i.e. 110 ha

(e) *Spring beans* to limit of arable area, i.e. 7.1 ha

The whole of Plan 1 is tested in the gang-work day chart in Fig. 10.1 and summarised thus:

		£
18.9 ha potatoes	@ GM £1 400/ha	26 460
24 ha sugar beet	@ GM £1 200/ha	28 800
40 ha w. wheat	@ GM £700/ha	28 000
110 ha s. barley	@ GM £600/ha	66 000
7.1 ha s. beans	@ GM £575/ha	4 082
	Total	153 342

Step 2

This step considers alternative ways of redeploying the labour force during these limiting periods, one day at a time, and calculates the net effect on cropping and gross margins.

One such marginal adjustment in the use of labour could be made in Period 1. In the present circumstances, one man must be left to plough, but the five men harvesting potatoes could be split into teams of three and two men preparing land and drilling wheat (i.e. 8 + 6 ha/day) which would replace spring barley. It would mean that five men withdrawn from potato harvesting for one day would reduce that enterprise by 0.9 ha, so the area of spring beans must be expanded to fulfil the break crop requirement. The net effect of this change for one day is

$$\begin{array}{cccc} \text{Potatoes} & \text{W. wheat} & \text{S. barley} & \text{S. beans} \\ (-0.9 \times £1400) + & (14 \times £700) - & (14 \times £600) + & (0.9 \times £575) \end{array}$$
$$= £657 \text{ per day}$$

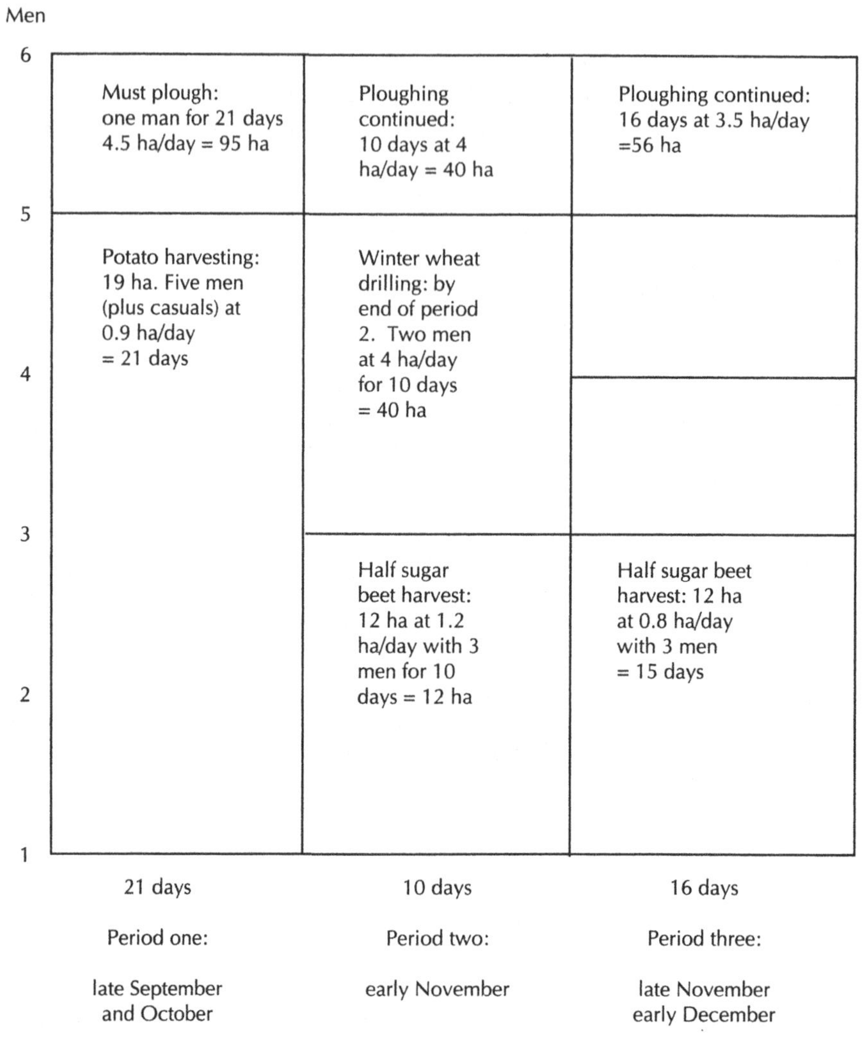

Men

6

Must plough: one man for 21 days 4.5 ha/day = 95 ha

Ploughing continued: 10 days at 4 ha/day = 40 ha

Ploughing continued: 16 days at 3.5 ha/day =56 ha

5

Potato harvesting: 19 ha. Five men (plus casuals) at 0.9 ha/day = 21 days

Winter wheat drilling: by end of period 2. Two men at 4 ha/day for 10 days = 40 ha

4

3

Half sugar beet harvest: 12 ha at 1.2 ha/day with 3 men for 10 days = 12 ha

Half sugar beet harvest: 12 ha at 0.8 ha/day with 3 men = 15 days

2

1

21 days	10 days	16 days
Period one:	Period two:	Period three:
late September and October	early November	late November early December

Fig. 10.1 Gang-work day chart for Plan 1.

It means that for each day this change is made, there will be a gain of £657. Now there are already 40 ha winter wheat and the maximum allowed is 100 ha. The change can therefore be made for only 60 ha/14 ha/day, i.e. 4.29 days.

Thus the maximum gain in gross margins by substituting wheat for potatoes is

$$4.29 \times £657 = £2820$$

In reality, this would be 'rounded-off' at either four or five days; but the resultant change is profitable and of course feasible.

Alternative changes can also be considered, some of which are listed in Table 10.7. The best suggestion is to replace potato harvesting by teams of three and two men drilling winter wheat and lifting sugar beet, respectively. Although the daily gain is only £1183, the number of days the change can be made increases to 5.3, giving a total gain of £6269.

Table 10.7 Alternative plans: marginal productivity of labour in alternative uses in Plan 1

Period	Substitution	Crop	Change/ day ha	Change in GM/day £	Gain max. No. of days	Total change £
1	Two gangs (3 + 2)	Pots	–0.9	–1 260	657	
	drilling w. wheat	Beet				
	in place of	Wheat	14.0	9 800	4.29	2 820
	potato	S. barley	–14.0	–8 400		
	harvest	S. beans	0.9	517		
1	Two men drill	Pots	–0.9	–1 260	1 357	
	w. wheat and	Beet	2.4	2 880		
	three men	Wheat	6.0	4 200	4.0	5 428
	lift beet instead	S. barley	–6.0	–3 600	(max. of	
	of potato	S. beans	–1.5	–862	beet)	
	harvest					
1	Three men drill	Pots	–0.9	–1 260	1 183	
	w. wheat and	Beet	1.8	2 160		
	two men	Wheat	8.0	5 600	5.3	6 269
	lift beet instead	S. barley	–8.0	–4 800	(above)	
	of potato	S. beans	–0.9	–517		
	harvest					

(It must be noted that any adjustments made to the area of sugar beet have a double effect as the area of late-harvested beet must not be more than half the total grown.)

Step 3

Select the best solution to date and substitute it in the farm system, which now becomes Plan 2.

Plan 2

		£
14.13 ha potatoes	@ GM £1 400/ha	19 782
33.54 ha sugar beet	@ GM £1 200/ha	40 248
82.4 ha w. wheat	@ GM £700/ha	57 680
67.6 ha s. barley	@ GM £600/ha	40 560
2.33 ha s. beans	@ GM £575/ha	1 340
200.0	Total	159 610

The gain over Plan 1 is £6269, as forecast.

Step 4

Steps 2 and 3 in the planning procedure can now be repeated several times, and the value of each marginal substitution calculated. The whole process can be repeated starting with a different initial plan, or varying the basis of selection at Step 3. If continued to the point where no further marginal substitution can profitably be made, it can be assumed that a nearly optimum solution has been found. In fact, this is not quite true, as shown by linear programming. The method does however yield a very good sub-optimum solution which itself is too precise for ordinary farm use. No practical farmers plan their farming system to one hundredth of a hectare, or consider only 2.33 ha of (say) spring beans. The method gives an indication of the best enterprise combination, which can now be adjusted manually to accommodate the actual farm problems of field size and other considerations.

Programme planning

'Creep budgeting' introduced formal selection by returns to a limiting resource and an arithmetic progression towards the optimum solution by marginal substitution on a mainly objective basis. Programme planning goes one step further. It is the ultimate 'hand' planning technique and a prelude to those which require a computer to resolve the complex mathematics involved. It is needless to say that the preparatory work for programme planning is also more refined.

In previous examples separate performance data have been collected for the activities involved in a single enterprise. For example, a wheat enterprise involves labour for harvest, baling, bale-carting (perhaps subsoiling), ploughing, cultivating and drilling, and maybe even an autumn application of chemicals. From Example 10.4 it can be seen that three men are needed to cultivate and drill 8 ha of winter wheat in an average day in Period 1. That is, three men for eight hours, i.e. 24 man hours to achieve eight ha or 3 man hours per ha.

Similarly, one man ploughs 4.5 ha per day in the same period, i.e. eight man hours to achieve 4.5 ha or 1.78 man hours per hectare. Thus the combined labour

Table 10.8 Table of ranks

Returns to:	Land		Late spring labour		Autumn labour	
	£/ha	Rank	£/hour	Rank	£/hour	Rank
Potatoes	1 400	1	140	4	15.6	5
Sugar beet	1 200	2	12	5	30	4
Clover	460	5	230	2		1
Winter wheat	700	3	234	1	70	2
Spring barley	600	4	200	3	67	3

The £/hour figures are calculated from data in Table 10.9.

requirement for wheat in Period 1 will be 4.78 man hours per hectare. The same can be done for other scarce labour periods and for other enterprises until a whole portfolio of information is gathered covering each enterprise that might be included in the eventual farm plan, this has been done in Table 10.8, which is based on data presented in Table 10.7.

The last stage of these preliminaries is to prepare a worksheet and write in the starting supply of each resource.

This allows the logic of the procedure to flow and removes some of the tedium of calculating.

Table 10.9 Labour requirements (hours/ha)

	Late spring hours	Autumn hours
Potatoes/ha	10	90
Sugar beet/ha	100	40
Winter wheat/ha	3	10
Clover/ha	2	—
Spring barley/ha	3	9

Step 1

The first step of formal planning is to select,by returns to a limiting resource and generally land in the first instance, the activity which shows the highest margin to that resource. Enter as many whole units as possible in the programme and deduct the required resources from the supply on the worksheet in Table 10.10. Thus potatoes are selected to their quota limit.

Step 2

Enter now as many whole units as possible of the activity which shows the next highest margin to the scarce resource (land) within the remaining available resources, i.e. sugar beet. Beet requires a lot of late spring labour and only 10 ha can be included in the plan, leaving a small balance of labour in columns three and four whilst only 22 ha of land have been used up.

Step 3

Continue to enter activities on this basis until one or more of the resources runs out. Wheat being the next in order of rank can be included to the limit of autumn labour, i.e. 8 ha.

Step 4

Determine which resource is most limiting for any possible additions to the plan. Calculate which activity shows the lowest return to this scarce resource and reduce the level of this activity by one whole unit from the worksheet. Potatoes rank last in this example so Step 4 returns 1 ha of this activity and 90 hours of autumn labour.

Step 5

Enter as many whole units as possible of the activity with the highest margin to the new limiting resource, followed by that with the next, until another resource becomes scarce. Thus clover, which has no autumn labour requirement and therefore ranks number 1, can be entered to the limit of late spring labour, i.e. 13 ha. (In fact some clover, 8 ha, could have been entered before the unit of potatoes was returned to the programme in Step 4. The result is the same whichever comes first.) The stage now reached shows in the worksheet (Fig. 10.2) that 42 ha of land are used up, yielding a gross margin to date of £38 980. Ninety hours of autumn labour remain but spring labour has run out. Steps 4 and 5 must now be repeated until all resources are exhausted or until the total gross margins can no longer be improved. At this stage, the arithmetic can be checked by orthodox budgeting and the near optimum plan declared. There are short cuts to programme planning as experience will show; indeed, it has been said that there are as many programme planning methods as there are programme planners. For the sake of completeness this example follows a formal procedure throughout.

Repeat (i) of Steps 4 and 5

One hectare of the enterprise which shows the lowest return to the limiting resource of late spring labour (namely sugar beet) has been returned to the 'balance', and as many hectares of the highest ranking activity as possible are entered (13 ha w. wheat). Clover can now be entered to its rotational limit as it has no need for the now limiting autumn labour.

Repeat (ii) and (iii)

The return of potatoes to the worksheet allows the entry of winter wheat until both late spring and autumn labour are limiting the consideration of further activities.

Fig. 10.2 Programme planning worksheet

		Land ha	GM £	Late spring labour hours	Autumn labour hours
	Start	120		1 160	1 560
Select initially by returns to land:					
Step 1 Potatoes to limit (quota)		12	16 800	120	1 080
	Balance	108	16 800	1 040	480
Step 2 Sugar beet to limit; late spring labour		10	12 000	1 000	400
	Balance	98	28 800	40	80
Step 3 W. wheat to limit; autumn labour		8	5 600	24	80
	Balance	90	34 400	16	—
Step 4 Clover to limit; late spring labour		13	5 980	26	—
	Balance	77	40 380	—	—
Step 5 Return 1 unit; potatoes		+1	–1 400	+10	+90
	Balance	78	38 980	10	90
Repeat 4 Return 1 unit; sugar beet (i)		+1	–1 200	+100	+40
	Balance	79	37 780	110	130
Repeat 5 W. wheat to limit; autumn labour		13	9 100	39	130
	Balance	66	46 880	71	—
Repeat 5 Clover to rotation limit		12	5 520	24	—
	Balance	54	52 400	47	—
Repeat 4 Return 1 unit; potatoes (ii)		+1	–1 400	+10	+90
	Balance	55	51 000	57	90
Repeat 5 W. wheat to limit; autumn labour		9	6 300	27	90
	Balance	46	57 300	30	—
Repeat 4 Return 1 unit; potatoes (iii)		+1	–1 400	+10	+90
	Balance	47	55 900	40	90
Repeat 5 W. wheat to limit; autumn labour		9	6 300	27	90
	Balance	38	62 200	13	—
Repeat 4 Return 1 unit; (iv) potatoes and sugar beet		+2	–2 600	+110	+130
	Balance	40	59 600	123	130
Repeat 5 W. wheat to limit; autumn labour		13	9 100	39	130
	Balance	27	68 700	84	—
Repeat 4 Return 1 unit; potatoes (v)		+1	–1 400	+10	+90
	Balance	28	67 300	94	90
Repeat 5 W. wheat to rotation limit		8	5 600	24	80

Fig. 10.2 *(continued)*

		Land ha	GM £	Late spring labour hours	Autumn labour hours
	Start	120		1 160	1 560
	Balance	20	72 900	70	10
Repeat 4 Return 1 unit; potatoes (vi)		+1	−1 400	+10	+90
	Balance	21	71 500	80	100
Repeat 5 S. barley to limit; autumn labour		11	6 600	33	99
	Balance	10	78 100	47	1
Repeat 4 Return 1 unit; potatoes (vii)		+1	−1 400	+10	+90
	Balance	11	76 700	57	91
Repeat 5 S. barley to limit; autumn labour		10	6 000	30	90
	Balance	1	82 700	27	1
Repeat 4 Return 1 unit; potatoes (viii)		+1	−1 400	+10	+90
	Balance	2	81 300	37	91
Repeat 5 Barley to limit of land		2	1 200	6	18
	Balance	—	82 500	31	73
Repeat 4 Return 7 units clover (ix)		+7	−3 220	+14	—
	Balance	7	79 280	45	73
Repeat 5 S. barley to limit of cereals		7	4 200	21	63
	Balance	—	83 480	24	10

Repeat (iv) and (v)

Sugar beet and potatoes are both returned to the plan, allowing wheat to be included to its rotational limit of 60 ha. (*NB:* it is important to keep an eye on the build-up of activities which have a specified maximum area. If space allows on the worksheet one or more columns could be set aside for that purpose.)

Repeat (vi), (vii) and (viii)

More potatoes returned to the plan allows spring barley to be entered until a situation arises where the resource, land, runs out whilst a surplus of labour remains.

Repeat (ix)

At this stage, the enterprise which shows the least return to land can be returned and replaced with a higher ranking alternative.

In this case, 7 ha of clover were returned to allow barley to be grown to its rotational limit of 30 ha. Only by splitting the units of activities into fractions can the plan be improved within the imposed constraints. The results of a linear programme

Table 10.10 Summary of the farm system developed by programme planning

	Area	Activity	Gross margin	Labour used	
				Late spring	Autumn
	ha		£	hours	hours
	4	Potatoes	5 600	40	360
	8	Sugar beet	9 600	800	320
	18	Clover	8 280	36	—
	60	W. wheat	42 000	180	600
	30	S. barley	18 000	90	270
Total	120		83 480	1 146	1 550
Supply	—			1 160	1 560
Surplus	—			14	10

show the optimum solution to be only marginally better than the one 'programme planned'.

Computers and linear programming

All the planning methods described require physical and financial data to be collected about the farm before planning procedures can commence. The solution can only be as good as the starting data allow, and as the planning techniques become more sophisticated, so does the need for reliable performance data become more significant. Data collection, therefore, is the crux of meaningful farm planning.

Again, as planning techniques become more sophisticated, so the derivation of a feasible solution depends more and more on tedious arithmetic. It is only common sense that a machine should ultimately be employed to do these sums.

Linear programming

Computers have been involved in problem-solving in agriculture for over 20 years, but have generally been inaccessible to, or too expensive for, most farmers. The computer techniques that have been developed are manifold and these have been described in several publications. The following examples of linear programming are intended to demonstrate the technique in its simplest form and may serve to whet the appetite of any budding enthusiast.

The data required is similar to that for 'creep budgeting' (Example 10.4) and almost identical to that for programme planning (Table 10.8), although some of the terminology is slightly different and the problem matrix takes on a new form (Fig. 10.3). The aim is the same, however; that is, to produce that combination of activities (enterprises or operations) which maximise the value of the objective function (gross margin or net revenue) within a predetermined list of opportunities and constraints (land, labour or capital). A standard layout of a linear programming matrix could take the form shown in Fig. 10.3.

Data entry

The first information a computer generally requires is the size of the matrix which it will ultimately have to handle. This refers to the number of variable activities which are shown as columns 1 to 6 in Fig. 10.3 and the number of resources and constraints (rows a to e) which have to be considered. Thus if the data used in programme planning (Table 10.10) were to be prepared for linear programming, the first consideration would be a matrix of six columns (potatoes, sugar beet, clover, wheat and total cereals; and the maximum labour available in the limiting periods of late spring and autumn). The activities can now be entered individually under their allocated column number.

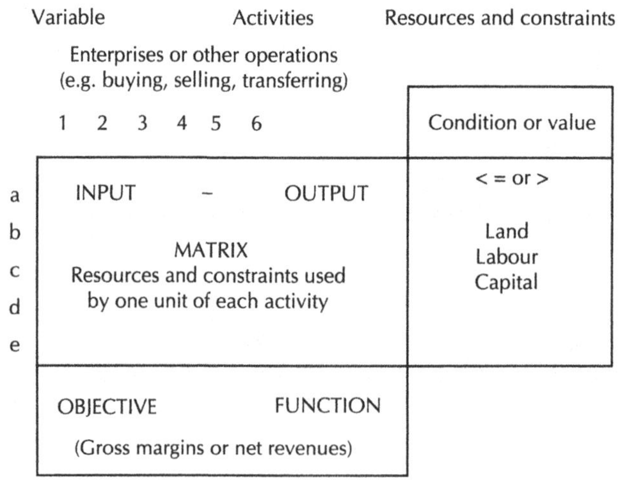

Variable Activities
Enterprises or other operations
(e.g. buying, selling, transferring)
1 2 3 4 5 6

Resources and constraints
Condition or value
< = or >
Land
Labour
Capital

a
b
c
d
e

INPUT – OUTPUT
MATRIX
Resources and constraints used
by one unit of each activity

OBJECTIVE FUNCTION
(Gross margins or net revenues)

Fig. 10.3 A linear programming matrix.

Constraints are now treated individually, and the level of each resource or constraint that is consumed by one unit of each of the activities is entered across the row of the main matrix. The right-hand-side column declares the maximum resource or constraint available and the condition that the sum of the inputs must obey. Figure 10.4 shows a complete linear programming matrix for the data first used in Table 10.6 (spring oats have not been manually discarded in these data). Row 'a' implies that each of the activities 1 to 6 consumes one unit of area and the total area must be less than or equal to 120 ha. The vertical column '4' shows the resources and constraints consumed by winter wheat and that one unit will contribute £700 towards the total objective function.

Fig. 10.4 Linear programming matrix with data entered

Variable / *Constraint*		1 Pots	2 Sugar beet	3 Clover	4 Wheat	5 Barley	6 Oats	*Condition value*
Maximum:								
Land	a	1	1	1	1	1	1	<120
Potatoes	b	1	0	0	0	0	0	<12
Sugar beet	c	0	1	0	0	0	0	<15
Clover	d	0	0	1	0	0	0	<25
Wheat	e	0	0	0	1	0	0	<60
Cereals	f	0	0	0	1	1	1	<90
Late spring labour	g	10	100	2	3	3	3	<1 160
Autumn labour	h	90	40	0	10	9	9	<1 560
Objective		1 400	1 200	460	700	600	540	*Maximum*

The computer is now in a position to solve the profit maximisation problem, the answer to which must be very similar to that achieved by programme planning in Table 10.10. The solution is shown in Fig. 10.5

Interpretation of the solution

One of the advantages of processing data by linear programming through a computer is that various other pieces of information are calculated as a 'spin-off'.

Item (1). Reveals a total net revenue of £83 629.176 in the optimum solution. Compare this with £83 480 in the solution achieved by programme planning and the gain can be seen as almost £150. The reasons for this will become obvious.

Item (2). Shows the levels of those activities included in the optimum solution. Programme planning operates in whole units of activities whereas the computer has broken these down, hence the slight increase in net revenue. The fact that it works to eight decimal places in some cases does tend to mock the expression 'optimum solution', because no farmer wants that level of precision. He can at least work back from the optimum to a practical cropping system.

Item (3). Introduces a new piece of information, which at first glance simply tells us that oats, as an enterprise, is not required. This is already known; but it also indicates the amount of change needed in net revenue before this enterprise would be included in the solution, i.e. if the net revenue for oats were improved by £58.59 per hectare, then it would be included in the solution. In more complex problems, especially where knowledge of the input data is not precise, information concerning the non-basic variables (unused enterprises) can be important. Minor changes to the input coefficients can be made in the matrix and the whole programme re-run.

Item (4). Is also new information. It shows which resources or constraints have been fully used up in solving the problem. Clearly the potato quota, the sugar beet

Fig. 10.5 Printout of a linear programming solution

(1) Value of objective	= 83629.176	
(2) basic variables:	Variable	Value
	5	30
	1	4.049 411 77
	3	17.811 764 7
	2	8.138 823 53
	4	60
(3) Non-basic variables:	Variable	Value
	6	60
(4) Binding constraints:	Constraint	Shadow price
	a	444.3789
	b	0
	c	0
	d	0
	e	104.8641
	f	24.9072
	g	2.4126
	h	12.5494
(5) Slack constraints:	Constraint	Slack
	b	7.950 588 23
	c	6.861 176 48
	d	7.188 235 3

quota and the clover limit have not been fully reached, but the remainder have. In addition, however, the programme has calculated the marginal value product (MVP) of each exhausted resource and constraint. For example, land is fully used, but the MVP indicates that if one additional hectare were available, then the total net revenue would increase by £444.38. Similarly, an extra hour of autumn labour would generate an additional £12.55 of cost.

Item (5): Shows those resources that have not been exhausted and the amount remaining.

Conclusion

There is still no substitute for sound husbandry practices! This statement has held good for so long that it is sometimes overlooked in the advent of new office technology.

This chapter has attempted to show the available methods for further precision in farm planning. Some of these are so obsolete or esoteric that they will never be applied to a farm situation but an understanding of their operation will assist the manager to appreciate how profitability, feasibility and worthwhileness can be accurately budgeted for.

11 Sources of Finance

Sources of capital for a business

Capital, along with land and labour, is a basic resource for any farm business. In many instances it can prove to be the ultimate limiting factor controlling the size of a business, preventing expansion and restricting profitability.

There are four basic forms in which a business can obtain capital. These are:

(1) The proprietor's own capital
(2) Profits retained in the business as a result of trading
(3) Capital provided by creditors
(4) Grants from the Government and European Union

The capital cycle

A business uses the proprietor's capital, grants and any creditor capital it can arrange to trade, with a view to making a trading profit. A trading profit itself does not necessarily mean that a business is prospering as there are a number of commitments to be met from this profit:

(1) *The servicing costs of any creditor capital.* These costs are deductible from the trading profit before assessing the taxable income
(2) *Taxation*
(3) *Repayment of creditor capital* This repayment is made out of income after tax has been paid. Repayment of loan capital will reduce the loan liability of the business as well as reducing 'retained profit'. It therefore follows that the two entries balance one another, leaving the overall level of proprietor's capital the same
(4) *The personal drawings of the sole trader or partner and the share dividends in the case of a limited company.* These drawings (dividends) in theory can only be taken after everybody else, including the taxman, has had his share of the returns

If there is any 'profit' left after these demands have been met by the business, then this 'net profit' is retained and will add to the proprietor's capital stake (Fig. 11.1).

There is always the possibility that a business is showing a trading profit but, because of the demands of creditor finance charges, taxation and drawings/dividends, total outgoings exceed total income, this will cause a drain on the proprietor's

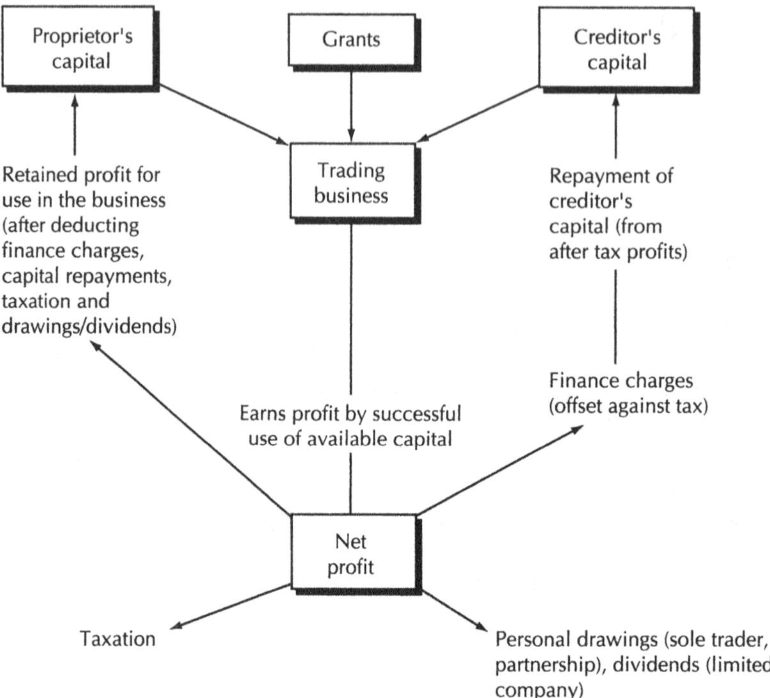

Fig. 11.1 The capital cycle.

capital which will progressively reduce the liquidity of the business. The test of liquidity is the ability to pay creditors' accounts as and when they fall due. If the trend of outgoings exceeding income continues it could result in the business becoming insolvent.

It is important to remember that a business can only provide its own capital, to fund expansion and the effects of inflation, from either retained profits or the injection of additional proprietor's capital.

The proprietor's own capital

This is the capital utilised in a business which belongs solely to the owner of the business. It is the money which will be available to the owner when all the assets of the business are sold and all the liabilities of the business paid out of the sale proceeds of the assets.

The proprietor's capital is assessed annually in the balance sheet of the business. The balance sheet is one of the financial accounts prepared for submission to the Inland Revenue for assessment of tax liability. In the balance sheet prepared by the accountant the assets may well be valued below their true sale price and so, to arrive at a more realistic assessment of proprietor's capital, revaluation is necessary (see interpretation of balance sheet ratios).

In the case of a partnership or sole trader, the proprietor's capital accumulates in the capital account which is recorded in the balance sheet. In the case of a company, the proprietor's capital is represented by the share capital of the company and various reserves belonging to the shareholders.

Whilst the proprietor does not have to pay interest on his own capital there is a definite cost to him of using his capital in a particular business venture. This cost is known as the opportunity cost.

There is a very high proportion of proprietor's capital in the national agricultural business (approximately 80%).

Profits retained in the business as a result of trading

To be able to retain funds in a business for future use means that sufficient trade profit must be generated to cover the commitments of the business (i.e. to pay tax, interest on creditor capital, repay creditor capital and to finance the living requirements of the proprietor). Any profit retained after meeting these commitments adds to the capital of the proprietor. It must be remembered that the repayment of creditor capital will not reduce capital; it does, however, reduce the liquid funds available to the business (see Chapter 5). Where a partnership or sole trader is concerned these profits add to the proprietor's capital account (see Fig. 11.2).

Fig. 11.2 Retention of profits in the capital account of a sole trader or partnership

	£	£
Capital account as at 31/5/1996	81 200	
Add profit for year end 31/5/1997	20 100	101 300
Less drawings	10 000	
Less tax	2 700	12 700
Capital account as at 31/5/1996		88 600

Capital provided by creditors

Creditor capital may take the form of trade creditors, merchant credit, family creditors or credit from financial organisations. Capital which is supplied to a business by creditors costs money in the form of interest and other charges. The amount of interest charged varies with the amount of capital borrowed and the source of capital.

The unit cost of creditor capital depends on a number of interrelated factors such as supply of and demand for capital, length of term of borrowing, the risk factors associated with a project and the stability and security of the individual farm business.

Creditor capital has now become a very important factor in many farm businesses as farmers have intensified and expanded their enterprises, with the help of outside finance.

As the demand for creditor capital has increased, so has the range of facilities made available by the financial institutions, in particular, as they have recognised the importance and stability of the industry and the market needs.

This increase in the complexity of the credit funds market demands that the farmer should alert himself to the opportunities available and at the same time he must be able to assess the relative advantages and disadvantages of the various sources of funds.

Creditor capital can be divided into three broad classes:

(1) *Long-term credit.* This is capital provided for a ten-year term or longer.
(2) *Medium-term credit.* This capital is provided for a term of between three and ten years.
(3) *Short-term credit.* This is provided for periods up to three years.

The main sources of credit funds and their uses are summarised in Table 11.1.

Table 11.1 Sources and uses of agricultural creditor capital from financial institutions

Period	Long-term credit (over 10 years)	Medium-term credit (3–10 years)	Short-term credit (up to 3 years)
Purpose	Land purchase Improvements Roads Houses Drainage	Plant and machinery Specialist buildings Breeding herds	Finance of arable and livestock production cycles Trading stock bridging finance
Source	AMC LIC Clearing banks Life assurance companies Merchant banks ICFC Private mortgages	Clearing banks Hire purchase companies Milk Marque Leasing companies AMC	Clearing banks Merchants' credit Hire purchase companies Leasing companies AMC

NB:

(a) Leasing companies do not lend money, but purchase equipment or machinery which they rent to customers. They can be regarded, however, as an alternative method of providing capital assets for use in the farm business
(b) AMC: Agricultural Mortgage Corporation
(c) LIC: Lands Improvement Company
(d) ICFC: Industrial and Commercial Finance Corporation

There is considerable overlap between the different classes of finance to cater for the needs of the individual farmer. However, it is prudent for the farmer and the lender to consider in broad terms the class of finance utilised for a particular project.

Financial institutions raise the funds they lend in differing ways and for different periods of time. Some raise long-term capital which they will not have to repay for many years; others will raise their funds on a shorter-term basis where they may have to be able to repay on demand or at short notice.

Clearly, if a financial institution is to remain viable, it must carefully consider the methods by which it has raised funds and when these moneys will have to be repaid. These facts will influence the period for which an institution will be prepared to 'on-lend' money to an individual farmer. If an institution borrows money on a short-term basis and lends it to a farmer for, say, 40 years, then it could experience difficulty in meeting its repayment commitment, or at best may have to raise additional funds in the market to cover its position. The cost of the new money could be very expensive and so reduce the profitability of the lending.

Alternatively, an institution may have raised long-term money at a fixed rate of interest and 'on-lent' on a short-term basis. In this case difficulties could arise if the borrower repaid early and the lender was unable to relend because of lack of new borrowers or because the cost of money in the market had dropped, making his long-term, fixed-rate interest look expensive.

Whilst this general premise holds true, financial institutions know that, in practice, they have some scope for flexibility. The major financiers now raise funds in a variety of ways, allowing them to increase the range of financial packages which they can provide.

A major reason for looking carefully at the class of finance used is to check (for both the farmer's and lender's benefit) that the capital is repaid within the effective life of a particular asset or project. In this way the farmer is not left paying for an asset that is not providing him with income to cover his commitments, and the financier is not left lending money for an asset which is not generating income for the farmer to assist him in servicing and repaying the credit taken.

How to assess the advantages and disadvantages of lines of credit

Apart from knowing the period of time for which a particular source of credit is available, it is important to compare the merits of the credit source in the following areas.

The purpose for which money will be lent

This is an obvious area for comparison and in most cases there is no problem in identifying which institutions are prepared to lend for what purposes.

The amount that will be lent

This will vary dramatically, but will depend on:

(1) *The amount of proprietor's capital a farmer has in his business.* If a loss situation arises in a business, then the first capital to be lost is that of the proprietor. Therefore, the more proprietor's capital there is in a business the greater protection against loss will be afforded to a supplier of credit finance. In general terms the more proprietor's capital there is in the business the more creditor capital will be available within the limits imposed by the profitability or otherwise of the business
(2) *The assessment by the financier of the farmer's management ability.* A farmer who can demonstrate that his technical performance is good and that he can translate this level of performance into good profits is likely to receive greater support than a farmer whose business performs modestly. The financier will be judging the ability of the farmer to achieve the performance levels outlined in his farm plan
(3) *The ability of a farm business to service and repay any borrowing.* In many respects the ability of a business to generate cash, to service and repay loans, is more important than ultimate profit. The 'profit' figure for a business may include valuation changes and will not have accounted for personal drawings and taxation. Thus the cash flow may be negative. Interest and capital repayments can only be met from a positive cash flow
(4) *The risk factor associated with a proposition.* The higher the risk factor associated with a project the greater will be the cushion of proprietor's capital the financier will want to see in a business to protect his lending. The level of proprietor's capital regarded as acceptable by the financier may be influenced by the amount of tangible security available

The cost of money

There are a number of ways that financial institutions quote interest rates and it is essential that the prospective borrower understands the basis of these calculations and ensures that all rates have been converted to a common denominator, before reaching a decision on which rate is best.

All interest rates should be converted to the true or annual percentage rate (APR). This rate takes account of the timing of interest charges and repayments reflecting the cash flow over the term of the loan.

There are two important factors to consider: how does a particular financier calculate the interest rate, and when will the farmer have to pay the interest?

How does the financier calculate interest rate?

There are two basic methods in use: nominal rate and flat rate.

The *annual nominal rate* is an annual rate applied to the daily balance outstanding. This means that if the daily balance goes down, then the annual nominal rate is applied to the reduced balance. If the balance outstanding increases, then the annual nominal rate is applied to the increased level of borrowing on a daily basis (see Example 11.1). This example ignores any impact of uncleared items.

Example 11.1 The basis of calculating the daily balance for calculating the annual nominal rate of interest

Date	Withdrawal £	Receipts £	Balance £	Balance on which interest charged £
10/1/1997	500	—	500 dr	500
11/1/1997	—	—	500 dr	500
12/1/1997	—	100	400 dr	400
13/1/1997	200	—	600 dr	600
14/1/1997	—	—	600 dr	600
15/1/1997	—	—	600 dr	600
16/1/1997	—	800	200 cr	—
17/1/1997	250	—	50 dr	50

The *period rate* expresses the annual nominal rate for periods of less than a year related to the particular charging period of a scheme. Interest is calculated on the balance outstanding on a specific day in the charging period. Thus the period rate is quoted as X% each charging period. For example, an annual nominal rate of 12% equates approximately to a monthly period rate of 1% per month, or 3% per quarter.

In the case of *flat rate* quotations the interest rate is expressed as a percentage of the original amount borrowed. The balance on which interest is charged does not reduce as the loan is repaid, this is shown in Example 11.2.

Example 11.2 Calculation of interest on a flat-rate basis

A farmer borrows £2000 to help purchase a forage harvester. The loan is to be repaid quarterly over two years. Interest will be charged on a flat-rate basis at 10%.

The interest charge is calculated as follows:

	£	£
Interest on £2000 at 10% pa over 2 years	= 400	
Total repayment of interest and capital	= 400 + 2 000	
	= 2 400	
Therefore quarterly repayments over 2 years	$= \frac{2\,400}{8}$	
	= 300	

When will the farmer have to pay interest?

If money is borrowed on the basis that interest is charged on a monthly basis, then it must be more expensive than borrowing when interest is charged on a quarterly, half-yearly or annual basis, because:

- each amount of interest is added to the borrowing. In this case there will be an interest-on-interest factor resulting in higher total interest charges; or
- new money has to be provided to pay off the interest, and this will mean the potential earning capacity of the new money is lost

The more frequent the charging period then, all other things being equal, the more expensive the credit becomes.

The conversion of annual nominal rates and flat rates to annual percentage rate equivalents is a complicated procedure and the farmer should ask for this conversion to be done for him. Tables 11.2 and 11.3 give an indication of approximate conversion rates.

Commission and other administrative costs

In addition to the interest rate charged it is important to establish any additional charges which may be associated with money that has been borrowed. In some cases

Table 11.2 Approximate APR equivalent of annual nominal rates for monthly, quarterly and half-yearly charging periods

	Approximate APR equivalent (%)		
Annual nominal rate	Monthly charging	Quarterly charging	Half yearly charging
5	5.1	5.0	5.0
6	6.1	6.1	6.0
7	7.2	7.1	7.1
8	8.2	8.2	8.1
9	9.3	9.3	9.2
10	10.4	10.3	10.2
11	11.5	11.4	11.3
12	12.6	12.5	12.3
13	13.8	13.6	13.4
14	14.9	14.7	14.4
15	16.0	15.8	15.5
16	17.2	16.9	16.6
17	18.3	18.1	17.7
18	19.5	19.2	18.8
19	20.7	20.3	19.9
20	21.9	21.5	21.0

Source: Consumer Credit Tables.

Table 11.3 Approximate APR equivalents of flat rates for various loan periods (equal monthly repayments of interest and capital)

	Approximate APR equivalent (%)			
Flat rate	6 month loan	12 month loan	18 month loan	24 month loan
3	5.2	5.6	5.7	5.8
4	7.0	7.5	7.7	7.7
5	8.8	9.4	9.6	9.7
6	10.7	11.4	11.6	11.7
7	12.5	13.4	13.6	13.7
8	14.4	15.4	15.6	15.7
9	16.3	17.4	17.7	17.7
10	18.3	19.5	19.7	19.7
11	20.2	21.6	21.8	21.7

Source: Consumer Credit Tables.

there may be an annual commission charge made, based on the size of the facility granted. Alternatively there may be an initial once and for all administrative charge or a charge for a manager's time. Banks charge for drawing cheques and paying-in credits but in some cases a small notional interest on any cleared credit balance is to be offset against these charges. However these charges are calculated they are bound to affect the overall cost of credit and should be established when the credit facility is agreed.

What factors affect the cost of money to farmers?

(1) *The cost of money to the financial institutions*. This is a complicated picture and a detailed commentary is beyond the scope of this book. However, in broad outline the factors affecting the cost of money are as follows.

Governmental factors. Governments have in the past manipulated interest rates, using them as a method of controlling the money supply. If they felt the money supply in the country was increasing too rapidly, resulting in inflationary trends, then they would raise the basic cost of money. This was achieved by raising the minimum lending rate (MLR) of the Bank of England. Minimum lending rate was abandoned on 28 August 1981, but the Government continued to influence interest rates through its activities in the short-term money market.

Sterling as an oil-based currency. With the advent of North Sea oil, sterling became a petro currency and is affected by oil prices. If prices rise sterling tends to strengthen and conversely if oil prices drop sterling tends to weaken. A government concerned with the level of sterling exchange rate might use the level of interest as a mechanism for stabilisation, hoping to encourage the buying of sterling when it was weak because the interest paid was attractive, or if sterling was riding high then reducing interest rates to reduce costs.

In June 1997 the responsibility for managing interest rates to control inflation was passed by the Chancellor, Gordon Brown, to a Bank of England Committee chaired by the Governor. They are charged with keeping inflation within set targets.

International interest rates. The rates of interest operating in other countries can influence our domestic rates. If rates are higher overseas then funds which are deposited in this country may be moved out (i.e. money deposited by the oil states). Where vast amounts of sterling are sold, then the international exchange rate of sterling can be adversely affected. As a result, our domestic interest rates can be forced up in an attempt to stabilise the outflow of deposits.

Supply and demand. There are occasions when the amount of money available to the financial institutions for on-lending is limited, for various reasons, at a time when demand is high. This can result in increased costs to the borrower as he competes for the limited funds available.

Inflation. High inflation is frequently associated with high borrowing and high wage rises, which result in the cost of goods increasing. This causes sterling to depreciate against other currencies because investors feel more secure with a stable currency. To encourage investors to retain sterling and thus maintain its international value, restrain borrowings and reduce inflation, interest rates are raised.

(2) *The methods by which financial institutions raise funds*. Some institutions raise the majority of their money directly from the general public through a branch network; others rely on raising money through the stock market and on the London money market.

The clearing banks (e.g. Barclays, NatWest), for example, raise a high, though reducing, proportion of their money in the form of current and deposit accounts held by their customers at branches throughout the country. The cost of these funds varies; deposit accounts attract interest and are therefore more expensive than current accounts, where very little or no interest is allowed to account holders at present. However, having large branch networks and providing money transmission services is costly and because the banks do not fully recover their overhead costs from commissions some of the interest margin is used for this purpose.

The Agricultural Mortgage Corporation (AMC), on the other hand, raises most of its funds through the stock market and the London money market, and the cost will vary according to the particular rates operating in the market at the time AMC wishes to borrow.

(3) *Whether the money raised by the borrower is on fixed or variable interest rate terms.*

Fixed rates of interest are determined at the commencement of the lending and the rate remains the same throughout the term of the facility granted.

When a farmer opts for fixed rates of interest, he knows precisely the cost of money for the duration of the loan and also the cash he will have to find at a particular time to finance and repay the borrowing. If a fixed rate loan is taken out when interest rates are high and there is no option to convert to a variable-rate loan, then when interest rates fall back the borrower could be committed to unnecessarily high costs over the term of the loan. The converse situation could also be true. If overall rates rise then interest costs may be stabilised. There are now options available whereby loans may be made with fixed rates to be reviewed every three to five years of a long-term loan.

Variable rates of interest are liable to change from time to time during the term of lending. The interest rate is linked to the cost of money to the lender. In the case of clearing banks the majority of their lending is on a variable-rate basis and linked to their base rates. The Agricultural Mortgage Corporation also has variable interest rate lendings which reflect the movement of the cost of money over short periods.

Where lendings are subject to variable rates of interest it is difficult, if not impossible, to predict the cost of borrowed money over the life of a project. The rates move in line with market interest rates and so will, over a period, reflect the average market cost of money. If a variable interest rate facility is taken when interest rates are high there is always the possibility of rates falling and so the cost of the facility might fall, whereas a fixed-rate loan would have remained at the high interest rate level. Where there is an option to convert from variable rate to fixed rate, then there is no doubt that in times of unusually high interest rates it is best to select the variable rate option initially and review the position as interest rates fall.

The ultimate decision of which option to select is not easy, and in the final analysis will be influenced by:

- The current level of interest rates in force at the time of the proposition
- An assessment of the likely trend in interest rates over the period for which finance is required
- The flexibility in any agreement with the lender to move from fixed to variable rates or vice versa

Later in this chapter various interest rate protection options are discussed.

(4) *The risk factors associated with a particular project and the term of the project.* The greater the risk factor (i.e. a weak security position or marginal viability), the higher the return the lender will be expecting from the investment. Similarly, longer-term rates tend to be higher than short-term rates, though this does not necessarily apply where one looks at future rates from a point in time when interest rates are historically high.

Supporting evidence required by the financier before he will lend

Most suppliers of credit will want the borrower to establish that his proposals are sensible and this will mean, in the case of an existing business, demonstrating that the present business is successful and the proposals for the future are viable.

To satisfy these requirements the farmer will almost certainly need to produce accounts, to establish past performance, with budgets and cash flows outlining the estimated future returns and cash requirements for the business.

Repayment terms of any borrowing

This is a very important area for consideration which is frequently neglected. The repayment terms should be matched to the cash flow of the business and to the effective life of the proposition. In recent years the suppliers of credit have become far more flexible in accepting fluctuating repayment patterns for their loans. It could well be that on a cereal farm it would be sensible to repay loans on an annual basis in October or November, whereas a mixed dairy and cereal unit might have a cash flow suggesting basic monthly repayments with a large reduction after harvest. Where a large capital item (e.g. land) is purchased and this is financed by a land loan, it might be sensible to arrange for the first repayment of capital to be delayed for two years to give the expanded business an opportunity to settle down and generate profits before committing the cash generated to capital repayments.

Security requirements

Security requirements vary tremendously, from completely unsecured lendings to loans where full legal mortgages are taken over specific assets such as land, houses, stocks and shares. Precisely what the security requirements are, will depend on:

(1) *The risk element of the project.* The higher the risk the higher the security arrangements are likely to be.
(2) *The proprietor's capital in the business.* The lower the gearing, the less concerned a supplier of credit will be to have tangible security.
(3) *The nature of the finance provided.* If, for example, a farmer is buying land, then it is likely and in some cases mandatory that the title to the land is taken as security.
(4) *The inherent profitability of the business.* The higher the profitability and by implication the higher the positive cash flow, then the less concerned a financier will be about tangible security.

Having pinpointed the factors a farmer and his advisor should review when considering using creditor finance, it is important for the farmer to select the source of finance most appropriate to his own situation. Some of the main sources of finance are discussed together with a number of the unusual methods of funding a farm business.

Long-term creditor finance (ten years plus)

The Agricultural Mortgage Corporation plc (AMC)

In the 1920s, farmers had no means of raising long-term funds from the public to help them finance land purchase and development in England and Wales. Entrepreneurs in other industries were able to form joint stock companies and so raise capital by issuing shares on the stock market. This form of finance was inappropriate for the agricultural industry where the individual units were very small.

Under the terms of the Agricultural Credits Act 1928, the Agricultural Mortgage Corporation Limited was born with a mandate to 'lend to farmers on the most favourable terms'. The share capital for the company came from the Bank of England and the London clearing banks.

Capital is raised on the London money market by means of the issue of debentures and, latterly, from short-term bonds and term loans from banks. To help keep the cost of this capital to a minimum, the 1928 Act enabled the Ministry of Agriculture to advance money to AMC, interest free, for 60 years. The money is invested in gilt-edged securities and forms the basis of a guarantee fund backing the debentures which have been issued. Since its inception in 1928, AMC has lent some £945 million on 2.39 million hectares. The amount outstanding in March 1987 was £606 million to 12 000 farmers on 1.01 million hectares, which is 8.5% of the agricultural land in England and Wales. AMC is now owned by Lloyds Bank plc.

For how long will AMC lend?

In the long-term market, AMC lends money from 10 to 40 years.

For what purpose will money be lent?

AMC lends the majority of its funds for land purchase and capital improvements. Table 11.4 shows the breakdown of the advances over the period 1980 to 1993.

How much will AMC lend in relation to the farmer's stake?

Loans will be considered for up to two-thirds of AMC's assessment of the value of the land which is offered as security, although normally no more than 50% of land value is lent.

The valuation of the land is based on an assessment of the worth of the land as a farming unit. Whilst it is possible that the valuation may equate to the current market price on occasions, the valuation may be pitched at a lower level.

It is feasible that AMC could provide 100% finance for a particular land purchase where they are offered additional mortgage-free land as security or a sitting tenant buys at 50–60% of vacant possession price.

Table 11.4 Purpose for which AMC loans were granted between 1980 and 1993

	1980 %	1987 %	1993 %
Purchase of additional land and new farms	50	41	48
Purchase of land by sitting tenants	8	4	4
Purchase of milk quota	—	—	2
Repayment of other loans	23	47	41
Capital improvements	10	3	3
Working capital	7	5	3

What does it cost to borrow from AMC?

Interest

Loans are available on fixed or variable interest rate terms or a combination of both.

The fixed interest rate option is based on the fixed rate which is ruling on the date the loan is actually completed. Loans taken out on a fixed interest rate basis cannot be converted to variable interest rate loans.

Variable interest rate loans have the interest rate reviewed on a quarterly basis. When a loan is taken out initially it will attract interest at the variable interest rate operating at the time. How the rate will vary will depend on the current cost of the fund from which AMC finances its variable rate loans.

In practice the variable interest rate has approximated to a rate equivalent to 1.5%–3% above the clearing banks' base rates. Variable-rate borrowers have the option to convert all or part of their loans to fixed rate at any time.

Interest rates are quoted on a nominal-rate basis.

The prevailing rates for all AMC loans are published in the national press.

Commissions and other administrative costs

A charge of 1% of the loan offered, with a minimum of £500 for new loans and £350 for further loans, is payable where a formal offer is made, whether or not the loan is taken up. Additional charges are payable for alterations to terms of loans. Interest penalties may be payable where loans are repaid early. AMC's solicitors' fees have to be met by the applicant.

What evidence of the viability of the proposition is required?

AMC will want to be satisfied that the borrower had adequate working capital available for the business to generate sufficient funds to service and repay the loan.

Firstly it wants to establish that the business is to be run by competent farmers. Assuming this is the case then where the borrower is already in business they will expect to see accounts for the past few years. In all cases they will want to see budgets and cash-flow projections for the future.

What repayment terms are available?

There are four repayment options available.

The annuity method

This method involves monthly, quarterly or half-yearly payments comprising both interest and capital. In the early years payments are largely made up of interest, which is eligible for tax relief. In later years the payments have a high proportion of capital repayment in their make-up which is not eligible for tax relief. In all cases interest will be calculated on a quarterly basis.

In overall terms the annuity method of repayment usually involves the lowest initial burden on the business (if the business can take advantage of tax concessions) and is consequently generally the best method to select for those wanting economy of cost in the early years of a loan.

The equal capital repayment method

Capital is repaid in equal half-yearly instalments with interest covered quarterly on the outstanding balance. This method is the cheapest but suffers from the disadvantage that the initial payments of capital, on which there is no tax relief, are high as well as having a high interest charge. This probably means that a better early cash flow is necessary if this method is to be adopted.

The endowment assurance method

This method allows the borrower to provide personal or family life cover to repay the capital at the end of the loan period from the maturity value of a life policy. The interest payments have to be met quarterly and will represent the interest on the initial capital borrowed. A suitable endowment assurance policy, maturing at the end of the term of the loan, is assigned to AMC. There are two schemes:

(1) A 'non-profits' endowment assurance policy can be taken out to cover the full amount of the loan. At maturity this policy will provide exactly the amount of capital to repay the loan
(2) A 'minimum cover with profits' endowment assurance policy which, if the insurance company's current rate of annual and intermediate bonus is maintained, will reach a maturity value equal to the capital necessary to repay the loan at the appropriate time. If the value of the policy at maturity is greater than the loan outstanding then the surplus is retained by the farmer. However, should the insurance company fail to maintain the bonuses predicted then it may be necessary to arrange a further loan from AMC or another source to fund any shortfall. As well as the interest on the loan being eligible for tax relief the premiums on life policies taken out before April 1981 attract tax relief at 15%. In certain circumstances this method can be cheaper than the annuity method

NB:

(a) An endowment assurance policy is one which is payable on death or after a given period of time. This differs from a whole life policy which only matures on death and is unacceptable to AMC
(b) Insurance cover must be on an adult life
(c) If an endowment policy is taken out when the assured is young then premiums will be low

The straight loan option method

This allows borrowers to take up to half of their loan on an interest-only basis. This part of the capital is repaid, in full, at the end of the term. The remaining half of the loan is repaid by one of the other repayment methods.

At first sight this alternative seems horrific if one considers a £200 000 loan of which £100 000 capital is left unpaid for 25 years. However, there may well be cases where individuals know they will receive capital sums at a future date which makes the straight loan option a viable proposition. There is also an argument in favour of having large amounts of capital repayment in the distant future where there is an inflationary economy.

If £100 000 is outstanding for 25 years then whilst in money terms it will always require £100 000 to repay it, in real terms the amount required to repay, assuming various annual inflation rates would be as follows:

Loan £	Annual inflation rate %	Repayment in money terms £	Repayment in real terms £
100 000	5	100 000	29 500
100 000	10	100 000	9 200
100 000	15	100 000	3 000

- The inflationary argument itself is not a justification for adopting this repayment method
- Before a farmer adopts a repayment method he should, without fail, consult his accountant to establish the most suitable method to fit his individual circumstances
- A fee is payable on early repayment of a loan but there is no fee for transferring a loan from one farm to another

What security is acceptable?

AMC can only lend against the security of a first mortgage over freehold land supported by the title deeds of that land.

The Lands Improvement Company Ltd (LIC)

Established and governed by successive Acts of Parliament between 1853 and 1969, the LIC can provide funds for agricultural businesses in England, Scotland and Wales.

For how long will LIC lend?

The LIC does not lay down specific periods for its long-term lending but loans up to 20 years have been sanctioned.

For what purpose will money be lent?

- Land purchase, including bridging advances
- Farm improvements (buildings, houses, fencing, drainage, roads and trees)
- Repayment of other borrowings
- Taxation, including inheritance tax
- Purchase of machinery and equipment

How much will LIC lend in relation to the farmer's stake?

There is a minimum loan figure of £25 000. The amount lent is limited to a maximum of two-thirds of the company's valuation of the security.

What does it cost to borrow from LIC?

Interest
Fixed and variable rate loans are available. The interest on fixed-rate loans is based on market interest rates at the time loans are taken up. Variable-rate loans attract an interest charge related to the company's base rate. The base rate is fixed half-yearly at a figure equivalent to the yield on Treasury 2.5% consolidated stock to the nearest 0.25% upwards. They also lend at a margin over London Interbank Offered Rate (LIBOR).

Commissions and other administrative costs

A fee of between 2 and 4% of the loan is charged and is payable when the loan is taken up. If a loan offer is accepted but not taken up within three months then a fee of 1% of the loan is payable. In addition, the farmer pays the legal fees of the company.

What evidence of the viability of a proposition is required?

In common with other lenders, LIC expects to see three years' accounts in the case of an existing farmer plus forward budgets and cash flows to demonstrate future profitability.

What repayment terms are available?

There are a number of options available. Borrowers can repay capital in full at the end of the term of the loan or by annual or half-yearly instalments. In some cases, with profit or non-profit endowment policies may be used to repay capital.

Where a farmer wishes to repay a loan early, he may be liable to a redemption fee of up to £500.

What security is acceptable?

The company will accept a first or subsequent mortgage on freehold agricultural land or woodland.

The clearing banks

The clearing banks have dabbled in long-term agricultural lending for some 20 years but it is only in the last decade that they have made any positive moves to obtain a significant share of the market. It has now become very popular to fund the purchase of land with bank loans at the expense of the Agricultural Mortgage Corporation. It is impossible to isolate the amount of money lent in this field but best estimates suggest that the clearers are lending some £400–500 million for land purchase.

For how long will clearing banks lend?

The maximum term available is 25 years.

For what purpose will money be lent?

The majority of 25-year money is lent for land purchase. Ten-year money is available for purchase of buildings and long-term installations such as roads. In general, the banks are flexible in this field subject to the duration of the loan being sensible in relation to the life of the asset being financed.

How much will the clearing banks lend in relation to the farmer's stake?

The relationship of overall bank lending to the farmer's stake in his business is discussed fully in the section on the clearing banks' operations under the heading Short-term Creditor Finance. As far as finance for land purchase is concerned the banks will lend up to two-thirds of their valuation of the land. Clearly, if additional land is offered as security or a sitting tenant is buying at 60% of vacant possession price, then 100% finance may be available.

What does it cost to borrow long-term from the clearing banks?

Interest rates
The clearing banks lend on both fixed and variable interest rate bases in this field. The normal rate of interest is 2–4% over the particular clearing bank's base rate where variable rates are concerned, although marginally finer rates may be available in certain circumstances. Interest, which is quoted on a nominal-rate basis, is charged quarterly in March, June, September and December. For larger loans, rates may be linked to a margin over LIBOR (London Interbank Offered Rate). Rates for fixed-rate lending are not based on bank base rates but on a view of likely rates over the long term. There are options to review fixed rates every three to five years. Some ten-year fixed-rate loans linked to LIBOR are available.

Commissions and other administrative costs
There is rarely a set commission charge, although the banks are probably looking for a fee of up to 1% of the money lent; negotiation is possible! In the majority of cases, a small administrative charge (up to £200) may be made to cover the work involved in perfecting the security taken.

What evidence of the viability of the proposition is required

The banks are looking to see if their customers can afford to finance the long-term loans which are granted. It is therefore important that forward budgets and cash flows are produced to demonstrate future viability of the business. With an existing business, the bank manager will have already established the 'track record' of the farmer from his accounts.

What repayment terms are available?

The banks are probably the ultimate in flexibility where repayment programmes are concerned. Repayment may be by conventional means such as monthly or three-monthly repayment programmes. Annuity repayment and endowment linked options on the same basis as AMC are acceptable. Some loan repayments are linked to commutation of pensions.

It is naturally best to gear the repayment programme to the particular farm business cash flow so that an arable farm might make repayments through the winter as cereals are sold. This is acceptable to the banker as is up to a two-year moratorium on capital repayments (capital holidays) in certain specialised circumstances. There are generally penalties for early repayment of long-term loans particularly if the lending is on a fixed interest rate basis.

What security is acceptable?

Most long-term loans require tangible assets as security. In the case of land purchase, there is generally a requirement that a first legal mortgage is taken over the land in question with the deeds being deposited at the bank. However, where there is only a very small first mortgage outstanding a second mortgage might be acceptable, particularly if the prior mortgage is at an attractive interest rate and it would be illogical for it to be repaid.

Insurance companies

Lendings to agricultural units by insurance companies are limited. The availability of finance will vary according to the general economic conditions and the attractions of alternative investment options. Insurance companies not already involved in agricultural investment are unlikely to provide funds.

For how long will they lend?

Contracts are essentially for 20 to 25 year periods.

For what purpose will money be lent?

The loans available are purely for land purchase.

How much will be lent in relation to the farmer's stake?

A maximum of 60% of the insurance company's valuation of the land. There is normally a minimum loan figure of between £30 000 and £50 000.

What does it cost to borrow?

Interest
Only fixed interest rate loans are available. The level of interest is based on current gilt-edged yields.

Commissions and other administrative costs
No commissions are charged in respect of the loan being agreed but the borrower has to pay the normal legal and other costs associated with the perfection of the company's mortgage.

What repayment terms are available?

The loan capital can be repaid either by annual instalments or, at the end of the loan period, from the proceeds of an endowment life policy.

What security is acceptable?

Insurance companies require a first mortgage on agricultural land.

Building societies

Although it is estimated that building societies, in total, lend some £30 million to agriculture in the form of long-term loans, it must be regarded as unusual for them to lend for the purchase of a farm. In the past, they have provided funds to finance the erection of houses on farms. Previously building societies have struggled to provide sufficient capital to fund conventional house purchase and with the problem of 'agricultural restrictions' on some farm houses there is little prospect of farmers obtaining funds from this quarter.

Private mortgages

These funds, if available, are for long-term finance normally secured on the deeds of a farm. Loans from solicitors' clients are a prime example. A significant amount of capital is provided by relatives of farmers and this is frequently available on an unsecured basis.

Sale and leaseback as a method of raising capital

Sale and leaseback involves the sale by the farmer of his freehold interest in the land. In return he retains the use of the land as a tenant. A market rent is paid for the tenancy.

The majority of sale and leaseback arrangements centre on the larger and better-quality farm units. The minimum size the institutions are normally interested in purchasing unless they own adjoining land is 120 ha. The availability of funds for this type of transaction will depend on money market conditions and the view the institutions take of land as an investment.

The important points to note about sale and leaseback agreements are:

(1) The farm is purchased by the institution at the tenanted value of the farm and not the vacant possession value of the freehold. (The tenanted value of a farm varies between 50 and 60% of its vacant possession value)
(2) Once the freehold is sold, there is no agreement for it to be repurchased by the farmer
(3) Capital gains tax may be payable on the proceeds of the sale which will reduce the cash value to the business
(4) The rentals charged normally represents 3 to 4% of the purchase price with provisions to review the rent every third year

There will be some cases where farmers will find it sensible to release money locked up in land by arranging a sale and leaseback agreement, but great care must be taken before any decision is reached. The rentals which have to be paid will be kept as high as possible by the institutions, and unless the farm is run effectively the farmer could find himself struggling to make the payments. Any decisions on entering a sale and leaseback agreement should only be made after close consultation with the farmer's accountant.

Example 11.3 The implication of a sale and leaseback

A farmer has a 290 h farm which he owns subject to an AMC mortgage of £125 000 giving rise to annual payments of £18 000.

The vacant possession value of the farm is £1 000 000 compared with the original purchase cost of £400 000. The farmer decides he needs to invest £100 000 in new buildings and equipment if his business is to generate acceptable profits in the future. To raise the money from the banks or AMC will result in an increase in annual commitments of around £11 500. He has been told that it should be possible for him to sell his farm to a pension fund on a sale and leaseback agreement so that he could release his equity locked up in land ownership and finance the investment in buildings and equipment from his own resources. The pension fund would expect to get a return of 4% on their investment. Before reaching a decision, the farmer asks his advisor to investigate the plan's economics:

	£
Vacant possession value of farm	1 000 000
Pension fund valuation for purchase on sale and leaseback at 55% of vacant posession valuation	550 000
Original purchase price	400 000
Potential capital gain	150 000
Rental to be charged by pension fund	= £550 000 × 4% = £22 000

Farmer's cash position after sale and investment:

	£	£
Sale proceeds		550 000
Less:		
Repayment AMC loan	125 000	
Investment in building, etc.	100 000	
Capital gains tax based on £50 000 at 40% not rolled over	20 000	245 000
Cash in hand		305 000

In this case, it can be seen that the farm business will be subject to a rental of £76 per hectare and this is likely to increase in line with inflation. This rental is significantly higher than the original finance charges of £18 000 per annum but much lower than if the money for investment had been raised through a financial institution. In addition, the farmer has £305 000 cash available for investment but has lost his main capital asset forever.

Having outlined the main sources of long-term credit for agriculture, it is important that the farmer and his advisor consider the advantages and disadvantages of each source and select one which will suit their own particular requirements. The various institutions have set out their wares to attract the borrower. It is for borrowers to be discerning in their selection of a credit source.

Land purchase on borrowed money

The returns from investment in land are discussed in Chapter 5 but it is appropriate when discussing long-term credit finance to consider, in general terms, the cost of financing land purchased on borrowed money.

Example 11.4 Land purchase by a tenant from his landlord

A farmer rents an 80 ha 'grade 1' farm which has a vacant possession value of £4940 per hectare. He is offered the freehold at 60% of the vacant possession valuation but will have to finance the purchase by borrowing. The interest rate payable on borrowed money is likely to be 10% over a maximum period of 40 years. His present rent is £90 per hectare.

Amount required to finance purchase:

$$80\,\text{h} \times £4940 \times \frac{60}{100} = £237\,120$$

Servicing and repayment cost of borrowing £237 120 repayable over 40 years

$$£102.00 \times £237.12 = £24\,186.24$$

Cost of borrowing per hectare farmed:

$$\frac{£24\,186.24}{80} = £302.33$$

In the majority of cases farmers would regard this as too onerous a burden to place on their business.

Example 11.5 Owner–occupier purchasing adjoining land

A farmer owns an 80 ha farm and has the opportunity to purchase an adjoining 20 ha at £4940 per hectare. His present farm is unencumbered. The money for the purchase will have to be borrowed at 10% over a maximum period of 40 years.

Amount required to finance the purchase:

$$20\,\text{ha} \times £4940 = £98\,800$$

Servicing and repayment cost of borrowing £98 800 repayable over 40 years:

$$£102.00 \times £98.9 = £10\,087.80$$

Cost of borrowing per hectare farmed:

$$\frac{£10\,087}{100} = £100.88$$

Most farmers would accept this as a price they would pay to purchase additional land.

Clearly it is vitally important that farmers analyse carefully the cost of land purchase and do not embark on a project which will prove a totally unacceptable burden on the business. Purchase of land purely on a basis that the opportunity to purchase locally may not occur again is a dangerous philosophy and the risks to the business of this approach should be carefully assessed before any decisions are taken.

Medium-term creditor finance (three to ten years)

Clearing banks' medium-term loans

Medium-term loans (MTLS) were introduced for two primary reasons. Firstly, they could be tailored to ensure assets were paid for within their working life and secondly, unlike bank overdrafts which are theoretically repayable on demand, medium-term loans cannot be recalled unless the borrower defaults on the repayment terms.

For how long will MTLS be granted?

Banks provide MTLs over the complete medium-term spectrum dependent on the asset which is being financed.

For what purposes are MTLS granted?

Loans are agreed for a wide variety of purposes ranging from the finance of breeding herds at the shorter end of the market (three to four years), through machinery and plant finance (three to eight years) to specialist buildings and land improvement (five to ten years). These periods are only generalisations and would vary on an individual farm basis.

How much will be lent in relation to the farmer's own stake?

The proportion of the money a bank will put up when financing the purchase of a medium-term asset is not necessarily governed by an specific deposit requirement. The bank will be looking to the overall percentage of the business which is funded by creditor money. Provided the farmer's stake in the business is acceptable then the governing factor will be the ability of the business to service and repay the loan from profits.

What will an MTL cost?

Interest
Interest rates vary significantly depending on the various factors discussed at the beginning of this chapter. The banks offer variable interest rates relating to their base rates. Typically rates vary from 1.5% to 5% above base rates. Interest rates are quoted on a nominal rate basis and debited to either the farmer's current account or to the loan account, depending on the type of loan agreed, on a quarterly basis (March, June, September, December). Fixed rate loans are also available, the farmer being quoted an 'all in' rate which includes the bank's margin. Fixed rate loans are expensive to repay early because the banks match funds, i.e. borrow money themselves on a fixed term basis.

Commissions and other administrative charges
There are a wide variety of charges made ranging from nominal initial administrative charges to a percentage commission (0.5 to 1%) of the balance remaining at the beginning of each year the loan is outstanding. It is important to define these charges at the commencement of the loan.

What evidence of the proposition's viability is required?

Normal bank requirements to test the servicing and repayment potential of the proposition, i.e. accounts, forward budgets and cash flows.

What repayment terms are available?

Flexibility is the key. The banks are prepared to adjust the timing of annual repayments to suit the cash flow of the business although interest has to be paid quarterly. There is generally no penalty for early repayment unless the loan is on a fixed interest rate basis. Regular monthly or quarterly repayments can be agreed which incorporate both the recovery of capital and interest.

What are the security requirements?

Medium-term loans vary from unsecured advances where the banks can see the farmer owns a high proportion of the assets utilised in his business to loans secured by legal mortgages, first or subsequent. It is impossible to define exactly which advance falls into which category of security requirement; the situation will depend on the weighting of the various factors discussed at the beginning of the chapter.

Hire purchase

Most of the major hire purchase companies are owned by the clearing banks (e.g. Barclays Mercantile is a wholly owned subsidiary of Barclays Bank plc and

Lombard North Central is owned by the NatWest Bank plc). They are complementary, in the banks' eyes, to the other services they provide.

For how long will they lend?

Hire purchase companies cover the complete spectrum of medium-term lending and indeed will lend some money in the short-term market. The majority of their lending falls in the three to seven year range.

For what purpose will money be lent?

Machinery and plant have traditionally represented the main agricultural hire purchase market. However, in recent times hire purchase agreements for the purchase of breeding stock have increased, as have agreements for purchasing buildings.

How much will hire purchase companies lend in relation to the farmer's stake?

Normally a significant deposit is required by the company, probably around 20% of the purchase price. Where, for example, machinery is being part-exchanged the trade-in value of the machine would count towards the required deposit. Indeed if the trade-in value is large enough it could cover the total deposit required by the hire purchase company.

What does it cost to borrow from a hire purchase company?

Interest
Hire purchase companies fund a high proportion of their operations through clearing bank finance and money raised on the money markets. The cost of this finance and the greater risks involved in hire purchase business usually lead to a significantly higher cost of money than borrowing from the clearers direct. Interest rates are generally quoted on a flat-rate basis. Variable interest rates available are related to the finance house base rate but the majority of business is on a fixed interest rate basis. (Finance house base rate is the base interest rate on which hire purchase companies fix their lending rates. It is calculated from averaging the two previous months' three-month London Interbank Offered Rate.)

Commission and other administrative charges
There are no separate charges with hire purchase agreements but in the event of a farmer wishing to repay a loan early he is likely to have to pay an interest penalty equivalent to up to three months' interest.

What evidence of the viability of a proposition is required?
The requirements of hire purchase companies vary considerably depending on the

proposition before them. When a farmer applies for a loan he will have to complete an application form requiring him to reveal information about his business, such as size of farm, enterprises, other financial commitments and his bankers. For some loans the hire purchase company might simply take up a reference from the farmer's bank and in conjunction with the information disclosed on the application form agree the loan. In more complex situations additional information might be called for, such as three years' accounts.

What repayment terms are available?

Repayment terms vary considerably but usually involve regular monthly payments of interest and capital.

What security is acceptable?

Hire purchase companies take as security the asset against which they have lent money. If a farmer fails to maintain the repayment programme for a loan then the hire purchase company is within its rights to call and repossess the asset, be it tractor or cow. It is not normal practice for additional security to be taken.

Leasing agreements

Although leasing does not involve the farmer in borrowing money it is sensible to discuss it as a form of medium-term finance of assets as it provides an alternative method of providing assets for use in the farm business. In essence a lease agreement allows a farmer to rent equipment of his choice from a company that wishes to own the equipment as a capital asset for tax purposes but does not want to work the equipment.

For how long will leasing companies rent out equipment?

Leases may be divided into two parts:

(1) *The primary lease period.* This is the period of time the farmer would normally wish to keep the equipment he is renting. In the case of a tractor it may be three years, or five years in the case of a combine harvester. During this period the leasing company (the lessor) will recover from the farmer (the lessee) the capital cost of the equipment purchased plus a profit for providing the equipment.

(2) *The secondary lease period.* Once the primary period has finished the agreement then enters the secondary period which normally extends the lease up to and beyond ten years. During this period a 'peppercorn' rent is charged for the machine (e.g. 0.5% of original cost up to a maximum £50).

For what purposes can leasing agreements be made?

Almost any piece of equipment can be leased. Leasing in agriculture, however, centres largely around machinery such as combine harvesters, tractors, forage harvesters and the like. Leasing of livestock is now uncommon.

What does it cost to lease equipment?

The cost of a lease hinges on the tax position of the lessor and his ability to utilise annual writing down (capital) allowances which are much lower today (25%) than used to be the case.

When a farmer buys a tractor either for cash or with the help of a bank loan or hire purchase finance he will have available to him writing down (AWD) allowances equal to the net cost of the tractor, which can be claimed at 25% pa on the outstanding balance. If a farmer rents a tractor on a lease then he cannot have these AWD allowances as he does not own the tractor. Instead the AWD allowances are taken by the lessor. The leasing companies which are generally owned by highly profitable parent companies (e.g. Highland Leasing plc, the major leasing company in agriculture, is owned by Barclays Bank plc) can utilise the AWD allowances to reduce their tax liability (tax shelter) and in return they offer attractive rentals which can be equated to a low 'interest rate equivalent'.

Comparing the cost of a lease with a medium-term bank loan or hire purchase agreement is complicated, involving the tax position of the farmer and the real value of money. Before embarking on a leasing agreement the farmer should consult his accountant or other professional advisor.

It is dangerous to generalise as to when it is best to lease and when to buy. Before the disbanding of the 100% first year (AWD) allowances farmers who had no tax liability, or were unlikely to have a tax liability in the foreseeable future, should have leased. If the farmer's marginal rate was 30% then there was little difference between leasing or buying. If his marginal rate of tax was 40% or higher he should have bought, this assumes that the farmer did not have the cash available to buy without borrowing! Now with AWD allowances reduced to 25% of purchase cost in the first year and then 25% of the outstanding balance in subsequent years the picture has changed. There is now a good case for higher-rate taxpayers using leasing because they can bring forward allowances by taking short-term leases, minimum two years (one year leases are not currently acceptable to the Inland Revenue), where the leasing company will have structured its rentals to cover the capital cost of the machinery in the primary lease period and these rentals are charged against the profit and loss account.

Example 11.6 illustrates how leasing can help the farmer to set the capital cost of machinery against profits over a shorter time span than if an outright purchase had been made. The example assumes the leasing or purchase of a tractor costing £22 000 over three years, a combine over four years and a farm vehicle over three years. Repayments or rentals are made half-yearly, annually and quarterly, respectively. At the end of five years, when all three leases would be on a peppercorn rent and in effect the farmer would have been able to set all his leasing

Example 11.6 Comparison of the time taken to recover the capital cost of machinery lease or purchase against tax

Machine	Value after trade-in	Period in years	Repayment interval	Total annual rental				
				Year 1	Year 2	Year 3	Year 4	Year 5
Tractor	22 000	3	6 months	9 500	9 500	9 500	—	—
Combine	42 000	4	annual		15 000	15 000	15 000	15 000
Farm vehicle	6 000	3	quarterly			2 600	2 600	2 600
Total allowable lease expense				9 500	24 500	27 100	17 600	17 600
Brought forward:					16 500	43 875	37 407	28 055
Additions:				22 000	42 000	6 000	—	—
Totals:				22 000	58 500	49 875	37 407	28 055
Total of 25% WDA				5 500	14 625	12 468	9 352	7 014
Carried forward				16 500	43 875	37 407	28 055	21 041
Interest on MTL at 13%				2 621	7 128	5 492	3 152	1 530
Total allowable purchasing expenses				8 121	21 753	17 960	12 504	8 544

costs against tax, he would still not have been able to recover all his AWD allowances against tax had he purchased. The example shows that of the original capital cost of £70 000 there is some £21 041 of allowances still unclaimed.

It is important to remember, however, that the total cost of the leases is £96 300 compared with the outright purchase which would have been £89 923. To decide which was the better option it would be necessary to know the detailed tax position of the farmer and then apply the discounting principle (see Chapter 9) to the payments.

What evidence of the proposition's viability is required?

The lessor will want to establish that the lessee can meet the rental for the primary period of the lease. Thus it will, in some cases, be necessary to provide the leasing company with audited accounts. At the other end of the range the leasing agreement application form will provide sufficient information for the company to agree to provide leasing facilities.

What rental terms are available?

Leasing companies are very flexible with regard to the timing of rental payments, a small penalty being incurred if rentals are paid in lump sums towards the end of a 'leasing year'. Rentals can be paid monthly, quarterly, half-yearly, annually, or be on a regular monthly rental with a higher rental charge when the harvest monies are to hand, or any other method which fits the farmer's cash flow. The lessor frequently requires a rental in advance which can take the form of a trade-in machine. The rentals are allowable against the farmer's tax. The rents are calculated on the cost net of VAT. VAT is added to the rental and can be reclaimed in the normal way.

What happens when the farmer wants to change his machine?

The farmer *never* owns the machine he is renting under a leasing agreement but he decides which machine he wants. Acting as the lessor's agent he will order the machine and the invoice will be sent to the leasing company. Similarly, when he wants a replacement machine he sells the old machine, as agent for the lessor, who pays the farmer a commission for his services, amounting to between 80 and 100% of the sale proceeds of the machine.

The farmer cannot buy the machine from the lessor or the tax advantages, on which the lessor bases the rental, disappear.

When and why a farmer may want to replace equipment will be governed by many factors but there is no restriction on his cancelling the rental of a machine which is leased to him subject to the lessor being able to recoup, through past rentals and the sale price of the machine, the capital cost of the machine plus a profit element.

What are the important points to consider when obtaining or comparing leasing quotations?

There are a number of areas in a leasing agreement that can affect the total cost. It is important to identify these areas to allow comparison of the costs of different leasing quotations and also to compare the total cost with other forms of finance. These areas are:

(1) *The rebate of rentals returned to the farmer when acting as the lessor's agent on the sale of a machine.* This can vary between 80 and 100% of the net sale proceeds. If a tractor is sold for £5000 and the rebate is only 90% of the proceeds then there is a 'loss' of £500 compared to the agreement which gives the farmer 100% rebate on sale. This loss adds to the cost of the lease
(2) *The continuation or peppercorn rental (i.e. rental charged in the secondary period of the lease).* These rentals can vary from £1 per annum to 2% of the original cost price of the machine. If a combine was purchased originally for £50 000 then the peppercorn rent could vary from £1 to £1000 per annum!
(3) *Payment structure.* The rental structure should be examined because if rentals are paid in advance they have the effect of shortening the leasing period and so increasing the true cost
(4) *Settlement.* If a machine needs to be changed or updated before the primary period has ended then it is important to establish at the outset if this can be done without penalty. If a penalty is to be charged then this should be ascertained, as should the position regarding 'unused rentals'
(5) *The type of lease.* Care should be taken to determine that the lease is a true operating lease and not a form of contract hire where there is a nil residual value. Similarly, the leasing discussed here is not leasepurchase, which is similar to a hire purchase agreement

Machinery contract hire

This method of acquiring up to date machinery for use on a farm is likely to be a growth area over the next few years. Basically the farmer rents the tractor or combine he wants from the range on offer from his dealer. It is preferable to ensure that the contract hire agreement is backed by the particular machinery manufacturer.

The machine is rented for two to four years at a set price per week or season based on the anticipated number of hours the machine will be used each year. If the machine is used for more hours then there will be an additional rental to pay.

The rental, which is tax deductible, is subject to VAT which can be recovered by the farmer. The rental agreement covers the following:

- *Maintenance.* The machine is fully maintained by the dealer, including major services in line with the manufacturer's recommendations. There will be a specific allowance for tyre replacement within the rental

- *Back-up machine.* If a machine needs repairing and cannot be repaired within 24 hours then either a back-up machine is provided or alternatively a refund of rentals will be made to help with replacement hire
- *Insurance.* Comprehensive insurance cover is provided
- *Road fund licence.* The road fund licence is covered for each year of the agreement

What are the benefits?

There are two main benefits to contract hire:

(1) *Fixed-price machinery costs.* With the possibility of fixed-price rentals for up to four years, budgeting is easy. The only costs associated with the machine not covered by the rental are the driver and the fuel. There is the possibility in some situations of an end of contract rebate if the net cost of the machine over the hire term is significantly less than anticipated.

(2) *Improved cash flow.* Minimal up-front rental payments are required (normally one month) and therefore the initial take up of a contract hire agreement could lead to release of equity in a 'trade-in' machine which could help the cash flow of the business.

What are the disadvantages?

It is possible that the overall cost could be higher than either leasing or machinery purchase. In the event of the farmer wanting to revert to purchasing or leasing, he could have to raise a deposit which traditionally had been available from the 'trade-in' value of the old machine.

Most of the major contract hire schemes are backed financially by the manufacturers supported by finance companies such as Highland Leasing.

Agricultural Credit Corporation Limited (ACC)

The Agricultural Credit Corporation, which operated throughout the UK, was established in 1964 and received government support under the provisions of the Agriculture and Horticulture Act 1964 and the Agriculture Act 1967.

The Corporation's function was 'to extend the availability of medium-term bank credit to efficient and progressive farmers, growers and their co-operatives'. This function was achieved by guaranteeing the bank borrowing required.

The guarantees were available for virtually any medium-term expenditure, including the purchase of farm buildings, machinery, livestock and working capital. The guarantee normally covered the whole of the bank borrowing and reduced annually in line with the repayment programme for the bank facility.

Medium-term loans were generally regarded as lasting from three to ten years, although in exceptional circumstances longer periods were considered.

Some assistance may now be available under the Small Firms Loan Guarantee Scheme, as the ACC no longer operates.

Table 11.5 compares medium-term loans with hire purchase and leasing options.

Table 11.5 Comparison of a medium-term loan, hire purchase and financial lease as methods of financing machinery purchase

Area of comparison	Medium-term loan	Hire purchase	Lease
Duration of agreement	Maximum period related to the effective life of the machine. Normally up to five years.	Maximum period related to the effective life of the machine. Normally up to five years.	Primary period of lease when lessor recovers capital cost and profit extends up to five years. This is followed by peppercorn rent extending the lease up to ten years or beyond.
Deposit required	No deposit is necessary provided the farmer has enough capital in his business to support the borrowing.	Normally 20% which can include any trade-in value of an old machine.	No deposit as such; the machine is purchased by the leasing company. Up to three months' rentals may be payable in advance.
Cost	Linked to the bank's base rate. The normal rates are 2–4% above the bank's base rate. Some fixed-rate lending.	Can be either fixed rates or variable rates linked to finance house base rate. The rates tend to be significantly higher than for a MTL.	The cost of a lease is fixed. The farmer pays the lessor a rental for the machine.
Repayment terms	An extremely flexible form of repayment can be agreed with a bank. No penalty for early repayment.	Tend towards regular monthly or quarterly repayments and do not have the flexibility of medium-term loans. Often a penalty for early repayment.	Can be very flexible to fit the cash flow of a particular business. Generally no penalty for early cancellation.

Table 11.5 *(continued)*

Area of comparison	Medium-term loan	Hire purchase	Lease
Security and criteria for lending	Loans can be unsecured. No special charge is taken on the machine. Banks require evidence that the business is trading profitably.	The hire purchase agreement gives a specific charge over the machine financed. The application form completed requires details of financial commitments and a bank reference is normally taken.	The lessor requires financial data to be sure the lessee can meet the payments. In some cases this may mean sight of audited accounts.
Tax	Capital (AWD) allowances are available to the farmer. The interest paid to the banks can be deducted from profit for tax purposes.	Capital (AWD) allowances are available to the farmer. The interest paid to the HP company can be deducted from profit for tax purposes.	Capital (AWD) allowances not available to the farmer. All of the rental can be deducted from profit for tax purposes.
Flexibility	A very flexible form of finance which is easily negotiated with the minimum of documentation.	HP does not have the flexibility of MTL in the areas of rerpayment terms and security. It is also generally more expensive.	Limitations on flexibility on replacement of a machine but the procedure is not complicated and can be cheap in certain cases.

Agricultural Mortgage Corporation five to ten year loans

AMC provide these medium-term loans on the basis of capital repayment in full at the end of the loan period with interest payments made every six months. The loan cannot exceed more than one half of the value of the property as valued by the Corporation's valuer.

In all other respects the terms are the same as those for long-term loans.

Medium-term point of sale finance

This is credit which is available at the source of supply (i.e. Barclays Farm Masterloan, NatWest (Growcash). The same credit terms can be obtained by farmers applying directly to the relevant section of the bank.

The main features of the loans are as follows:

(1) They are available on an unsecured basis for periods of one to five years for the purchase of new or second-hand machinery
(2) Interest rates may be fixed or variable and tend to be very competitive
(3) Finance will be provided for up to 80% of the cost of the machine including VAT
(4) There is usually no obligation to conduct other business with the bank which is providing the finance, allowing a farmer additional flexibility

It is important to remember that to obtain finance is relatively easy, and where an additional source of finance is introduced and no one lender has overall knowledge of a farmer's commitments then close financial management is vital to ensure that a level of borrowing is not reached at which the business is placed at risk. Equally, because credit is readily available, a farmer should not be induced to make an unprofitable purchase.

Machinery Syndicate Credit Schemes

The high cost of buying machinery can mean that a farmer has to forego replacing vital equipment at the optimum time because the capital cost of replacement would adversely affect profitability and cash flow. Clearly if a number of farmers were able to group together and purchase some machinery between them this would cut down the capital outlay and provide each farm with the use of up to date equipment. The Machinery Syndicate Credit Schemes were set up for members of the National Farmers' Union (NFU) precisely for this purpose. The schemes were administered by the NFU and financed by the clearing banks.

To some extent farmers are loath to share and show dependence on others. In some cases syndication would never work because the individuals within a syndicate would not be able (or prepared) to compromise, which is the vital ingredient for successful syndication. However, there are many successful syndicates in operation today which have been in existence for a decade or more and syndication, in the form of Machinery Rings, has a role to play in the future, not only in the purchase of equipment but in the sharing of labour resources which has become associated with some machinery syndicates.

Eurocurrency loans

With the abolition of the exchange control regulations in the 1970s, Eurocurrency loans have attracted considerable interest when farmers look at the low rates of interest available to European farmers and compare them with the rates prevailing in the United Kingdom. Whilst the attraction to a farmer of opting for low cost Eurocurrency money to finance his business appears clear cut on the surface the potential dangers are high. To counter these dangers is costly and eliminates the interest rate advantage which attracted the borrower in the first place.

The disadvantages of borrowing in foreign currencies

The exchange rate risk

If the farmer has no income in the foreign currency he proposes to borrow, then when the time comes to repay the loan he will have to convert sterling into that currency. Unless the farmer takes out forward cover to eliminate this forward exchange exposure then the conversion of sterling to the foreign currency will be subject to the prevailing exchange rates at the time of the loan repayment. Therefore, if the pound sterling has weakened in relation to the foreign currency it will take more sterling to repay the loan.

Example 11.7 Borrowing by way of Eurocurrency loan

A farmer decides to take out a one-year Swiss franc loan equivalent to £100 000 sterling. The interest rate on the loan will be 5% payable quarterly compared with 10% for the equivalent sterling loan. The exchange rate at the time the loan is taken out is 2.4 Swiss francs per pound sterling. Assume the exchange rate remains the same for the first three quarters:

Swiss franc equivalent of £100 000 sterling:

$$£100\,000 \times 2.4 = 240\,000 \text{ Swiss francs}$$

Interest payable in first three quarters:

$$240\,000 \text{ Swiss francs} \times \frac{5}{100} \times \frac{3}{4} = 9000 \text{ Swiss francs}$$

$$\frac{9000 \text{ Swiss francs}}{2.4} = £3750$$

If it is assumed that the pound sterling weakens so that a pound sterling only buys 2.0 Swiss francs, then:

Repayment of 240 000 Swiss francs from sterling funds:

$$\frac{240\,000 \text{ Swiss francs}}{2.0} = £120\,000$$

Interest payable in final quarter:

$$240\,000 \text{ Swiss francs} \times \frac{5}{100} \times \frac{1}{4} = 3000 \text{ Swiss francs}$$

$$\frac{3000 \text{ Swiss francs}}{2.0} = £1500$$

Therefore the total cost of the Swiss franc loan in pounds sterling is:

	£
Capital cost	120 000
Interest for first three quarters	3 750
Interest for final quarter	1 500
Total cost	125 250

The equivalent cost of a loan in sterling at 10% interest would have been:

	£
Capital repayment	100 000
Interest at 10%	10 000
Total cost	110 000

Fluctuations in exchange rate of this magnitude are well within the bounds of possibility as Fig. 11.3 illustrates. Between 1976 and 1996 the Swiss franc varied from 6.1 Swiss francs down to 2.2 Swiss francs to the pound sterling.

If farmers were gamblers, or clairvoyants, then they might wish to take the exchange risk because if they happened to hold a foreign currency loan over a stable exchange rate period they would gain the appropriate interest rate advantage. Indeed, if they were lucky enough to be repaying with a strong pound sterling they

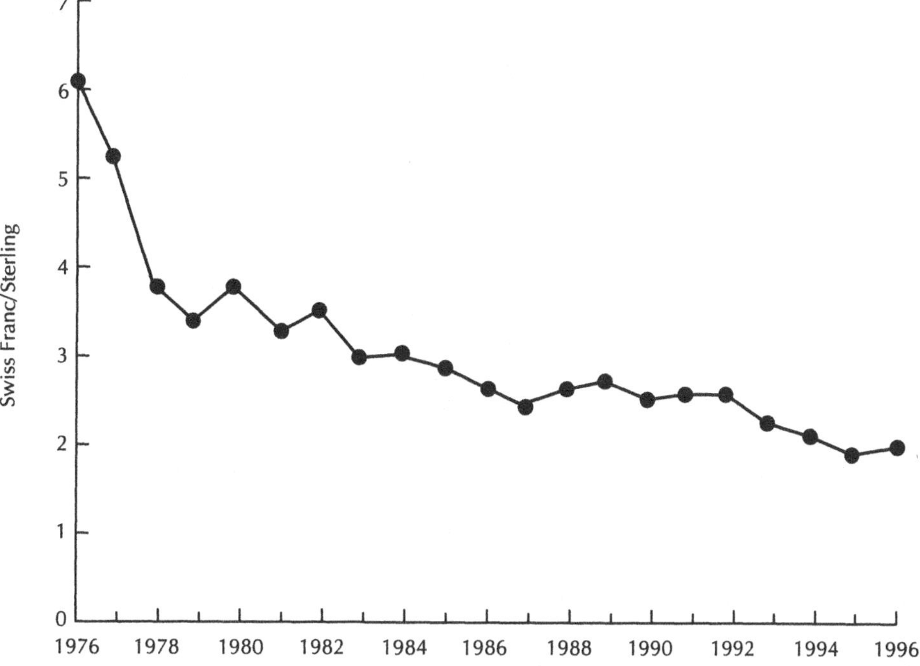

Fig. 11.3 Sterling–Swiss franc exchange rate movements 1976–1996.

would make a capital gain as well. No financier would recommend a working farmer to expose his business to the exchange rate risk.

If a farmer wants to fix the total sterling cost of his foreign currency borrowing he must buy forward the necessary foreign currency to repay his currency borrowing. Unfortunately, as a general rule currencies which are cheaper to borrow tend to be more expensive to buy forward than to buy for immediate settlement and so the potential interest cost advantage is eliminated. Commission costs must also be considered.

There is usually a minimum borrowing figure
The minimum foreign currency loan a bank would consider is likely to be in the region of £5000.

Taxation implications of foreign exchange gains or losses

(1) *Capital taxation.* If a capital asset is purchased with the foreign currency loan and a profit is made on the foreign exchange then this profit is not liable to capital gains tax; but if a foreign exchange loss occurs this loss is not allowable for capital gains tax relief purposes.
(2) *Revenue taxation.* If the foreign currency loan has been used for revenue purposes such as the purchase of trading stock then any foreign exchange profits are taxable as trading profit. Foreign exchange losses are deductible against trading profits.

Interest rate protection

The importance of interest rate protection

Between 1973 and 1993 interest rates in the United Kingdom were high in relation to those in some other countries, for example, Germany, United States, Japan and France. Average annual bank base rates only fell below 10% on three occasions and then only marginally, although in October 1977 interest rates did reach a low of 6% for one month as shown in Fig. 11.4. The first year when a significant fall occurred was 1993, with the average annual base rate at 6%, a level still being held at the end of 1996. The volatility of rates shown in Fig. 11.4 can and does create problems for businesses in budgeting and ultimately in achieving acceptable levels of profit.

The effect on borrowers, who have suffered interest rates as high as 20% (base rate plus margin) and depositors who have enjoyed interest income as high as 14% has in many cases been significant.

If you have a business and are borrowing a substantial amount of money the impact of a big increase in interest rates can be quite dramatic, as shown in Example 11.10.

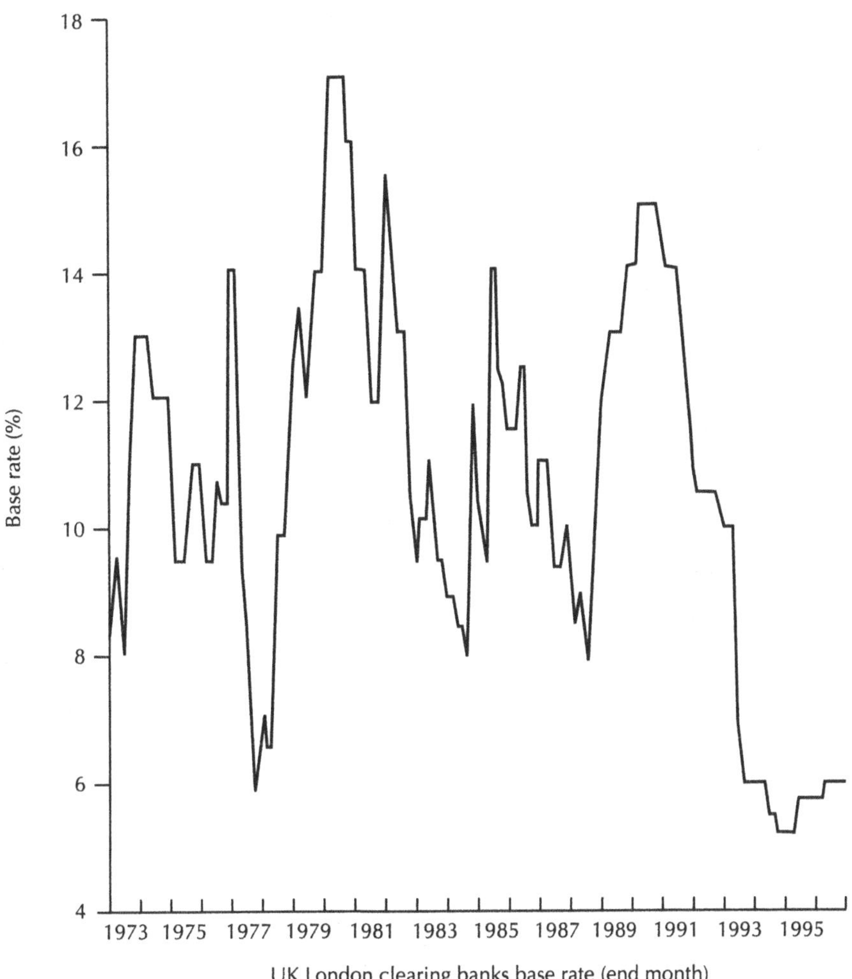

Fig. 11.4

It is therefore important for a business to consider interest rate protection to:

- ensure that reduction in income is minimal
- ensure that borrowing commitments can be met
- help to determine, in advance, financial costs

How important it is to protect a business against interest rate movements will depend on the level of exposure the business has and the length of time over which the exposure is likely to exist. The potential effect on a business of interest rate movements, be the business a depositor or borrower, is highlighted in Figure 11.5.

Example 11.10 The impact of interest rate movements on a business.

John Fordson has a 120-cow dairy herd with followers on a 60 ha farm he bought in December 1987 when base rate was 8.5%. The farm was bought for £350,000 including the stock. John had £200,000 cash available and took out a £150,000 20-year variable-rate endowment loan at 2.5% over base rate. The business performed well initially with profits retained in the business:

Year ending 31/12/1988	£	£
Profit before interest and drawings		43 000
Interest at 11% for 12 months	16 500	
Drawings (including tax payments)	20 000	
		36 500
Retained profit		6 500

Base rate remained within 1% for seven months but then increased steadily until in 1990 base rate averaged 15%. The impact on the business was as follows:

Year ending 31/12/1990	£	£
Profit before interest and drawings		43 000
Interest at 17.5% for 12 months	26 250	
Drawings (including tax payments)	20 000	
		46 250
Retained profit		(3 250)

The decision-making process

There are a number of alternative solutions available in the marketplace to help with interest rate risk management but there are a number of factors which the farmer will need to have considered before he can decide on the best solution, be he borrower or depositor:

(1) Is there a 'natural hedge', i.e. are there borrowings and deposits either both fixed or both variable? If this is the case then no additional protection may be necessary
(2) At what interest rate does the business require protection? It is logical that the farmer should calculate the maximum interest rate he believes his business can stand based on the level of borrowing anticipated
(3) The farmer's own view compared to the market's view of likely interest rate movements

Nobody can predict precisely what will happen to interest rates and when! However, most farmers and businessmen will hold a view based on current interest rate levels, their feeling about the economy, inflation and exchange rates.

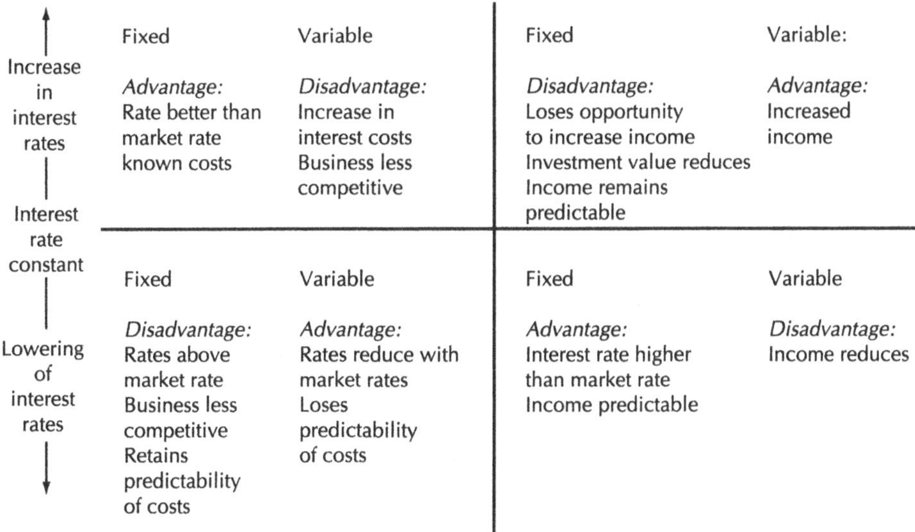

Fig. 11.5 The fixed and variable interest rate guide.

Information is also available in the marketplace which may help with the decision-making process (Fig. 11.5).

Yield curves

There are a series of interest rate markets where predictions are made of what the market expects to happen to interest rates over time (Fig. 11.6). The fact that a market expects a particular trend to be followed does not guarantee that this will be the case. However, it provides additional information which can be used in decision making.

Farmers who are borrowing should determine the peak interest rate they are prepared to accept on the money they borrow. Farmers who hold cash deposits need to decide the minimum interest rate return they require from their deposits.

A decision tree can be prepared based on the current level of interest rates (Fig. 11.7).

How to protect against potentially adverse interest rate movements

There are basically two types of product in the market place:

(1) Fixed interest rate products which fix rates over a specific period.
(2) Option-based interest rate products which allow for favourable interest rate movements but limit unfavourable movements. They act as an insurance policy, usually with an up-front premium.

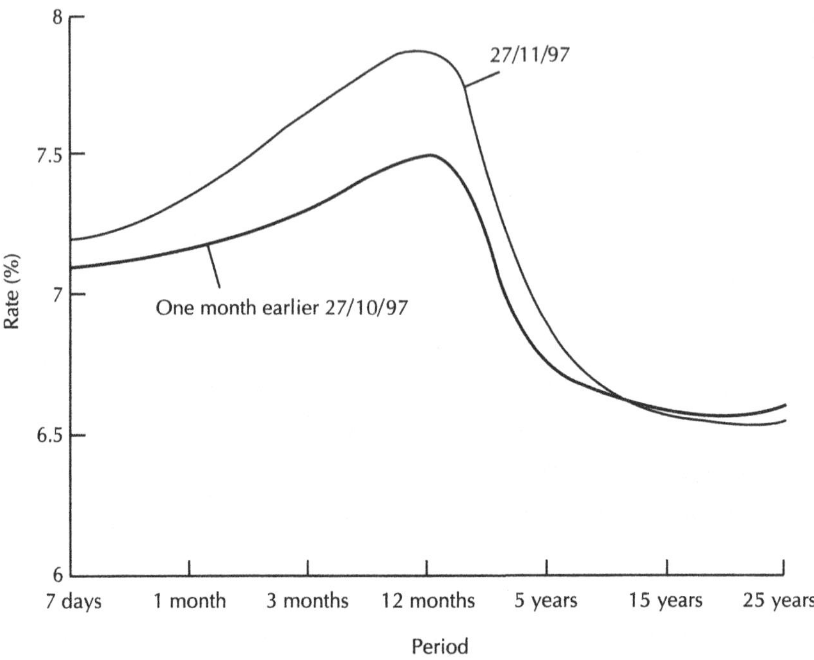

Fig. 11.6 Sterling yield curves 27/10/97 vs. 27/11/97.

Fixed-rate products

Fixed-rate products give an interest rate which is constant for the duration of the agreement. They also eliminate the possibility of taking advantage of beneficial interest rate changes so in the case of borrowers the possibility of benefiting from lower interest is eliminated, and the possibility of depositors benefiting from higher rates is stopped.

There is an exception to this rule in the case of interest rate SWAPS which can change fixed-rate agreements to variable rates.

Examples of fixed-rate products are fixed-rate loans, forward rate agreements and interest rate SWAPS.

Fixed-rate loans

These are loans which are fixed throughout the term of the loan. The rate charged on these loans will reflect the market view of rates over the term of the loans. Generally there is a penalty for early repayment of these loans unless there is an advantage to the lender at the time repayment is made.

Fixed-rate Agreements (FRA)

This is a contract up to a maximum maturity date of three years, between the

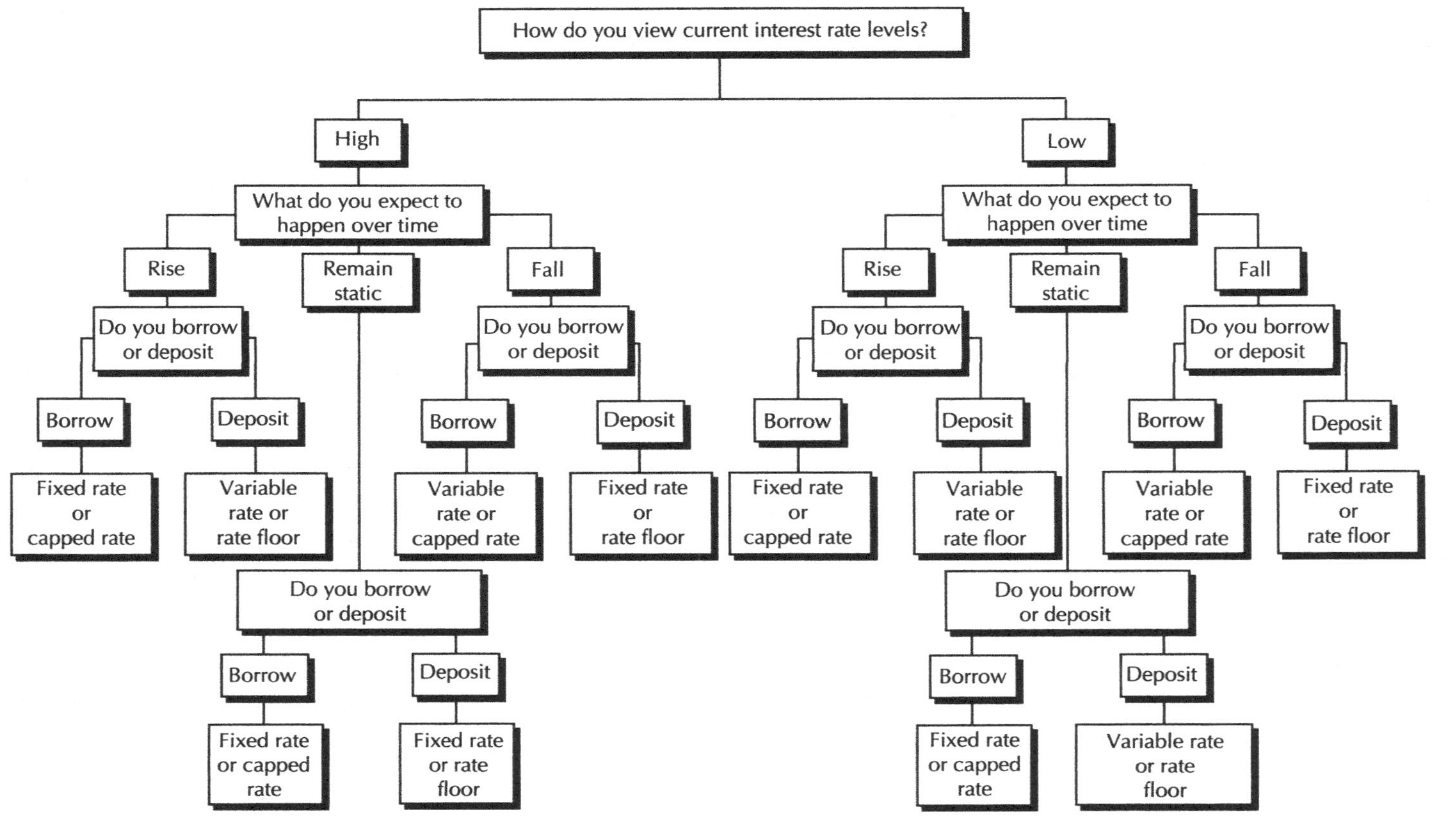

Fig. 11.7 Interest rate protection decision tree.

borrower or depositor and their bank to pay the other party if interest rates in the future move either higher or lower than the level agreed in the contract, e.g. in the case of a borrower, if the future market interest rate is above the agreed FRA rate then the bank pays the borrower the excess interest cost, if the rate is lower the borrower pays the bank. This is calculated on a 'notional' loan for a defined period.

The agreements may be in almost any of the major currencies (.e.g. sterling, US dollars, Swiss francs, German marks, French francs) and have to have a minimum value of £1 m or US $1 m.

A disadvantage with fixed-rate agreements is that if rates move against the business it may be disadvantaged to a greater extent than if it had done nothing.

There is no up-front payment for a fixed-rate agreement.

Interest rate SWAP

An interest rate SWAP allows one type of interest to be exchanged for another. Thus, if a business had a variable-rate loan they could 'swap' the variable rate interest for a fixed rate, the reverse also applies. Most SWAPS are for periods from one to ten years.

If loan interest payments are being exchanged then they are called liability SWAPS, if interest income is being exchanged then they are called asset SWAPS.

In SWAPS deals the bank involved acts as a principal, thus any credit risk is against the bank.

As with FRAs, businesses cannot benefit from advantageous interest rate movements and there is no upfront premium to pay.

SWAPS can be done in most major currencies with minimum deals normally £250 000 or the currency equivalent.

Both forward-rate agreements and SWAPS require a bank to agree a credit risk facility.

Option-based products

Option-based interest rate protection differs from fixed-rate products in requiring the payment of up-front premiums (but see zero premium collar). They enable interest rates to be fixed at a maximum (borrower) or minimum (investor) whilst still enabling the business to benefit from what to them is a beneficial interest rate movement.

Examples of option-based products are interest-rate CAPS, interest-rate FLOORS and interest-rate COLLARS.

Interest rate CAPS

An interest-rate CAP enables a business to fix the peak interest rate it would have to pay in a particular period whilst at the same time retaining the ability to benefit if interest rates drop.

CAPS can be arranged for up to five years in sterling, deutschmarks or dollars with a minimum figure normally of £10 k or other currency equivalent.

The level of interest at which the CAP is set and for how long it is taken out is decided by the business and in this way the level of the up-front premium can be controlled.

CAPS can be bought and sold independently of any borrowing facility and because the cost is paid up front banks do not have to record a credit risk against the business.

Interest rate FLOOR

An interest-rate FLOOR is the opposite of a CAP. It enables a business with deposit funds to guarantee a minimum return, whilst retaining the ability to benefit from higher interest rates.

Interest rate COLLAR

An interest-rate COLLAR is a combination of CAP and FLOOR with the object of limiting within a specific range the movement of interest rates. Thus, a borrower will ensure that he has a maximum interest rate that he will have to pay but his ability to benefit from reducing interest rates will be limited by the FLOOR.

It is possible by selecting the 'right' spread of CAP and FLOOR to have a COLLAR with no up-front premium.

It is important to remember that if a bank overdraft and loan has a minimum lending rate then this should be taken into account when setting a floor.

When considering any of the interest rate protection products there can be tax and accounting implications and it is therefore important for farmers to talk to their accountant before entering into a contract.

Most banks deal in these products.

How much does interest rate protection cost?

Earlier in this chapter mention was made of yield curves and in the case of forward rate agreements and SWAPS it is these yield curves which reflect the market's view of the future movement of interest rates which determines the price of these products.

Whilst yield curves also influence the price of option-based products, other factors are also involved:

- The interest rate level chosen. The market will take a view on the likelihood of the ceiling or floor being breached during the period of the contract and price accordingly
- The term of the contract. In general terms, the longer the contract the bigger will be the premium.

Prices in the market place are not subject to negotiation, a cost is quoted and that is final.

How to decide whether a fixed-rate product and/or option-based product is appropriate

It is the view of the individuals within the business against the market's view that determines whether a business takes a fixed-rate product or an option-based product.

When a business is less optimistic than the market or it wants to be absolutely sure of cost or income, then it should use a fixed-rate product.

If the business' view of the future is more optimistic than the market then it should use an option-based product.

Summary

For some businesses protection of borrowing costs or interest income will be regarded as being of little consequence because the exposure to borrowing or the level of deposits in relation top overall income stream is small.

There will be other businesses where a 5% movement in interest rate in the wrong direction could be critical in terms of profitability, for these businesses protection should be seriously considered and, in some borrowing situations, the lender may make it a condition of granting a loan.

Short-term creditor finance (up to three years)

The clearing banks

Apart from the farmer himself the clearing banks are the largest single suppliers of finance to the agricultural industry. A high proportion of their lending has been at the short-term end of the market. Total bank lending in December 1997 stood at £7104 million.

Generally speaking the banks borrow their money from depositors on a short-term basis. Current accounts represent 40% and deposit accounts 30%. Current accounts are repayable on demand and deposit accounts can vary from demand deposits to requiring the giving of three months' notice. As prudent businessmen the bank's general managers do not want to commit themselves to too much long-term lending when they are on lending funds where they have a potential commitment to repay in the short term. Thus the majority of bank finance for agriculture is for working capital purposes plus stock and debtor finance. (Working capital is the capital required to finance the production cycle.)

The bank overdraft

This is the traditional method of financing the working capital requirements of a farm business. The farmer agrees with his bank an overdraft limit he may go up to on his current account. The overdrawn balance on this account will vary as the farmer pays cleared credits into the account from debtors or issues cheques in payment of bills to creditors. The farmer will only pay interest on his daily balance, when overdrawn, and not on the overdraft limit which he has agreed with the bank.

The bank overdraft is theoretically repayable on demand but in practice overdraft limits are agreed for up to 12 months and unless the farmer deliberately misled the bank or his business collapses the facility is unlikely to be withdrawn at short notice. Indeed history indicates that many short-term bank overdrafts have been in existence on some farms for 30 years or more, financing successive production cycles.

How much will be lent in relation to the farmer's own stake?

Traditionally, banks have lent on a maximum pound for pound basis. If the farmer has a pound in his business then, in principle, a bank has been prepared to provide up to another pound to help finance working capital requirements. This is not a hard and fast rule and is governed by a number of factors, such as the ability of the farmer, the profitability of the business and the risks associated with the project.

What will a bank overdraft cost?

Interest

Interest rates are always variable and linked to the particular bank's base rate. The margins above the base rates vary from 1.5% to about 4%, depending on the individual and the particular proposition. The finer rates are associated with the overdraft which is fully fluctuating, that is, it swings from being overdrawn to having credit balances on the account on a regular basis. The higher rates can be associated with higher risk or with hardcore borrowings where the current account has a proportion of the overdraft limit permanently in use, year in, year out. It is quite common for an overdraft facility to have split interest rates, the fluctuating portion at a finer rate than the hardcore element. Overdraft interest rates are quoted on a nominal-rate basis and interest is debited to the current account on a quarterly basis in arrears.

Commissions and other administrative charges

All commission charges for loans are debited to the customer's current account in addition to specific charges associated with the current account itself.

Entry count

All current accounts whether or not they are overdrawn attract a charge based on the number of cheques drawn against the account. This charge is based on the cost to the banks of processing these cheques. In addition, credits paid into the account attract an entry cost on the components of those credits, namely cash, house cheques and cheques from other banks. Where a current account carries credit balances from time to time, an allowance may be made for the value of these balances to the bank and this value is offset against the entry count charges (see Example 11.8). However, if the value of the credit balances exceeds the entry count charge, the farmer does not receive a credit for the difference.

Example 11.8 Calculation of commission costs

Number of cheques drawn in the quarter	50
Cost per cheque	52 p
Number of credits paid in the quarter	15
Cost per credit	65 p

The account was in credit on a number of occasions and the value of these credit balances at interest worked at 3% was £5.

Calculated entry cost:

	£
50 entries × 52 p =	26.00
15 entries × 65 p =	9.75
Less value of credit balances	5.00
Net entry cost	30.75

There is some cost saving to be made if the bank's 'electronic' products are used, these cut out paper transactions and in some cases enable customers to move their funds around electronically

Commissions and lending fees

These relate to the amount of work involved in providing services which are not covered by the entry count costs, such as the time spent processing requests for finance, management time or perfecting the bank's title to security which the farmer has offered to support his borrowing.

Generally speaking, the charges the banks make in this field are considerably lower than those recovered by other financial organisations.

What supporting evidence of the viability of the proposition is required?

The banks require evidence that the proposition is able to demonstrate that any overdraft facility can be serviced and that, within reason, the level of borrowing will show good fluctuations on a monthly or seasonal basis. In some cases where the customer is well known as a successful and profitable farmer, the bank may well agree facilities on the basics of a track record shown in the accounts for the past few years. Where a new customer is asking for facilities or the facilities required by an existing farmer are materially different from those given in the past, then the bank will probably request forward budgets and cash flows to support the request in addition to seeing accounts.

What are the security requirements?

Overdrafts may be fully secured by security such as land, stocks and shares or life policies with surrender values which cover the facility. Alternatively an agricultural charge may be taken over the business. At the other extreme, they may be unsecured, with the bank relying on the proprietor's stake in his business as the ultimate cushion against the bank losing money. Where no tangible security is available the bank may insist on a cheap form of life cover on the farmer as the success of the business will depend on him.

Short-term point of sale finance

As with machinery purchases, the banks provide finance at the point of sale for the variable inputs to a farm business such as fertilisers, feeds, seeds and agrochemicals. The leading names in this field are Farm Masterloan (Barclays) and Growcash (Nat West).

Loans range from three months to two years, depending on what is being purchased and how quickly the input will show a cash return. The main features are:

(1) Loans are unsecured
(2) Interest rates may be fixed or variable
(3) Up to 100% finance will be provided
(4) The repayment programme can be tailored to the cash flow of the business

Merchants' credit

Merchants have traditionally provided a significant amount of short-term credit to their farmer customers. It was quite common in the 1960s and early 1970s to find merchants providing 6 to 12 months' credit. The credit charge made was not always clear but some firms quoted a monthly interest charge for failure to pay on a monthly basis whilst others quoted a discount for paying each month.

In the past few years, merchants have become far stricter on credit terms as they, along with every other business, having experienced pressure on profits and cash flow. The general philosophy seems to be to minimise the finance they provide to agriculture and hand this credit-giving function to the clearing banks. Symptomatic of this change are the fortnightly trading terms which are now common and the large discounts offered to encourage farmers to pay on time, or high interest penalties for late payment.

The taking of discounts can be very lucrative from the farmer's point of view and expensive for the merchant to give as shown in Example 11.9. Sometimes it will be necessary for a farmer to borrow to take advantage of a discount. There is a simple formula for evaluating the merits of a particular set of circumstances:

$$y = \frac{100X \times 365 \times 100}{(100 - X)t}$$

where y = % interest rate the farmer will pay if he fails to take the discount, X = % discount offered and t = time in days by which the farmer has to shorten his normal payment terms to take advantage of the discount offered.

Example 11.9 An assessment of the saving in taking discounts

A farmer who normally pays his bills 90 days after receiving his invoice is offered a 5% discount if he will pay one month from the invoice date. What rate of interest could the farmer afford to pay on a bank overdraft before he would be out of pocket by taking the discount?

X is the discount offered, i.e. 5%
t is the time by which the normal credit period is shortened, which equals:

$$90 \text{ days} - 30 \text{ days} = 60 \text{ days}$$

$$\text{interest percentage } y = \frac{100 \times 0.05}{(100 - 5)} \times \frac{365}{60} \times 100$$

$$= \frac{5}{95} \times \frac{365}{60} \times 100$$

$$= 32\%$$

In this case, if the farmer can borrow from his bank at less than 32% interest, which he can, then he should take the discount he is offered. Similarly, it would be expensive for a farmer to give a discount of this nature.

How to approach your bank for finance

A farmer's bank manager is an ordinary businessman and human being. He deals on a daily basis with a wide range of businesses and in some cases he will have a very intimate knowledge of the trade or industry a particular business represents. In the vast majority of cases, he will have relatively little knowledge of the industry he is dealing with but will be well versed in financial matters and will have considerable experience of the factors which are necessary to make a successful business.

Most bank managers deal in common sense and are keen to help in the development of their customers' businesses. Assuming that the farmer has a manager who falls into this category, he need not be concerned about the reception he will get when he approaches his bank for finance. If his proposition is sensible then he will get the support he asks for even if the bank manager does not know the front end of a cow from the back! In practice, a significant proportion of country bank managers do know a little about the agricultural industry and are generally keen to learn more.

Indeed, some banks have appointed agricultural managers to deal purely with agricultural business.

The path to success can be smoothed by understanding how a banker approaches a lending proposition and what information he requires to reach a decision. There are many factors to be considered when trying to assess the acceptability of a lending proposition. Clearly, profitability, to both the farmer and the bank, must play an important part but it must be remembered that social, management and, indeed, political considerations can influence the end result.

If the main objectives of the farmer and the bank are the same, that is to make money, farmers with the bank's money, banks with their customers' money, then it seems reasonable that there should be some basic requirements which are common to both parties' approaches in establishing the viability of a proposition. Whilst it may not always be possible to agree in detail the interpretation of a particular situation, the basic facts required to determine the wisdom of borrowing money are the same. If this analysis is correct, and the writers believe this is the case, what factors should be present in a lending proposition and what evidence will the banker need to help him determine:

(1) If it would pay his farmer customer to borrow money from the bank?
(2) If it would pay him, the banker, to lend his bank's money to the farmer?

A borrower of integrity and ability

The most important factor in any business proposition is the integrity and ability of the individual farmer. The banker is looking for a man or woman who in his estimation is trustworthy and capable of managing the money he or she is lent profitably. Having put forward a proposition, the farmer must be capable of carrying it out successfully. The critical factor is the management ability of the individual farmer. Unfortunately it is extremely difficult to upgrade management and in the past many propositions have been put to bankers for financial assistance where the standard of management required to achieve the predicted performance level has been far beyond the management capabilities of the farmer concerned in terms of both husbandry and financial acumen.

A good deal of money in past years has been lent, to owner–occupier farmers in particular, which has not been used to increase output but purely to allow the farmer to maintain his standard of living.

The banks had been relying solely on the security of the farm and looking forward to the day when the farmer died or took out an AMC mortgage. This type of lending is not really banking, it is pawnbroking!

The capital of the farmer

The capital of the farmer should be in sensible ratio to the finance provided by the bank. (The farmer should provide 60% or more of the capital.) Banks do not want to be in a position of having a greater involvement in the farm business than the farmer himself.

In general terms the farmer's capital will largely finance the fixed assets on the farm with the bank overdraft supporting the current assets and the breeding stock. There are of course longer-term loans for machinery, plant and land purchase.

In the case of a tenant farmer, where there is no supporting security in the form of freehold deeds, bankers will consider providing assistance up to a maximum of one pound for every pound which the farmer has in his business. This capital stake is normally assessed using a farmer's balance sheet, shown in Figure 11.8.

The precise level of finance will vary considerably according to the judgement which the individual banker makes of the farmer's management ability and the quality of the proposition.

The viability of the proposition

The proposition must contain within itself the means to repay on demand or within a reasonable time or on a seasonal basis; or if considering the medium or long-term loan market, then the proposition must demonstrate that the loan repayments can be met on the due date and within the life span of the particular asset being financed without prejudicing the survival of the business.

If there is the correct mix of these three factors, banks are usually content to review facilities on an annual basis unless pressure is exerted from an outside source, knowing that their money will be assisting the profitable expansion of their customer's business while making a profit for the bank. It is not in the interest of either party for the bank to recall this money because:

(1) The farmer will have to find somebody else to lend him the money for expansion
(2) The bank manager will have to find a new farmer with equal (or better!) ability and integrity to borrow his money

Having broadly outlined the requirements for an acceptable banking proposition it is necessary to establish where the information is found to support or condemn the acceptability of an individual case.

Whether there is a change in policy, be it to keep more cows, pigs or sheep, or a farmer is simply contemplating continuing his present farming system, it is essential that he prepare the appropriate budgets and cash flows. It is vital that wherever possible the figures and performance levels used in these calculations are those achieved by the farmer and not performance standards attained by a selected sample of recorded farms in a farm management handbook. Such figures have their uses but only in a comparative role.

Where the figures for returns and inputs are historic in nature they must be updated without improved performance levels unless there is justification for such a move.

FARMER'S BALANCE SHEET **BARCLAYS**

Branch	CIRENCESTER	Date	29/11/97
Customer	Turner and Taylor Farms		
Farm	Hollybush Farm, Cambridge		
Landlord	o/o	Total Milk quota available	800,000 L
Insurance	Buildings £ 750,000	Livestock £ 180 000	
	Implements £200,000	Deadstock £ 75,000	
Other charges	Drainage rate, Tithes, etc		

Area in acres/ha	Owned	Rented Agricultural Tenancy	Rented Seasonal
Arable	365		
Grass	85		
Other (specify)			
Wood and roughland	91		
Total	541		
Number full time men employed: 5 — Rent per annum			
When due			

LIABILITIES

Rent owing and accrued from / /19		£	
Due to Bank	Loan account 1	£	
	Loan account 2	£	
	Cheque account	£	93,182
Trade creditors	Merchants	£	16,400
	Auctioneers	£	
Other creditors	Tax	£	10,000
		£	
		£	
Hire purchase	Total amount outstanding carried from E on reverse **E**	£	24,000
Leasing	Total amount outstanding carried from F on reverse **F**	£	
Sub Total		£	143,682
Mortgages	1. Barclays Agri	£	151,000
	2.	£	
	Interest owing	£	
Other borrowed money	Relatives	£	
		£	
Dilapidations		£	
Other liabilities		£	
Total Liabilities		£	294,582
Farmer's net worth		£	2,932,241
Contingent liabilities such as guarantees Name		£	
Name		£	
Total		£	3,226,823

ASSETS

Cash at bank		£	
Milk cheque owing		£	16443
Book debts	Good	£	21100
Stocks of	Fertilizer and spray	£	12000
	Feeding stuffs	£	3500
Value of livestock carried from A on reverse **A**		£	158150
Value of growing crops carried from B on reverse **B**		£	32130
Value of produce in store carried from C on reverse **C**		£	91500
Sub Total		£	334823
Land and buildings		£	
	Area 450ha @£5000	£	2,250,000
	Area 91ha @£1500	£	136,500
Quota owned		£	
	Milk 800,000 litres @ 35	£	280,000
	Suckler cow premium cows @	£	
	Sheep annual premium ewes @	£	
Value of landlord approved improvements to land and buildings (Tenant Farmer)		£	
Tenant right		£	
Value of machinery carried from D on reverse **D**		£	156,500
Other farming assets (Shares in co-operatives, etc)		£	
Sub Total		£	3,157,823
Non farming assets			
	Stocks and shares	£	50,000
	Life policies		
	Face value £250 Surrender value	£	
Other personal assets	Vehicles Discovery	£	19,000
		£	
Total Assets		£	3,226,823

I/We do hereby declare that the above is a true statement of my/our financial position; that the assets shown are all my/our property (except those that are leased) and that no other party has any interest therein.

I/We further declare that all my/our liabilities are shown in the above statement and that I/we have no indebtedness outstanding except as stated above.

Signature(s) [signature] Signature(s) Martin Taylor

83 (4/95)

Fig. 11.8 A farmer's balance sheet.

LIVESTOCK

Proposed annual stocking	Number today			Current market price per head	Total value today
150	135	Dairy cows	non CI/CI	@£600	£ 81000
	39	Dairy heifers	Over 2 years	@£600	£ 23400
	35		1-2 years	@£375	£ 13125
	37		under 1 year	@£125	£ 4625
		Beef cows	Hill/Lowland	@£	£
		Other beef cattle	over 2 years	@£	£
			1-2 years	@£	£
			under 1 year	@£	£
250	200	Pigs	Sows and Gilts	@£110	£22,000
			Boars	@£	£
	400		Weaners	@£ 35	£ 14,000
		Other	Pork/Cutter/Bacon	@£	£
		Sheep Ewes	Hill/Lowland	@£	£
			Lambs	@£	£
			Rams	@£	£
		Poultry		@£	£
				@£	£
				@£	£
				@£	£
				@£	£
				@£	£
		Value of Livestock	**A**	£	158150

GROWING CROPS

Proposed annual cropping	Present area planted		Estimated value per acre/ha	Total value today
	217	Winter Wheat	@£90	£19530
	140	Winter Barley	@£90	£12600
		Spring Barley	@£	£
			@£	£
			@£	£
			@£	£
			@£	£
			@£	£
		Grass	@£	£
		Value of Growing Crops	**B** £	32130

NORMAL YIELDS

		Per acre/ha		
Dairy cow yield (calving) Ayr, Spring, Autumn	6000 litres	Winter Wheat	7.5	tons
Lambs reared per ewe		Winter Barley	6.5	tons
Weaners per sow	22.5	Spring Barley		tons
				tons
				tons
				tons

PRODUCE IN STORE

	Quantity			Value
Wheat	800	@£80	per Tonne	£64,000
Barley		@£	per	£
		@£	per	£
		@£	per	£
		@£	per	£
Silage	1500t	@£15	per tonne	£22500
Hay		@£	per	£
Straw	500t	@£10	per	£ 5,000
Forage roots		@£	per	£
Value of Produce in Store		**C**	£	91500

MACHINERY ON FARM (Including Leased)

Type	Age	Present value
Combine Harvester Claas		£35,000
Tractors		£
John Deere		£19,000
Matbro TR250-110		£17,000
New Holland 8560		£23,000
Hardy LY Sprayer		£ 8,000
Horsch co/p 6.25		£ 5,500
HKK Disc		£ 3,000
		£
Baler		£
Forage Harvester		£ 5,000
Milking equipment		£15,000
Trailers 4		£ 6,000
Other Equipment		£20,000
		£
Fixed equipment		£
Value of Machinery	**D**	£ 156500

HIRE PURCHASE

Details of Hire purchase debts	Annual payments	Date of final payment	Amount outstanding on agreement
Highland Leasing	£8,000	15/12/99	£24,000
	£		£
	£		£
Total	**E** £		24,000

LEASING

Details of Leasing agreements	Annual payments	Date of final payment	Amount outstanding on agreement
	£		£
	£		£
	£		£
Total	**F** £		

Fig. 11.8 *(continued)*

What documentation will the bank manager require to aid him in reaching a decision?

Audited accounts

Where there is an existing business the bank manager will be able to look at the past record through accounts. These will reveal:

(1) Past profitability
(2) Farmer's net worth at historic values (see Chapter 5)
(3) The level of fixed costs on the farm
(4) The liquidity of the business (see Chapter 5)
(5) The farmer's 'track record'

A forward budget

The track record of a business is an important starting point in assessing a business. To test the future potential profitability of the farm business and its ability to service and repay any borrowing, however, a farm budget is essential, particularly where there is a change of emphasis or enterprise. It is important to record the assumptions behind the budget as the bank manager will want to assess the likelihood of these assumptions being achieved in practice. It is also helpful if a sensitivity analysis is carried out, to gauge the implications for the business of failure to reach budget.

The use of yardsticks

Experience has shown the banker that the financial pressure on a business can be gauged using relatively crude yardsticks. It must be remembered that these yardsticks are not absolute measures and must not be used as such. There are two popular yardsticks:

(1) *Rental equivalent*. There are a number of 'definitions' of rental equivalent but basically it takes into account all the financial commitments of the business and the cost of owning or renting the land. This total is then apportioned on a per hectare basis. The calculation of the rent equivalent should include:
    (a) The rent and rates of the business
    (b) Bank interest
    (c) Bank and other loan interest and annual capital repayment
    (d) Hire purchase and leasing charges.

    Table 11.6 gives an indication of average and high gross margins for particular farm types.

    The level at which a rent equivalent is feasible on an individual farm will depend on the management expertise and technical performance of the farmer. A low rent equivalent does not necessarily mean the farm will be profitable! Unfortunately rent equivalents are not as helpful if there are significant non-land-using enterprises on the farm, for example,

Table 11.6 Maximum rental equivalents associated with different farm businesses and different performance levels

Farm type	Maximum rent equivalent (£/ha)	
	Average performance	High performance
Dairy (rearing replacements)	150	270
Dairy (flying herd)	190	350
Cereals	95	200
Cereals/dairy	140	220
Arable (including roots and pulses)	180	350
Beef/sheep	90	150

housed pigs or poultry. Rent equivalents do not take into account the output from a business and so a more sophisticated measure can be used as detailed below.

(2) *Rental equivalent as a percentage of gross output.* Relating rental equivalent to gross output is fine but there are practical problems in calculating a future gross output with any sort of accuracy, and the extra sophistication has little benefit unless non-land-using enterprises are present on the farm. Between 8 and 15% is the normal range and 20% is considered high.

A forward budget which gives rental equivalents or rent equivalents as a percentage of gross output above the guidelines gives the bank manager a clear warning that further detailed investigation is required!

Measures used along with rental equivalents

(1) *Cash flow.* This is required to establish that there is sufficient cash available to fund the proposed budget and to establish that the overdraft/loan facilities requested are sufficient (see Chapter 8). It is also possible to calculate rental equivalent from cash flows.

(2) *Farmer's balance sheet.* This will establish the proprietor's capital in the business at current valuations (see Chapter 5). It must be remembered, however, that there could be a capital taxation liability. Knowing proprietor's capital, it is then possible to assess the gearing of the business by taking the proposed level of borrowing from the cash flow.

Exactly how much of the above information will be required will be governed to a large extent by the bank manager's knowledge of the individual case. However, the good farmer will usually have the information readily available from his normal management data.

12 Taxation

The importance of husbandry to a successful farm business was stressed at the beginning of this book and subsequently a series of business management techniques have been explained which should aid the reader in developing the management skills to ensure that management profits are maximised.

It is, however, impossible to consider business management and maximisation of profit retention without making some reference to taxation and how it impinges on the farm business. In Chapters 5, 9 and 11 the taxation treatment of various decisions was briefly discussed.

The subject of taxation is a discipline in itself and it is not therefore within the scope of this book to discuss taxation in depth. Indeed, with the rapid changes which take place in the taxation field, it is essential that all taxation matters are discussed with the business's accountant. In an ideal world, the farmer would be reviewing the tax treatment of his annual profits two months before the end of his financial year, based on ten months' management accounts!

The remainder of this chapter outlines a few of the basic principles of our taxation system, hopefully giving the reader a flavour of the potential complexity of the subject. For those who wish to learn more, there are a number of specialised text books.

In what form do farm businesses trade?

There are three basic types of farm business:

(1) The sole trader
(2) The partnership
(3) The limited liability company

Currently in the UK around 93% of all farm businesses trade as either sole traders or partnerships with the remaining 7% as limited companies. It is important to understand that the 'trading vehicle' of the business can change so that a business which began as a sole trader might turn into a partnership and later even become a limited company. It is critical that the business's accountant is involved in any proposed change at a very early stage.

Who is subject to tax?

Any business resident in the UK will be subject to taxation of its net income, irrespective of whether it operates as a sole trader, partnership or limited company.

The income of a UK resident business will be taxable whether it is earned in this country or abroad. In addition, income received by non-resident businesses is highly likely to be subject to UK taxation.

What income is subject to tax?

To pay tax a profit, net of expenses, must be made. This profit may be made as a result of trading or from interest received on investments, in which case it would be subject to income tax in the case of a sole trader or partnership; a limited company would pay corporation tax. Alternatively, a profit may be made from the sale of a capital asset, in which case there might be a liability to capital gains tax.

How to distinguish between revenue and capital income

If a capital asset is sold, such as a farm or stocks and shares, and it can be demonstrated that it is not the normal practice of the business to 'trade' in these assets, then they will be subject to capital gains tax, not income or corporation tax. However, if these assets were a normal trading asset of the business, then the profits would be subject to income or corporation tax. Example 12.1 illustrates the different treatment of asset sales depending on who is the seller.

Capital and revenue expenditure is distinguished in the same way.

Not all income/profit is subject to tax and there are a number of well-known exceptions, such as the profit on the sale of an individual's main residence (provided it is used for the whole time as such and no part is used solely for business purposes), winnings on Premium Bonds, interest on National Savings Certificates, or the lottery!

Example 12.1 An illustration of the different treatments of assets sold as 'income or capital'

Asset sold	Capital profit	Revenue profit
farm	farmer	land agent
shares	farmer	broker

Why is it important to distinguish between revenue and capital expenditure?

For tax purposes, revenue expenditure can only be set off against revenue profits, and similarly capital expenditure can only be offset against capital profits. There is an exception in the case of limited companies where a trading loss in a year can be set against a capital gain in the same year.

How are different sources of income assessed for tax?

Sole traders, partnerships and companies have income from a variety of sources and to cater for this the Inland Revenue breaks the income down into classes or schedules for taxation purposes. The major schedules are:

- *Schedule A*. This covers income from land and buildings, including rents
- *Schedule D*. This schedule is broken down into a series of cases of which I, III and VI are of most significance to the farmer.

  Case I: Profits of a trade
  Case III: Interest receivable
  Case VI: Any income which doesn't fit elsewhere!

- *Schedule E*. Covers taxation of employees including directors of farming companies. The cases depend on where the employee and employer are resident and domiciled.

When do businesses have to pay tax?

Income tax (paid by sole traders and partnerships) is payable on an annual basis for each year of assessment (6 April to 5 April). Businesses do not necessarily have their accounting years corresponding to the year of assessment and so the assessment for tax is based on the profits for the accounting year ending in the year of assessment. There are special rules for opening or closing years of a business.

Sole traders and partnerships have to pay tax in two equal instalments on 1 January of the year in question, and the following 1 July. Late payment attracts interest!

Companies pay corporation tax. The tax year for companies is from 1 April to 31 March. This is known as the financial year. Clearly different companies will have different accounting years and in farming, for example, a popular accounting year is 1 October to 30 September. Companies pay corporation tax nine months after the end of their accounting period. If the tax is not paid on time, then interest is charged on the amount overdue.

Income tax

Income tax is chargeable on taxable income (after personal allowances) and the rate will depend on the level of taxable income for the year of assessment. The tax rates and the bands within which they operate are set in the government's annual budget. The rates and bands for 1997/8 are shown in Table 12.1. The rate structure has been greatly simplified over the past 10 years with the elimination of higher tax rate bands. As a general rule, the level of the tax bands moves in line with inflation.

Table 12.1 Income tax returns and bands for fiscal year 1997/8

Taxable income £	Rate %	Tax £	Slice £	Total tax £
Up to £4 100	20	820	4 100	820
Next £22 000	23	5 060	26 100	5 880
Over 26 100	40			

Self assessment

From 6 April 1996 a new system of tax assessment was introduced called self assessment and it applies to the tax returns of individuals. From 6 April 1997 partners in a partnership will be assessed as individual partners on their profit share in the partnership. This will be based on the partnership accounts prepared by a nominated partner.

Companies are not affected by these changes.

Individuals should discuss the requirements with their accountant but the main features are:

(1) Since 5 April 1997 farmers have been taxed on the 'current year' basis of assessment. This means that they are taxed on the profits shown in their accounts ending in the same tax year, i.e. between the 6 April and 5 April the following year. For example, if the farm business accounting year ended on 31 October 1998 then it will be taxed for that year in the tax year 6 April 1998 to 5 April 1999.

(2) Taxpayers will work out the amount payable for each period and in normal circumstances the inspector of taxes will not need to issue a notice of assessment. However, if the information is submitted early enough there is the option of getting the inspector to do the calculation!

(3) The returns will not be checked as they are received but at a later date. The Inland Revenue will still make enquiries where they believe the returns are inaccurate. The Revenue can raise queries up to 12 months after the deadline for submission of returns, and charge interest on underpayments. The assessment is fixed if no queries are raised in this period, unless fraud is discovered.

(4) Returns must be submitted by 31 January after the end of the tax year, or by the previous September if you require the Revenue to work out the tax.

(5) There are penalties when returns are submitted late, ranging from £100 when up to 6 months late (this is automatic) to up to £60 per day if they are more than 12 months overdue.

Corporation tax

Corporation tax was introduced as a separate business tax by the Finance Act of 1965. The schedules under which companies are assessed for corporation tax are the same as for sole traders or partnerships. However, the financial year runs from 1 April to the following 31 March.

There are two basic rates of corporation tax. First there is the 'standard rate' which for the 1997/98 year stands at 33% on profits in excess of £1 500 000. Secondly, there is the 'small company' rate for companies with profits of less than £300 000. This rate for 1997/98 stands at 23%.

For companies with profits between these two figures there is a 'marginal' relief on the higher rate which is effectively 33.5%.

Where a company's accounting year is different from the corporation tax financial year the profits are taxed at the rates applicable in the two years. The profits are assumed to be made evenly over the year!

Capital gains tax

Capital gains tax is payable by sole traders and partnerships, but not companies. The rate equates to the income tax rates, with individuals being allowed an annual exemption which, for 1997/98, is £6500. In the past the annual exemption was increased by inflation but from 6 April 1998 the allowance will be frozen. Gains beyond this level will be taxed subject to tapering reliefs depending on the type of asset and the length of time the asset is held after 5 April 1998.

Companies pay corporation tax on gains. There are no exemptions.

With the variations in land prices the significance of capital gains has changed depending on when it was purchased.

The gain which results from the disposal (this could be as a gift or a sale) of a relevant asset attracts capital gains tax. This gain is the difference between the cost (purchase price plus costs) and the sale price (sale price less costs).

For disposals of a business before 6 April 1982 there was no tax relief for inflation but since that date there has been an indexation allowance based on the 31 March 1982 base.

There are a number of exemptions to capital gains tax, for example, disposals of a business on retirement for gains up to £250 000 attract full relief with gains between £250 000 and £1 m attracting 50% relief. Whilst there is no capital gains tax charged when a gift is made between husband and wife, this is not exempt. The recipient simply takes the donor's base cost. It is effectively a holdover increasing the value of the recipient's estate.

Capital gains tax may be deferred on the sale of business assets if the proceeds of the sale are completely reinvested in other eligible business assets. These 'new' assets must have been acquired no more than 12 months before or three years after the sale in question. It is important to discuss this with an accountant.

Personal allowances

Private individuals are eligible for allowances to set against their income and so the sole trader, partners in a partnership or employees of companies (including directors) can reduce their taxable income by the amount of these allowances.

It is not within the scope of this book to detail all these allowances but Table 12.2 outlines a few of the more notable ones.

Table 12.2 Some personal allowances and reliefs, 1997/98 fiscal year

Type	Circumstances	Relief £
Personal allowance	Single	4 045
	Married couple*	1 830
Age allowance relief		
65–74	Single	5 220
75 and over	Married couple*	3 185
	Single	5 400
	Married couple*	3 225

* Relief restricted to lower rate 15%.

The allowances are supposed to be index linked on an annual basis unless the Treasury orders otherwise.

Capital allowances

When a profit and loss account is prepared for a farmer by his accountant it will have, as an expense item, depreciation. This depreciation figure represents the write-off of capital expenditure on fixed assets over the life of the asset. Thus a tractor purchased for £20 000 might show depreciation in the profit and loss account of, say, £4000.

However, for taxation purposes depreciation is not an allowable expense. Instead, capital allowances are available both for corporate and non-corporate businesses. The allowances are available for a 'chargeable period', i.e. in the case of a non-corporate body the year of assessment and for a corporate body the accounting period.

Capital allowances are applicable to new and second-hand assets purchased at any time after 26 October 1976.

Plant and machinery

Until 13 March 1984 it was possible to charge 100% of the cost of plant and machinery against the profits in the year of purchase. These allowances were called first year allowances (FYA). Since that date the Chancellor of the Exchequer has changed the rules so that first year allowances were gradually phased out and from 1 April 1986 replaced by the writing down allowance.

Writing down allowance for plant and machinery has been in existence since 1946 (with alterations in the 1971 Finance Act), currently the Capital Allowances Act 1990 applies. Frequently where first year allowances were available, profitable farmers made use of these and charged the cost of a purchase in one year's accounts, thus reducing tax.

When FYA were withdrawn all capital expenditure on plant and machinery had to be put into a 'pool' of capital expenditure which could be set off against profits at the rate of 25% per annum of the reducing balance, i.e. in the first year £1000 capital expenditure would give allowances of £250 to be charged against the profit. In the following year, there would be £750 left in the pool of which £187.50 could be charged against the profit. It would therefore take seven to eight years to recover the bulk of the capital expenditure against profits. It is because of this that leasing has become attractive to tax-paying farmers (see Chapter 11).

If machinery is sold for more than its written-down value in the accounts this will give rise to a 'balancing charge' in the pool and in effect create a capital gain which will increase the tax liability.

In the July 1997 budget the Chancellor allowed a 50% FYA and this may be the case in subsequent years.

Agricultural building

Until 1 April 1986, it was possible to claim 30% of the cost of a farm building, fencing, roads or drainage against profits in the first year: 20% represented an 'initial allowance' and 10% a straight-line allowance, which meant that the total cost could be recovered against profits over eight years. It is now only possible to claim 4% per annum of this capital expenditure against profits, which means it is 25 years before all the cost is recovered!

The valuation of livestock

What is regarded as farm stock for tax purposes?

Included in farm stocks are crops in store, growing crops, tillages and livestock.

The trading stock basis of valuation

The general principle is that stock should be valued at the lower of cost or net realisable value.

It is not a function of this book to deal in detail with valuation, which is a subject in its own right, but in practical terms the cost attributed to harvested crops is the variable costs of production (i.e. seeds, fertilisers and sprays) plus some allowance for rent and cultivation costs. The valuation of crops in the ground is usually based on the variable costs plus the actual cultivation work completed.

Livestock are valued on the basis of the lower of cost or market value but a problem area arises with home-grown livestock because it is frequently very difficult to establish the cost of rearing them. The Inland Revenue have acknowledged the fact and are prepared to accept a percentage of market value as an estimate of the cost of rearing.

The arrangements for acceptance of this method are generally made with the local Inspector of Taxes and the normal figures are 60% for home-reared cattle and 75% for sheep and pigs. If the stock are pedigree or this percentage would not give a realistic value of cost then the equation is different.

The mechanics of the trading stock basis of accounting for stock is for the cost of the animals to be charged in the trading account as an expense and the proceeds of any sale to be included as a receipt. The trading account is also adjusted for any change in the stock valuation over the period. Thus, the opening and closing stock valuations are included in the accounts.

If over the year the stock valuation has increased (i.e. closing stock valuation is greater than opening stock valuation), then the accounts will show a paper profit which has not yet been realised. However, this profit will be subject to tax in the appropriate year of assessment.

There is clearly a disadvantage in having to pay tax on paper profits and this has resulted in the introduction of the herd basis of assessment.

Herd basis of assessment for production herds

In the Taxes Act 1970 it was recognised that the purchase of production herds which were kept solely to produce income from products such as milk, eggs and the sale of offspring was really capital expenditure and so should not enter into the trading account for the purpose of assessing profit.

The definition of a production herd for the purpose of this Act is a herd of animals of the same species, irrespective of breed, kept wholly or mainly for the sake of products for sale, such products being from the living animal. That is:

(1) The young of the animal
(2) Any other product that does not involve the slaughter of the production animal (e.g. milk). Production herds include dairy herds, suckler herds, sheep flocks, pig herds, laying bird flocks, stud mares

The herd basis of assessment will not be applied to production herds unless the farmer elects for this form of assessment. The election will only be valid if:

(1) It is made in writing
(2) It is made within two years of the end of the year of assessment in which the

farmer first kept the particular class of production herd (financial year end 25 March 1996 election must be made by 5 April 1998).

(3) It is made within two years of receiving compensation for the slaughter of the whole or a substantial part of the herd by order under animal diseases laws

(4) It specifies the class of herd covered. Herds are deemed to be of the same class if they:
   (a) Are of the same species (breed is irrelevant).
   (b) Produce the same products for sale (e.g. cows producing milk for sale would form a production herd, cows producing suckler calves for sale as a separate herd)

The election once made is irrevocable for a particular class of livestock.

In general only mature animals can be accepted as qualifying for inclusion in a production herd because products for sale would not come from the young or immature animal. The female of a species is regarded as mature when it produces its first young.

However, immature animals will qualify for inclusion in a production herd where they are solely reared for replacement and the land on which the herd is kept is such that replacement for draft animals can only be replaced by animals bred and reared on that land, e.g. acclimatised hill sheep on Welsh mountains.

Animals which are kept for a short term are not eligible even if they produce a saleable product (e.g. flying herd). Similarly, working animals, racing animals, animals kept for competition or exhibition purposes are excluded.

How does the herd basis of assessment work?

(1) The herd is treated as a capital asset and the initial cost of purchasing the herd cannot be deducted from the trading profit. If the herd is expanded then the cost of the additional animals is a non-deductible expense

(2) The value of the herd is never brought into the trading account

Treatment of additions or reductions to the production herd

If an animal is part of the trading stock on a farm and it is put into the production herd as an addition (not a replacement), then the trading account is credited with the purchase price of the animal, plus any cost there might have been in bringing the animal to maturity. If the animal is a home-grown product, then the cost of breeding or a specific percentage of market value is credited to the trading account and the cost of the replacement animal debited to the trading account (if home reared, then this cost would already be in the trading account).

Where the original animal is replaced by an animal of different quality then the trading account is credited with the sale price of the animal and debited with the cost of replacing with an animal of the same quality. The herd asset account would be increased by the difference between the actual cost of replacement and the cost of replacing with an animal of the same quality.

Treatment or replacement of the whole herd

Where the whole of a herd is sold and replaced by a herd of the same class then the ordinary replacement rules apply.

If over a 12-month period the whole herd is sold or a substantial part of the herd is sold (20% +) then any profit or loss on the transaction does not pass through the trading account provided another herd of the same class is not purchased within five years or there is no move to build up numbers in the five-year period with a view to replacement. Insurance or compensation receipts for a herd are treated in the same way as a sale. The herd basis of assessment demands that a more detailed record is kept so that it is possible to identify individual animals and their value.

What are the advantages of the herd basis of assessment?

How much advantage can be gained from the herd basis of assessment will be determined by, firstly, the size of the herd; secondly, the difference between the market value and cost of the animals (this is a measure of the quality of the stock); thirdly, the situation of the individual farmer.

The main advantage on an annual basis is that since the value of the herd and any additions to it are excluded from the trading account there will be no paper profit on which tax might be payable. How advantageous this is will depend on whether the stock are home reared or purchased in the open market. The greatest potential advantage of the herd basis arises when the herd is sold and there is a large profit between book value and sale price. This profit is free of income tax and capital gains tax.

Inheritance tax

The taxation of distribution of an individual's wealth is governed by the Inheritance Tax Act 1984 which has been amended by subsequent Finance Acts. Inheritance tax is payable on transfers of value though there are some transfers which are exempt and others which are potentially exempt.

Transfers of value

A transfer of value is deemed to have taken place when an individual's net assets are reduced in value as a result of a gift or disposition. The value of the transfer is the reduction in value of the net assets as a result of the gift or disposition.

Where the transfer is made as a result of death then the value of the transfer equates to the value of the estate just prior to the death of the individual.

Exemptions and reliefs from inheritance tax

A number of transfers of value are either exempt or have relief from the full impact of inheritance tax and a number of these have a direct bearing on agricultural businesses. These may be transfers in an individual's lifetime or upon death. Table 12.3 highlights some of these exemptions and reliefs.

Levels of inheritance tax

Inheritance tax becomes penal as the level of the chargeable transfer increases. Table 12.4 shows the impact of this tax on estates.

Apart from the exemptions and reliefs shown in Table 12.3 there are potentially exempt transfers which require the survival of the transferor for seven years for complete exemption or they will attract inheritance tax at a proportion of the death rate.

It can be seen from Table 12.3 that inheritance tax can have an enormous impact on the successful transfer of a farm business from generation to generation and it is therefore of paramount importance that this aspect of taxation is carefully examined and appropriate action taken to avoid or reduce its impact as far as is sensible.

Inheritance tax is regarded as an optional tax but this means properly drawn wills and, where appropriate, lifetime transfers. Consultation with the appropriate specialists is essential to prepare the best plan, which may well need updating on a regular basis if it is to remain effective.

Table 12.3 A sample of some exemptions and reliefs to inheritance tax

Exemption/relief	Amount	Lifetime	Death
Transfers between husband and wife		X	X
Annual transfers up to £3 000 per donor	3 000	X	
Marriage gifts up to:			
Parents	5 000	X	
Grandparents	2 500	X	
Agricultural property reliefs:			
Where there is a tenancy	100%	X	X
Where there is vacant possession within 12 months	100%	X	X
Business property relief:			
Transfers controlling shareholdings in company > 25%	100%	X	X
Transfer of business assets	100%	X	X
Transfers of minority shareholdings < 25%	100%	X	X
Growing timber (deferral)	All	X	X

NB: This list is not exhaustive.

Table 12.4 The effect on death of inheritance tax on chargeable transfers inheritance tax deaths after 6 April 1997

Transfer £'000s	Slice £'000s	Rate %	Tax/slice £	Total tax £
Up to £215 000	215 000	Nil	Nil	Nil
Over £215 000		40		

Professional advice

This chapter has hardly scratched the surface of the taxation minefield!

Tax planning is a very important part of running a farm business and it requires close consultation with the accountant.

Remember, a little knowledge is a dangerous thing!

13 Marketing

Food consumption

Total household expenditure on food and meals out was around £33 billion in 1982, rising to £76.3 billion in 1992. Spending on meals out in this period rose from £6.5 billion to £21.1 billion. The pattern of food consumption has altered, with calorie intake declining a little while volume has fallen more rapidly. The public are generally more conscious of healthy eating habits and some are able to pay more for additive-free, natural and consumer-ready foods. The sector appears more susceptible to 'food scares' than in the past, with greater media attention being focused on disease and residue-free food. Consumer concern over salmonella, especially in poultry, and Bovine Spongiform Encephalopathy (BSE) in cattle have had a significant effect on sales and though conclusive proof has not been found to substantiate consumer fears, neither has full confidence been restored, despite efforts to show that the industry has food safety as its highest priority.

The proportion of total expenditure spent on household food is falling, as shown in Table 13.1. This trend means that farmers must pay more attention to the specific wants of customers and consumers in order to maintain (and increase) market shares.

Table 13.1 Expenditure on food as a percentage of total UK expenditure

	%
1976	21
1985	16
1995	11

The demand trend will be towards better quality and consumers want products which compete on price; are prepared with due diligence to food safety and are free of perceived harmful residues and ingredients; exhibit freshness, health, variety, naturalness and are well prepared (e.g. French meat which has the collagenous tissue removed pre-sale) but not over-processed, and convenient. The food industry must also respond positively to increasing consumer awareness of the effect of production systems on the environment and of the conditions in which animals live and die.

Competition

The static volume of food consumption must be compensated for by individual producers increasing their emphasis on market segmentation so that they can compete effectively with other producers, especially from Europe and overseas. Britain is now the fifth largest food and drink exporter in the world. Exports were 25% of imports in the 1970s but now comprise 60%. Twelve of the top twenty European food and drink manufacturers are British companies and prepared poultry, soft drinks, ice cream and beer are among the fastest-growing export products (see Fig. 13.1). The value of UK market segmentation is demonstrated by the success of existing producer retailers, farm shops, organic producers and healthfood stores showing that opportunities do exist in the market place for individuals, albeit low volume at present. Competition in export markets will be particularly strong and we must fulfil the market needs more specifically than our competitors to maintain sales. In cereals, for example, it is vital that we convince our buyers (by creating an identifiable product with a unique selling point USP) that we can supply high-quality produce.

Negotiations

The pressure of competition within the UK and in export markets plus a real effort on behalf of the agricultural supply trade to cut costs will mean that farmers are going to have to improve their negotiating skills with regard to buying inputs and selling outputs. It seems likely that in many areas marketing will increasingly be done in package deals so that the sale of grain or livestock products will be linked to the purchase of feed, seed, fertiliser and sprays in particular from the merchants; while the provision of advice by the merchants seems likely to diminish (as a cost-cutting exercise) and will be left to specialists whom farmers will pay for directly. These packages are already in existence so that, for example, it is possible to supply cereals to a feed compounder or merchant in return for animal feed, fertiliser or sprays (contra trading). No money changes hands for individual transactions but balancing payments will occur periodically. In some cases farmers are given inputs 'on credit' before delivering any grain. Thus both parties are benefiting from assured outlets and negotiated trading which remove uncertainty from the management of the business.

In the UK there are still market opportunities for home producers, especially in bacon and ham, cheese, lamb, fruit and vegetables (Table 13.2), although some EU production restrictions may make it difficult to expand production. In the EU there are fewer opportunities (Table 13.3).

Pressures from competition, changing consumption patterns, surplus controls, cost cutting (including the effect of economies of scale) and public concern over animal welfare and the environment are likely to continue into the next decade. These pressures are already influencing the structure of farming and the pattern is being set for the foreseeable future.

The number of holdings in the dairy sector is falling. The change is almost entirely

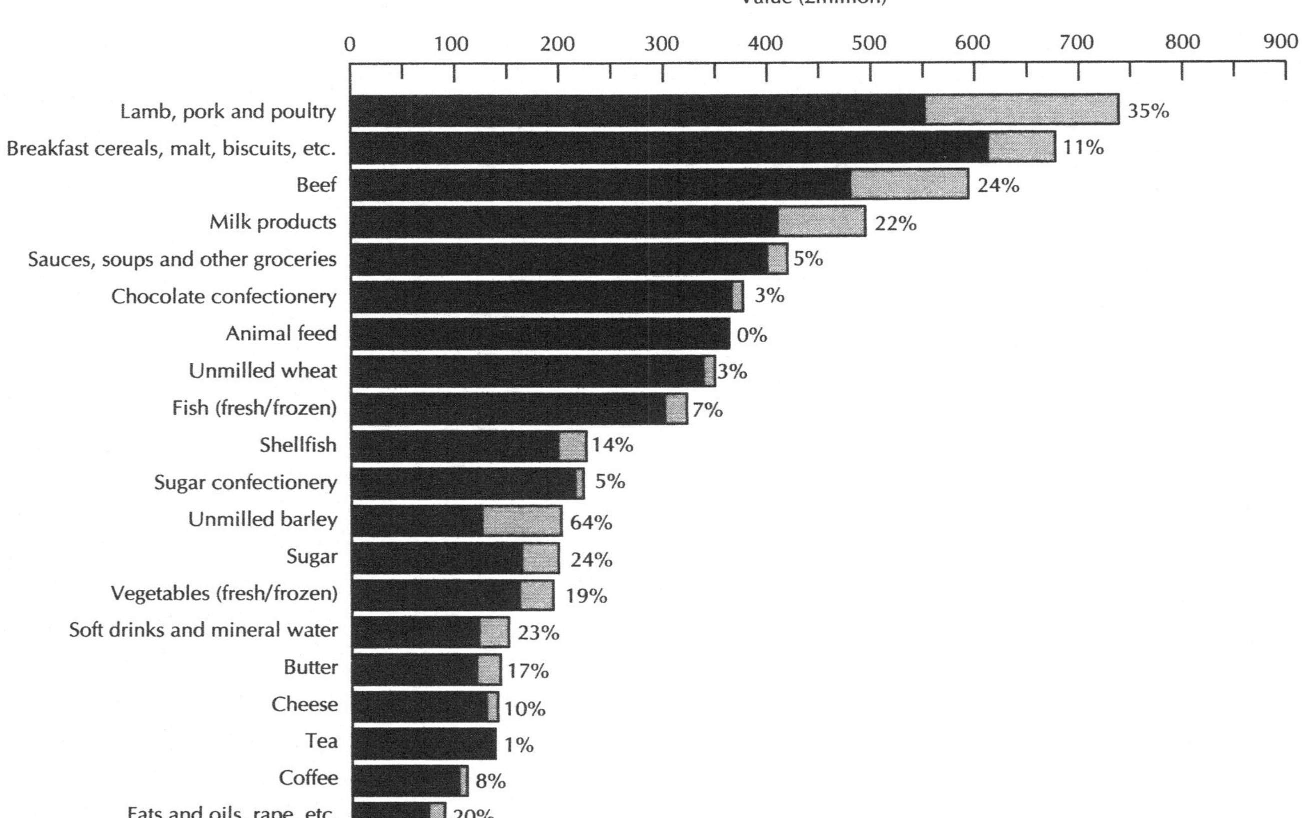

Fig. 13.1 Top 20 food and drink exports in 1995 (with percentage growth over 1994). Source: Food from Britain.

Table 13.2 UK self sufficiency in selected products (%)

	1966	1976	1986	1991	1992	1993	1994	1995	1996
Wheat	47	56	117	130	124	120	120	117	121
Barley	109	94	164	130	127	115	125	135	122
Oats	100	94	97	102	118	108	102	153	129
Rye, mixed corn, Triticale	62*	44*	71	93	93	92	91	103	100
Oilseed rape	n/a	49	123	101	93	86	80	72	101
Potatoes	92	75	87	90	91	91	89	88	90
Sugar	33	29	55	54	59	65	62	62	60
Apples	58	46	42	41	47	43	42	37	36
Cauliflowers	89	86	83	89	92	85	81	78	74
Tomatoes	28	41	32	34	32	34	32	29	28
Beef and veal	78	89	96	95	95	100	112	113	86
Mutton and lamb	44	57	84	98	95	116	113	112	99
Pork	100	100	104	103	103	100	104	102	98
Bacon and ham	35	47	45	41	42	44	46	48	44
Poultry meat	98	102	94	96	92	94	93	96	95
Butter	8	20	66	63	61	72	70	68	67
Cheese	43	60	65	68	65	72	68	70	69
Milk powder, full cream	59	116	562	467	960	357	327	426	526
Milk powder, skimmed	—	157	252	179	123	152	161	157	113
Linseed	—	—	—	144	141	136	93	68	80
Eggs	—	—	—	96	97	98	96	97	96
Wool	—	—	—	60	56	61	47	63	54

* Rye only.
Source: MAFF.

in small herds and it is likely that this trend will continue. Average herd size continues to increase. The pattern is repeated in the pig sector, with the number of holdings falling and the smaller herds disappearing.

The message is clear. Farmers' markets are becoming much more sophisticated and more attention to correct, consistent weight and quality to meet customer and consumer demands must be forthcoming from all those involved in the food chain.

This chapter now concentrates on the objectives and concepts of marketing, and on the effects which EU and UK organisations and policy have on food production and marketing.

The marketing concept

In the days of the feudal system people in local areas produced all the food which they consumed. This subsistence farming altered when migration from the land occurred during the industrial revolution (of the 1800s) and people began to specialise in particular jobs. They earned a wage and were able to buy the food which they required. Thus it became necessary for food products to be channelled to the places where demand existed, that is in population centres. By channelling

Table 13.3 Self sufficiency in selected products in the EU (%)

	1991	1992	1993	1994
Rye	—	134	113	125
Maize	—	97	111	99
Barley	—	133	123	116
Durum wheat	—	221	134	95
Common wheat	—	138	136	119
Sugar	110	115	120	113
Peaches (processed)	114	132	125	155
Tomatoes (processed)	119	117	134	124
Wine	123	112	133	113
Potatoes	101	100	103	101
Beef and veal	115	112	104	101
Pig meat	104	103	106	107
Olive oil	79	117	93	98
Soya seed	7	8	5	6
Sunflower	91	87	79	72
Oilseed rape and colza seed	96	103	100	85
Rice	109	94	86	

Source: Euro Commission.

food, selling was easily achieved and producers simply grew as much food as they could without much regard to specific consumer demand. By the early 1900s, however, technological advances meant that production had increased dramatically both in the UK and overseas, and with improved transportation aided by refrigeration, excess supplies of food had arisen. Selling was no longer an automatic result of production, even when foods were taken to the right place, as producers in the UK and abroad began to compete for the privilege of supplying the consumer.

Under these conditions the requirements of the consumer become vitally important, and production has to be orientated towards their specific needs. Thus farmers must become market orientated rather than production orientated. This means that as far as possible farm resources should be managed to produce specifically what the market demands so that a profitable outlet can be satisfied. Indeed, the outlet should be identified first, and then farm resources manipulated to fulfil the requirements of the outlet. Marketing therefore is the planned production and disposal of goods, whereas selling is preoccupied with converting a product into cash – the unplanned disposal of something that has been produced, with no predetermined outlet.

The marketing concept demands that producers put all their energy into producing specifically what people want and are able to buy, rather than trying to sell to them something which has been produced which may not satisfy market requirements. Adoption of this concept naturally leads to a discussion of market segmentation.

Market segmentation

In order to segment any market the producer has to identify the specific requirements of each segment. The market for potatoes, for example, can be segmented into:

- boiling
- baking
- canning
- chipping
- dehydrating
- crisping
- roasting

In order to make the best return the producer must know which segment to aim for, and this will often result in contractual agreements between the producer and the processor. This will then allow selection of the variety which suits the chosen segment and concentration on the husbandry techniques which will ensure that the crop meets the minimum acceptable quality standards. Thus the producer has determined the specific segment to aim for before planting the crop, and has a market outlet ready. The processor is more assured of the quantity and quality which is required and therefore both have organised and planned their marketing to mutual benefit. This planned marketing should mean that the occurrence of over-supply or shortage of one particular variety is minimised, and if the whole sector could be organised in the same way it would benefit all producers. This has been accomplished in several local areas in the UK either by potato co-operatives, working on behalf of producers, or by large individual growers.

Market segmentation can be carried out for almost all agricultural products and should be given more emphasis by producers if they are to improve total market returns in the face of increased competition at home and from overseas.

Importance of market orientation

Disposing of products is largely influenced by the attitude of the producer. The traditional (and thankfully changing) attitude of the UK farmer has been, with few exceptions, to concentrate on production efficiency and play a passive role in marketing. Until comparatively recently little energy and enthusiasm have been put into the disposal of farm output, either in the UK market or abroad. In mitigation of the poor performance of producers in marketing it must be said that a great deal of initiative was taken from them by the setting up of the marketing boards and other para-statal organisations (in the first half of this century, see Chapter 15), when they were actively dissuaded from marketing their own products. In addition, many farmers had no time or inclination to become involved in the very specialised and complex process of marketing food. The recent significant change in market awareness has been brought about by a series of events which have increased the pressures on farmers (especially those who are paying rents or have a mortgage). These are:

(1) *Over-supply and increased competition amongst producers.* This is caused by increased productivity and a wider choice of goods in farming, especially in developing economies.

Competition has particularly been felt by UK producers since joining the EU because many of the member states have nationally organised policies of exporting food into the UK, particularly horticultural products and pig meat

(2) *Higher investment risks in production.* Capital investments are increasing and, in order to finance borrowing and create an acceptable return on capital, producers are looking for better returns; these can more easily be obtained by further advances in production efficiency and marketing expertise

(3) *Rising operating costs*

(4) *Increased consumer awareness of the food products available, their quality,* any residual medication in food, and the perception of a 'healthy' diet

(5) *Increase in concentration of buyers.* The firms buying from farmers are becoming larger by expansion and amalgamation and thus control more of the market. Farmers therefore have to negotiate with experienced, specialist buyers who have a strong position from which to determine quality and price. As a result farmers remain as pricetakers while the pricemaking is done by the buyers. (Market price support arrangements do of course remove some of the incentive for farmers to extend their marketing efforts when their total returns may not be significantly affected)

The marketing process

Production on the farm is the first stage in the process of getting food in an edible state to the final consumer. Farmers at the first stage of this process need to recognise that they are only one part of the total process (albeit a vital part) otherwise they will not get the best out of their resources or the market. In fact serious consequences have resulted in the past for those who have not recognised the true nature of the business they are in and the wider implications this has for them. Some passenger shipping lines failed because they thought they were still in the transport business, whereas they were in the floating hotel business. Similarly canal owners failed because they did not develop railways themselves, thinking that they were in the inland waterway business, whereas they were in the transportation business. Governments too have a tendency to suffer, by isolating parts of a larger system. The rationalisation of the railways was done in isolation rather than in conjunction with the whole system of transportation. (Also governments until recently could not decide whether transport was to be a social service or a profit-making organisation.)

Bearing this in mind the marketing chain for the food sector can be simplified into several main functions and integrated into the process are the important elements of storage, financing, insurance and transport which facilitate the channelling of food to the consumer. A simple marketing chain (or marketing column) could be as follows:

(1) Production	Farms
(2) Assembly (grading and sorting)	Auction markets, private sales
(3) Processing/packaging	Abattoirs, packing plants
(4) Distribution	Wholesalers
(5) Final sale	Retailers
(6) Consumption	Householders

These stages in the chain are no longer clearly defined and neither are all the functions carried out by separate firms. The role of wholesalers has certainly diminished in the last ten years and some products do not need to pass through all of the links in the chain, whilst some, being sold directly to the consumer from the farm, bypass the intermediate steps in the chain.

The majority of products do, however, need processing. Marketing involves the co-ordinating of the processes which are necessary to equate market supply and market demand.

Probably the best co-ordinated of all is broiler production where the whole process, from parent stock at breeding farms, through chick production at hatcheriers, to broiler production and processing, is controlled within one firm.

Other products, such as 18-months-old beef, are more difficult to co-ordinate since the producer is usually dependent on someone else for his calf supply and therefore cannot control the time when an animal will be ready for market, except by altering feeding and extending the finishing period in order to get the animal to market at a time of peak market demand and highest prices.

Despite the fact that more producers are trying to orientate production towards the market, there are many fundamental difficulties which they face.

Marketing difficulties

The time-lag is the biggest problem faced by producers. It is characteristic of agricultural production. The time-lag relates to the time period which farmers need in order to react to market prices. This leads to cycles in production, particularly for pigs and potatoes, which several governments have endeavoured to control using price and quantity measures; thus pig production fluctuated from minus 31% to plus 41% from year to year between 1921 and 1932.

The cycle occurs when production is either dramatically reduced or increased in any year by weather or disease changes. There is disequilibrium in the market and if, for example, the weather was so bad that potato supplies were cut to 5.5 m/t, when these were offered for sale the price would be bid-up by competition amongst buyers. The producers, seeing a high price (for example, £140/t in 1994–1995) will plan their next crop in relation to the price of £140/t (based upon their market supply curve). This will mean that, despite quotas on area, more potatoes will be planted and producers will accept the levy on excess area in order to raise their profitability.

Since many producers will react in this way, given a normal growing season there will be a surplus of potatoes flooding the market during the next marketing period and prices will fall. The lower price will cause fewer potatoes to be planted in the next production season and the market supply, given an average growing season, will be lower and closer to equilibrium between supply and demand.

The cycle thus occurs because the farmer responds slowly to market price (because he can only alter supply significantly in his next growing season) but the market price responds immediately to the supply of potatoes currently available. The 'cobweb cycle' as this is called is shown in Fig. 13.2.

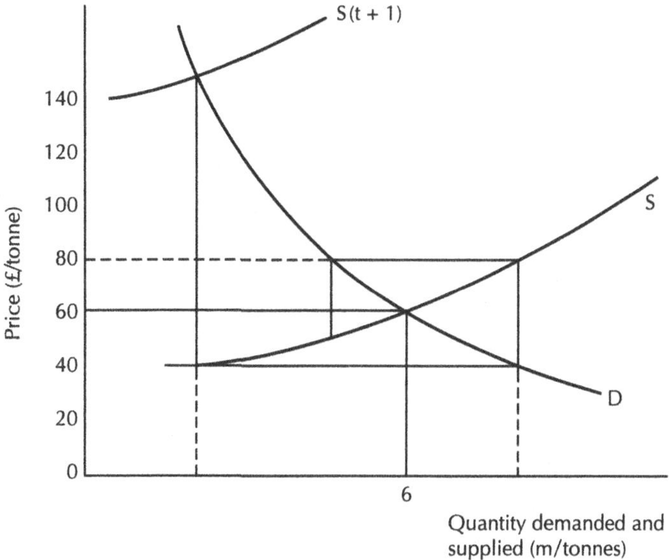

Fig. 13.2 The cobweb cycle.

The magnitude of the cycle depends upon the relative elasticities of supply and demand, that is, the responsiveness of supply and demand to a change in price (see Fig. 13.3). The normal cycle in agriculture is a converging one, that is, one which tends to return towards equilibrium.

The control which government and the EU exert on agricultural production now prevents the worst effects of the cycle being felt for all products which have a guaranteed minimum price as part of a support scheme for the producer.

Minimum prices therefore protect the producer from market forces so that he does not respond to actual prices if they are below the minimum but will try to do so if they are above it.

Figure 13.3 shows the market supply in period 1 moving from the 'normal market supply curve S to supply curve S(t + 1), as a result of disease or bad weather, so that the quantity OQ3 is produced. The price will rise to OP2 and the next season farmers will aim to produce OQ4. Given an average season, when this amount is presented on the market the price will fall to OP3, but the difference between the market price OP3 and guaranteed price OP1 is made up by the UK Government or

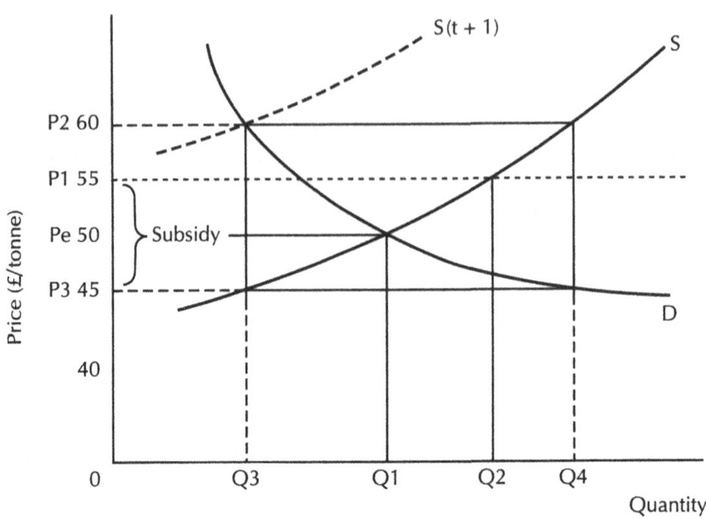

Fig. 13.3 The effect of guaranteed prices on the cobweb cycle.

EU. This is a deficiency payment or variable premium. Thus the producer will respond by planning next season's crop to produce OQ2 just slightly above the equilibrium level of production. *In this case the producer gets his signals from the guarantee price, not the market price, and therefore plans production accordingly.* Thus the market is stabilised but surpluses have arisen.

This simplified example shows the principles behind guarantee price schemes. There are, however, other factors to consider in the schemes, some of which create buffer stocks rather than making-up the price. These other factors will be discussed later. The difficulty with them is how to set the guarantee price at the correct level to equate supply and demand. It is a fact, however, that the guarantees have been set too high in the past to create equilibrium of supply and demand, and have encouraged producers to create over-supply which is one reason why the CAP schemes have been reformed. This has been a determined policy of several EU members who do not wish to risk food shortages because many people on the continent of Europe can still remember times of starvation, especially in and after World War II.

Associated with the time-lag problem are *green rates*.

Effect of green rates in the agri-monetary system

Prices in the EU Support Schemes have been significantly influenced by the existence of 'green' (representative) rates. Each year when the annual review of farm prices occurs, the Council of Ministers agrees the prices for agricultural products. If the ministers do not agree the Commission usually introduces interim measures until agreement is reached. These prices are quoted in European Currency Units (ecus), the common currency of the EU. These prices have then to be converted into national currencies using the 'green' rate. In May 1995 the green pound sterling was equivalent to 1.048685 ecus. The intervention price for feed wheat and feed barley in May 1995 was quoted at 115.00 ecus/tonne, the UK intervention price was calculated as:

$$\frac{115.00}{1.048685}\ \text{ecus/t} = £109.66/\text{t}$$

The obvious implication of this conversion process is that it was possible for a national government to use the green rate as a means of benefiting farmers or, on the other hand, keeping food prices comparatively low. In the UK, where the farmers form an insignificant part of the voting public, a cheap food policy has been maintained while in Germany, where far more are involved in agriculture, it has been possible to keep farm guaranteed prices relatively high compared to the expected 'common' price level.

How does this happen?

The original intention for the financial system of the EU was that ecus would be converted for each member state at a rate which reflected the value of the currency on foreign exchange markets. As currency levels altered there would therefore be changes in the representative 'green' rate to reflect an increase or decrease in the value of any currency and maintain a parity between its market value and its 'green' value which is used to convert agricultural prices.

In the UK this parity was certainly not maintained and between 1973 and 1980 the green pound was over-valued. This meant that UK farmers were offered lower guaranteed intervention prices than would have been the case if the green pound and market value of sterling were the same (at parity).

This had obvious implications for the farming community and the same situation arose in 1986 when, with the sharp fall in sterling, farm prices in the UK fell to 20% below those in other EU countries. The strengthening pound in 1987 eased the situation, although a large gap remained.

Thus the level of farm prices was influenced by the strength of the pound on foreign exchange and if the pound devalued (as it did between 1973 and 1979). Then to remain at the common level farmers should have received more pounds, which could have been achieved by devaluing the green rate. This is a simple procedure since other member states do not normally oppose green rate changes. The UK Government of the day was, however, not prepared to go to the extent of raising food prices by this means during this period. In Germany, on the other hand, the green deutschmark was under-valued, meaning that German farmers were receiving intervention prices above the common price level.

Example 13.1 illustrates the impact of changing agri-monetary values on farm prices.

Example 13.1 UK intervention price for feed wheat and feed barley, in April 1981. The official rate used for conversion in the UK was

$$£1 \text{ green} = 1.70148 \text{ ecus}$$

$$\text{parity } £1 \text{ (foreign exchange market value, FEMV)} = £1 \text{ green} = 1.53133 \text{ ecus}$$

Thus the intervention price offered was

$$\frac{163.49}{1.70148} \text{ ecus (i.e. guaranteed price)} = £96.08/\text{tonne}$$

when it should have been

$$\frac{163.49}{1.53133} \text{ ecus/t} = £106.76$$

if ecus were converted at the market rate of pounds sterling.

'Cheap' food was produced in the UK compared to the EU common price level. Why not export it to the continental member states and at least benefit from the higher prices there while at the same time expanding our markets?

Under the system of monetary compensatory amounts (mcas) begun by Germany in 1969, UK exports to Germany, for example, had levies imposed upon them to compensate for representative (green) rate distortions. At the same time imports from Germany had subsidies paid to them; this prevented unfair competition arising purely out of green rate disparities.

The UK situation changed in 1980 when oil resources and better economic results raised the value of sterling on foreign exchanges.

In 1981 the intervention price offered was

$$\frac{165.24}{1.61641} \text{ ecus/t} = £102.23/\text{tonne}$$

Parity value would have been

$$\text{i.e. } £1 \text{ (FEMV) equal to } £1 \text{ (green) } \frac{165.24}{1.89139} \text{ ecus/t} = £87.36/\text{tonne}$$

The October 1982 situation was

$$\text{price offered } \frac{184.16}{1.83075} \text{ ecus/t} = £100.59/\text{tonne}$$

This clearly demonstrates that for a period farmers enjoyed higher than common level prices.

The predominant situation in the UK has been that the green rate is over-valued (as shown in Table 13.4 which refers to December 1986) with the effect of low prices

Table 13.4 Grain intervention prices 1986 to 1987 (£/tonne)

	At December 1986 £1 green = 1.59491 ecus	*At parity rate* £1 green = 1.35501 ecus
Bread milling wheat	120.19	141.47
Feed wheat and barley	106.88	134.85
Oilseed rape	276.72	325.71
Butter	1 990.78	2 365.84
Skim milk powder	1 106.24	1 332.80

Table 13.5 The effect on UK trade with EU members and on monetary compensatory amounts of an overvalued 'green' rate

	Monetary compensatory amounts (mcas): export tax or import subsidy (£/t or p/kg)	
	July 1986	Dec 1986
Exports:		
Wheat to Germany	£24.75	£34.80
Barley to Belgium	£20.30	£29.60
Beef to France	25.65p	7.64p
Imports		
Pig meat from Denmark	6.52p	—
Beef from Eire	17.72p	–2.82p

and negative mcas (see Table 13.5). These were more likely to occur for sterling since it was not a full European Monetary System (EMS) currency.

The aim of the Commission was to abolish (or at least control) all green rate disparities but rapid currency movements and national interests make this very difficult to accomplish. The pound was stabilised within the Exchange Rate Mechanism (ERM) for a period up to September 1992, when it left the ERM. CAP reforms in 1992 abolished green rate disparities but the switchover mechanism was retained so that German farmers did not suffer reduced support prices as national currencies were revalued. Another difficulty for traders was that the mca changed quite rapidly (Table 13.5).

The relationship between the pound on foreign exchange markets and the green pound is shown in Fig. 13.4 for the period 1973 to 1987 (when many problems arose in the UK) and in Table 13.6 for 1991 to 1997.

The disparity between exchange rates in member states causes guaranteed prices to vary quite markedly from the 'common' level which the European Commission would like to have in existence throughout the EU. Table 13.7 illustrates this by comparing the intervention prices prevalent in the UK and Germany at both the 'green' and the foreign exchange (spot) rates.

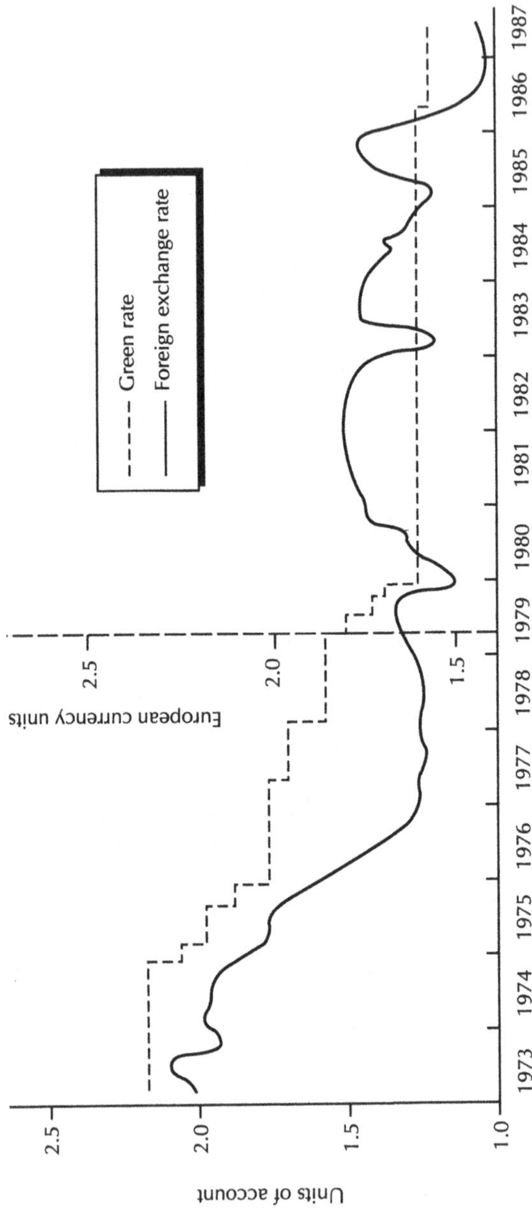

Fig. 13.4 Green and foreign exchange rates of the pound.

Table 13.6 The green and market exchange rates of sterling 1991–1997

	Market rate £ Value of 1 ecu	Green rate £ value of 1 ecu
1991	0.69	0.79
1992	0.71	0.81
1993	0.75	0.95
1994	0.79	0.93
1995	0.84	0.84
1996	0.811	0.83
1997 (March)	0.71	0.74

Table 13.7 Variance between intervention prices due to 'green' rate disparities

Current intervention price feed barley	UK value with green rate at £1 = 1.59491 ecus spot rate is: £1 = 1.35501	German value with green rate 1DM = 0.417028 ecus spot rate is 1DM = 0.473747	UK value converting DM to pounds at the spot rate of: £1 = 2.83DM
ecus/t	£/t	DM	£/t
182.72	114.56 (134.85)	438.15 (385.69)	154.82 (136.29)
		French value with green rate 1FF = 0.140852	UK value
		spot rate is: 1FF = 0.145493	spot rate of: £1 = 9.27FF
182.72	114.56 (134.85)	1 297.25 (1 255.87)	139.94 (135.48)
		Italian value with green rate 1L = 0.000649773	UK value
		spot rate is: 1L = 0.000691023	spot rate of: £1 = 1963L
182.72	114.56 (134.85)	281 205.9 (264 419.6)	134.25 (134.70)

Summary on green rates

(1) An over-valued green rate caused lower farm prices. Agricultural exports in this situation attracted levies, while imports gained subsidies

(2) An under-valued green rate caused higher farm prices. Exports gained subsidies while imports attracted levies

(3) Under the green rate and mca system it was possible to artificially influence prices and distort member trade by not keeping green rates in line with foreign exchange rates

(4) The existence of green rate disparities in the UK provided some protection for farmers when basic ecu support prices were only increased slightly (or kept static or reduced), because simultaneous devaluations of the green rate (to bring it back into line with market currency rates) meant that UK support prices increased and 'cushioned' UK producers from the real price changes which they would otherwise have been subject to

(5) The switchover mechanism. The switchover mechanism was first used in 1984 to protect German farmers from falling prices as the deutschmark (DM) strengthened. The agri-monetary system, first used in 1969 when France devalued the franc by 11%, was simplified with the abolition of monetary compensatory amounts (mcas) on January 1993. The switchover mechanism, however, was retained and had the immediate effect of maintaining farm support prices at about 19.5% higher than they should have been. German farmers were protected by the process of 'ratcheting-up' support prices as the DM strengthened or ERM member currencies weakened in relation to other currencies. This proved very expensive (about 83 million ecu in 1992–1993) and so the switchover system ended in December 1994.

With the pound sterling floating outside the exchange rate mechanism (ERM), along with the Irish punt, large gaps between the green rate of the pound and the agricultural market rate of the pound became possible, with no mcas to compensate for price differences when goods moved between member states. If the combined monetary gap between two floating currencies is greater than 5%, a green rate revaluation is triggered. The combined monetary gap is monitored within reference periods as shown in Table 13.8.

The agricultural market rate is the daily European Currency Unit (ecu) rate (which alters as member currencies fluctuate) multiplied by the switchover coefficient. The coefficient is calculated by referring only to currencies within the ERM (but may in future include floating currencies too), but it is then applied to all member currencies for the purpose of calculating farm prices. It is adjusted to ensure support prices in Germany do not fall as the DM increases in value. If the pound sterling strengthens, however, farm support prices in the UK will fall (as green pound revaluations occur) and there will be no compensatory switchover increase to farm support prices because the pound is outside the ERM.

Table 13.8 Reference periods for the combined monetary gap

	1	2	3
Reference period	between the 21st and month end	any 3 consecutive days	1st to 10th, 11th to 20th, 21st to month end
Combined memory gap	> 2% but 4%	> 6%	< 4% but < 6%
Green rate action	halved so it is within 2%	immediate revaluation to within 2%	reduced to within 2%

This occurs because the switchover converts the ecu into the green ecu in the following way. The system is arranged so that all green rates should be within 2% of the agricultural market rate, at the end of every month. Adjustments could, however, be made every three days if the combined monetary gap exceeds 6% (see Table 13.9). The instability of currencies in the second half of 1993 led to the freezing of green rates, while agreement was sought regarding the future operation of the monetary system and the switchover. The currency re-adjustments in 1992 and 1993 added a net 83 million ecu to the cost of guarantee payments up to the end of July 1993. Extra payments of this kind are provided for in the 400 million ecu margin built into the annual budget. When payments (or savings) exceed 400 million ecu, the reserve fund of 1 billion ecu (set up in 1988 to provide for unforeseen changes between the budgeted $: ecu rate and the actual market $:ecu rate) is brought into play. The budget rate was based on the average market rate for the first three months of the year preceding the budget.

Table 13.9 Value of the switchover coefficient

Date	Value
12 September 1992	1.145109
16 September 1992	1.154338
22 November 1992	1.157346
31 January 1993	1.195066
13 May 1993	1.205454
31 July 1993 to February 1995	1.207509

Agri-monetary arrangements from February 1995

(1) The switchover mechanism was abandoned on 1 February 1995. All ecu prices and aids, however, were then multiplied by 1.207509 and all green rates (expressed at national currency/ecu) divided by 1.207509, effectively raising basic prices by 20.7%, the exact amount that abolishing the switch over would have reduced them by.

(2) Upper limits to monetary gaps between currencies (the so-called franchises) were set at +5% and – 2% with the maximum spread of gaps between positive and negative at 5%.
Currencies with a common border may have a smaller gap enforced to prevent trade distortions.

(3) The Council has to decide on any revaluation deemed to be 'appreciable', that is, greater than the difference between the current green rate of a currency and its lowest rate during the previous three years. Agreement will be reached by qualified majority on revaluations and any appeals to proposed revaluations which must be lodged within three days of publication.

(4) Any green rate devaluations and 'non-appreciable' revaluations, triggered by gaps exceeding the franchise, will take the form of halving the monetary gap. If this does not prove sufficient to meet the rule on five point spreads for negative gaps, then it will have to be halved again.

(5) Revaluations must be carried out before the negative gaps are assessed.

(6) Appreciable revaluations can be delayed (and maybe more than halved) up to 60 days, that is, six reference periods, to confirm the trend, whereas non-appreciable ones are only delayed for one reference period.

(7) The mini switchover was introduced in February 1995 to protect (especially German farmers). As a result there were pan EU increases in the ecu amounts of flat rate aids per hectare or per animal of sheep and goat compensatory premiums and amounts of a structural or environmental nature, wherever a green rate revaluation was implemented. The increase was calculated to neutralise price falls in the revaluing country and applied to all currencies. In June 1995 a revaluation of the German mark should have reduced the area payments to German farmers based on the 1 July 1995 green mark rate. The Germans resisted this move and got an agreement to freeze green conversion rates at the June 1995 level, thus protecting their farmers from appreciable green rate revaluations for 1995. This protection was extended for the first six months of 1996 to protect Swedish and Finnish farmers from appreciable green rate revaluations and it seems likely that it will be used again as a future measure for protection against further appreciable revaluations if requested by a member state. The strengthening of the pound in 1996–1997 has meant a reduction in farm support prices. The value of 1 ecu in March 1998 was 69 p and the intervention price for feed wheat was £86.40 per tonne.

The inelastic demand for food

It is a simple fact that as the price of a product rises the market demand (with few exceptions) will fall. Attempts are made in agriculture to measure and therefore to forecast the magnitude of responses of both supply and demand to changes in price.

It can be seen from Fig. 13.5 that where demand is elastic the response of quantity (ΔQ) is much greater for a given price change (ΔP) than where demand is inelastic.

In agricultural production demand is generally inelastic.

It therefore follows that when supply is related to demand for food for small changes in supply there are resulting large changes in price which affect the income of producers dramatically from one year to the next causing some uncertainty and instability in the industry (another reason why guarantees are given and supply controls attempted).

Other problems faced by farmers and associated businesses involved in agricultural marketing are *fluctuation in supply*, which can be caused by weather and disease at any time during a production cycle. However, there are always seasonality problems which mean that the ability to store produce is critical. The perishable nature of most food products means that shortage is usually expensive, and high losses can be incurred. *The fragmented nature of production units* makes co-ordination difficult and *lack of conformity in quality* makes standardisation difficult to carry out on a consistent basis, at either market or retail outlet level.

The EU has a great influence on UK agricultural marketing and to understand the operation of the community schemes it is necessary to look briefly at the background, aims and objectives of the EU.

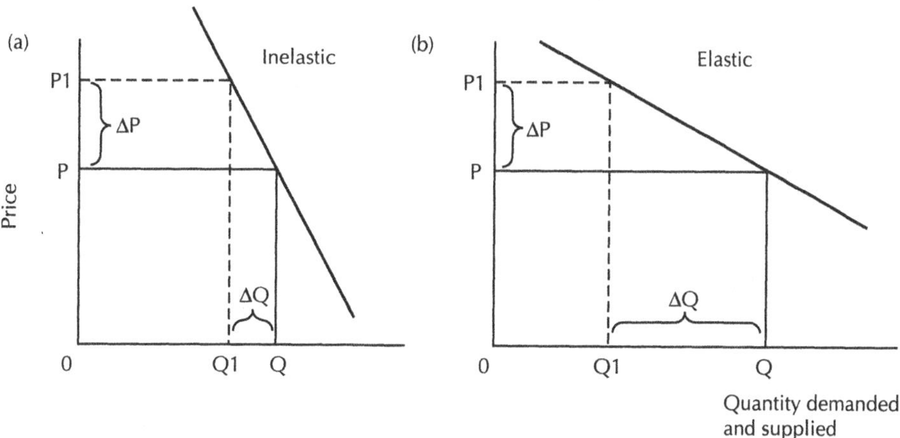

Fig. 13.5 The effect of price elasticities of demand on the responsiveness of quantity to price changes.

14 The European Union (EU)

The EU and Common Agricultural Policy (CAP)

The background to the formation of the European Union

1946 A United Europe was advocated by Churchill in an effort to prevent a third world war arising from conflict in Europe.

1951 The European Coal and Steel Community was set up by the Treaty of Paris. This was initiated by Franco-German co-operation in an attempt to combat rising Soviet power. It was signed by the 'Six' (West Germany, France, Italy, the Netherlands, Belgium and Luxembourg).

1957 The Treaty of Rome set up the European Atomic Energy Community and European Economic Community.

1961 In Bonn leaders of the 'Six' were discussing political union and Macmillan (the British Prime Minister), realising that the UK would be isolated in Europe, announced our intention to join. (De Gaulle, the French President, vetoed our entry. When he died the way was clear.)

1973 UK, Denmark, and Eire join with effect from 1 January.

1975 A referendum (5 June) in the UK endorsed the decision to join with a 2:1 majority in favour of staying in.

1981 Greece joins.

1986 Spain and Portugal join and the Single European Act is passed.

1992 The Single Market comes into being.

1993 European Union (formed 1 November; resulting from the Maastricht Treaty).

1994 European Economic Area (EEA) formed, EU links with EFTA.

1995 Austria, Sweden and Finland join (1 January).

It is important to understand that the EU was set up only 12 years after the end of World War II when furniture and food shortages in Europe were still remembered. The founders were determined that no-one in the original 'Six' would be subjected to food shortages again. The policy at the outset was aimed at solving the problem which these nations had in 1958 and was particularly biased towards the improve-

Table 14.1 EU countries: area, population and GDP compared (1993)

	Area ('000 km²)	(%)	Population ('000)	(%)	GDP ('000 ecu)	(%)	GDP (ecu/head)	GDP ranking (ecu/head)		Purchasing power standards	
								EU 12	EU 15	(GDP/head) EU 12	EU 12 Rank
Belgium	31	1.0	10 085	2.7	180.0	3.0	17 849	5	7	108.3	3
Denmark	43	1.3	5 189	1.4	115.5	2.0	22 253	2	2	108.0	4
Germany	357	11.0	81 180	21.9	1 631.5	27.6	20 097	3	3	107.2	5
Greece	132	4.1	10 362	2.8	76.7	1.3	7 406	11	14	60.8	12
Spain	505	15.6	39 141	10.6	408.4	6.9	10 434	10	13	76.7	10
France	544	16.8	57 327	15.5	1 068.6	18.1	18 640	4	5	110.7	2
Ireland	69	2.1	3 561	1.0	40.4	0.7	11 334	9	12	81.0	9
Italy	301	9.3	58 098	15.7	847.3	14.3	14 584	7	9	106.1	6
Luxembourg	3	0.1	398	0.1	10.7	0.2	26 859	1	1	133.6	1
The Netherlands	42	1.3	15 290	4.1	264.0	4.5	17 268	6	8	102.3	7
Portugal	92	2.8	9 877	2.7	72.3	1.2	7 323	12	15	61.1	11
UK	244	7.5	58 168	15.7	807.8	13.7	13 887	8	11	100.4	8
EU of 12	2 363	—	348 676	—	5 523.2	—	15 840			100.0	
Austria	84	2.6	7 991	2.2	155.5	2.6	19 453		4	—	
Finland	338	10.5	5 066	1.4	71.5	1.2	14 110		10	—	
Sweden	450	13.9	8 719	2.4	159.2	2.7	18 256		6	—	
EU of 15	3 235	100.0	370 452	100.0	5 909.3	100.0	15 951				
USA	9 373	289.7	258 311	69.7	5 367.4	90.8	20 779			147.7	
Japan	378	11.7	124 674	33.7	3 600.6	60.9	28 880			123.6	

Source: Eurostat.

ment of French agriculture. This occurred because the first area of focus was agriculture and De Gaulle negotiated excellent terms for his country. As a result, and because the UK (pre 1973) used a very different approach in supporting agricultural production to ensure food supplies, many conflicts have arisen between UK policy and the objectives and implementation of pricing and support policy for the original six. Table 14.1 sets the scene on the current EU members, showing their size and relative prosperity, it also provides a comparison with the USA and Japan. The EU was initially intended as a political expedient to prevent further conflict in Europe, and it brought about a series of economic commitments from the signatories, culminating in the European Monetary System (EMS) in December 1978. The UK did not join the EMS until October 1990.

European Monetary System (EMS)

The objective was to create exchange rate stability in the European Union, within predetermined limits, by having close monetary co-operation and economic policies which were to converge. The co-operation involved the European Currency Unit (ecu), the Exchange Rate Mechanism (ERM), and the European Monetary Co-operation Fund (EMCF).

Table 14.2 The ECU basket

	%		%
Deutschmark	30.4	Peseta	5.2
French franc	19.3	Kroner	2.4
Sterling (£)	12.6	Punt	1.1
Lira	9.9	Drachma	0.7
Guilder	9.5	Escudo	0.8
Belgian/Lux franc	8.1		

The European Currency Unit (ecu)

Is made up of a 'basket' of currencies, weighted to reflect the relative strength of EMS members' economy and share of intra-community trade. The basket structure is reviewed every five years, but more frequent adjustments occur with short-term movements of the exchange rates. The ecu is intended to lead to a 'common currency' throughout the European Union (EU), indeed Union-wide farm prices are quoted in ecu.

The Exchange Rate Mechanism (ERM)

Each country in the EMS has a fixed central rate with the ecu. It began in March 1979, with Belgium, Denmark, France Eire, the Netherlands, Luxembourg and

Germany committed to keeping their currencies operating within a 2.25% band of the central rate while Italy used a 6% band. The UK was not a member. Greece joined the EMS and the ecu in 1984 but not the ERM. In 1989 the escudo and peseta joined the ecu; only the peseta entered the ERM, at a 6% band, while the lira moved towards a 2.25% band in January 1990. The pound sterling finally entered the ERM, in the 6% band in October 1990, and was joined by the escudo in April 1992. The process of exchange rate maintenance was arranged to prevent all the burden falling on members with a weakening currency, a traditional failing of fixed-rate systems. A divergence indicator was triggered when a currency had moved 75% of the way towards its limit against the ecu. The formula used for sterling was

$$75\% \text{ of } 2.25 \times (1 - 0.121)\%$$

giving an indicator of + or −1.48%. For the peseta it was

$$75\% \text{ of } 2.25 \times (1 - 0.052)\%$$

giving an indicator of +/−1.6%.

When a currency reached its indicator the other members were expected to take action that would resolve the problem. Movement towards full economic and monetary union, and the high interest rates kept by Germany in order to combat inflation fuelled by re-unification, increased pressure on the member currencies and in September 1992 the UK and Italy suspended their membership of the ERM after spending millions trying to keep their currencies within the agreed bands. There was realignment of the other currencies at this stage and the unstable relationships led to an official acceptance of a much wider 15% band, in July 1993, except the deutschmark and guilder which maintained the 2.25% band. In March 1995 the peseta and escudo were devalued by 7% and 3.5% respectively. Since the bands were widened there has been a tendency towards decreased interest rate volatility and increased exchange rate volatility.

Economic and Monetary Union (EMU)

The process towards EMU began in earnest on 1 July 1990 with the elimination of restrictions on capital movements, (as part of the Single Market programme) and the start of convergence programmes. In January 1994 the European Monetary Institute (EMI) was established; the organisation which precedes the European Central Bank (ECB). The process leading to the independence of the ECB has begun and also procedures for co-ordinating economic policies and measuring economic convergence. The convergence criteria, which should be met before any currency can enter the final phase of EMU – adopting the single currency – relate to the following:

(1) Inflation – a rate within 1.5% of the best three performing member states.
(2) Deficit – a deficit which should not to be more than 3% of GDP.

(3) Debt – Government debt should not exceed 60% of GDP.
(4) Exchange rate – a currency should stay within normal ranges of fluctuation provided by the ERM of the EMS, for at least two years, without devaluing against the currency of any other member states.
(5) Long-term interest rates – should be within 2% of the three best performing members. It is likely that the single currency (the Euro) will be launched on 1 January 1999, although this could be altered. In order to meet this deadline, a decision will be made, in May 1998, on which currencies have met the criteria. This decision will be based on data relating to the end of 1997 for each member. At this point (May 1998) the list of participating states will be agreed, the start date of EMU announced (or confirmed as 1 January 1999), and a deadline set for the final changeover to the Euro (probably 2002). To reach this point the fixing of the conversion rates between member currencies and the Euro will be agreed upon, and the changeover process to be accomplished in each state. For those not joining the Euro in the first wave, there will be an agreement on the exchange rate between the Euro and the national currency. If the pound continues to strengthen, this will have the effect of reducing support prices and difficulty in exporting.

Benefits of EMU/single currency

EMU will consolidate the single market by eliminating the effects of monetary disturbance, and the need for national protective measures. There will be a reduction in transaction costs for trade and travel because currency conversion costs will be eliminated, as will the costs attached to hedging the risks of currency exposure. There will be more economic stability because exchange rate variation will be abolished, and intra-EU trade will be less vulnerable to volatility as the impact of monetary fluctuations is reduced. The monetary system will be managed by the ECB, an organisation independent of national political control. This should ensure that decisions are motivated by economic criteria and not by political dogma or survival strategy.

Costs of EMU/single currency

To achieve EMU member states must meet the agreed criteria. Although the 'costs' of doing this will have been incurred before EMU they could nevertheless have a considerable impact on economic growth, interest rates and employment in EU states. The major issue, concerning UK politicians in particular, is the loss of independence relating to the decisions regarding interest rates and exchange rates. The debate continues to focus on whether the financial and economic rewards of joining a system controlled by a central, independent European bank outweigh the risks associated with it. This is often described as the loss of sovereignty and freedom of Parliament to make decisions relating to the particular needs of the UK, and is 'rolled' into the debate on the evils of federation. Many businessmen have declared their support for a single currency, although some politicians and financiers appear sceptical. There will also be substantial costs associated with the changeover to using the Euro in accounting and information systems, cash systems and recording equipment. It is estimated that these costs may be as high as 2.3% of turnover for up to three years. No doubt part (or all) of these costs will be passed to consumers.

Most of these attempts at Economic Union have foundered upon the 'rock' of national interest, preferred over and above EU interests, by individual member states. Obvious examples of this are the Social Chapter and the free movement of people to which the UK Government has not agreed.

The 'unofficial' aims of the EU (in 1958) were to form one trading, farming and industrial system. This was to be achieved by adopting the four freedoms: free movement of goods, persons, services and capital, and having:

(1) Economic union
(2) A common currency
(3) A common foreign policy

The movement towards completion of these aims has been slow, especially since the Union has been expanded in 1973, 1981, 1986 and 1995; but progress continues to be made. The EU–EFTA link to form the European Economic Area (EEA) in 1994 created the biggest 'free trade' area, embracing 372 million consumers. Further expansion is expected since associate membership has been agreed with Poland, Hungary, the Czech Republic and Slovakia (1991), Bulgaria and Romania (1993).

Who controls the Community?

There are three main institutions of which it has been said 'the Commission proposes; the Parliament opposes; and the Council of Ministers disposes'.

The Council of Ministers

The Council is the supreme decision-making body and each country is represented. It was intended that decisions should be taken by a qualified majority (66 out of a total of 87) in the Council voting in favour; voting being weighted by the size of the country. The principle of 'overriding national interest' was introduced in the late 1960s, which in effect gave each member state a veto in the Council of Ministers. Since the Maastricht Treaty, however, many more issues are settled by qualified majority voting (see Table 14.3) which requires 70% of the votes to be in favour of a motion. Anyone wanting to block legislation must therefore get at least 26 votes. The Council is assisted by a Committee of permanent representatives (COREPER), who are 'ambassadors' of the member states and who 'prepare the ground' for the Council meetings, and a general secretariat. The chairmanship of the Council rotates around the members at six-monthly intervals.

Table 14.3 The Council, the Commission and the Parliament

	Presidency	Votes in Council	Commission members	Euro Parliament members
Austria	11	4	1	21
Belgium	1	5	1	25
Denmark	2	3	1	16
Finland	13	3	1	16
France	6	10	2	87
Germany	3	10	2	99
Greece	4	5	1	25
Ireland	7	3	1	15
Italy	8	10	2	87
Luxembourg	9	2	1	6
The Netherlands	10	5	1	31
Portugal	12	5	1	25
Spain	5	8	2	64
Sweden	14	4	1	22
UK	15	10	2	87
Total		87	20	626

Voting: voting by 'qualified majority' means that 26 votes are needed to block a proposal. The UK has proposed that 24 or 25 should be enough.

Presidency of Council: the order of presidency is altered slightly so that members do not always get the same half of the year because January–June is more important. Denmark therefore came before Belgium in 1993, Greece came before Germany in 1994, France preceded Spain in 1995, Italy preceded Ireland in 1996, Netherlands came before Luxembourg in 1997 and Portugal will be before Austria in 1998.

Languages: there are now 11 languages with the addition of Swedish and Finnish.

The European Commission

The Commission initiates and subsequently manages EU policy particularly with regard to the Community budget, which it controls. Before any proposal reaches the Council of Ministers, from the Commission, a great deal of consultation and refinement occurs, carried out by various advisory, management and drafting committees, and working parties.

The Commission consists of 20 members who are engaged in serving the interests of the Union and not national interests. France, Germany, Italy, Spain and the UK provide two members each, whilst Belgium, Luxembourg, the Netherlands, Eire, Denmark, Greece, Austria, Sweden, Finland and Portugal provide one. These members are all appointed by their national governments, and can serve for five years. The President of the Commission is appointed by the heads of government and each member is given an area of EU policy to oversee. The Commission employs about 12 000 people in 22 Directorates General; these are shown below:

DG I	Foreign relations
DG II	Economic affairs
DG III	Industrial relations
DG IV	Competition
DG V	Employment, industrial relations, social affairs
DG VI	Agriculture and rural development
DG VII	Transport
DG VIII	Development
DG IX	Personnel and administration
DG X	Audio–visual, information, communications, culture
DG XI	Environment
DG XII	Science, research and development
DG XIII	Telecommunications
DG XIV	Fisheries
DG XV	Internal market and financial services
DG XVI	Regional affairs
DG XVII	Energy
DG XIX	Budget
DG XX	Financial control
DG XXI	Customs and indirect taxes
DG XXII	Small and medium-size enterprises, commerce, tourism and social economy

DG VI (Agriculture) has 10 divisions and a total of about 600 personnel.

The European Parliament

The Parliament consists of 626 members and elections are held in member states every five years, 1979 being the first. The Parliament has two main powers:

(1) It has to approve the budget proposals made by the Commission
(2) It has to approve the members of the Commission en bloc

Parliament therefore has little influence on the decisions made by the Council of Ministers and usually only offers opinions on matters which require action. However this does enable the Parliament to 'slow down' decisions and therefore exert some influence over the Council.

The EU budget

The money to finance EU expenditure is obtained from several sources. These are shown in Tables 14.4 and 14.5.

Table 14.6 shows a breakdown of budget expenditure in the Union. One problem with the budget is that the agricultural burden is so high that development of other common policies has been neglected. Money for agricultural support and devel-

Table 14.4 Financing the EU budget (%) 1982–1995

	1982	1986	1990	1991	1992	1993	1994	1995
GNP-based own resources	—	—	—	12.9	13.6	24.7	25.9	28.4
Customs duties	32.5	27.8	23.9	22.1	20.6	18.2	17.5	16.9
Agricultural levies	9.8	4.5	2.4	2.8	2.0	1.5	1.3	1.1
Sugar levies	2.9	3.2	1.9	2.0	1.6	1.6	1.9	1.5
VAT	53.9	63.8	57.6	55.5	56.7	53.4	51.0	51.4
Other	0.9	0.7	14.2	4.7	4.5	0.8	2.4	0.7

Source: Compiled from European Commission information.

Table 14.5 Community revenue 1982–1995 (million ecus)

	1982	1986	1990	1993	1994	1995
GNP-based own resources	—	—	94.3	16 555.7	18 456.3	21 726.0
Customs duties	6 815.3	9 700.5	11 427.9	12 259.6	12 419.9	12 928.5
Ordinary levies and sugar levies	2 227.8	2 698.7	2 084.0	2 147.4	2 304.6	1 989.0
VAT	12 000.0	20 468.2	27 440.1	35 560.0	36 313.5	39 321.0
Financial contribution	197.0	189.9	—	—	—	—
Miscellaneous	219.1	257.3	6 595.2	1 504.0	8 685.1	535.5
All resources	21 459.2	33 314.6	46 290.9	68 026.7	65 759.5	76 500.0

Source: European Commission.

Table 14.6 How the budget is spent (%): 1973–1996

	1973	1988	1993	1994	1995	1996*
EAGGF (FEOGA) Guarantee	80.6	58.4	48.8	49.3	46.2	47.2
Structural operations	—	17.4	30.9	30.5	32.9	33.7
Research	1.6	2.3	3.6	3.6	3.7	3.7
Other internal policies	12.3	11.3	2.2	2.1	2.6	2.4
External action	1.4	2.3	6.0	6.3	6.1	5.9
Administration	5.5	4.2	4.7	4.8	5.0	4.8
Other (including Euro development)	—	3.7	4.0	3.4	3.5	2.3
Total (%)	100	100	100	100	100	100

Source: Compiled from European Commission information.
* Draft.
— Part of other internal policies.

opment is kept in the European Agricultural Guarantee and Guidance Fund, known as FEOGA (a French acronym).

The allocation of money from the budget is shown in Tables 14.6 and 14.7. The distribution of funds within the agricultural sector, from the FEOGA fund, is shown in Tables 14.8 and 14.9.

Table 14.7 FEOGA share of the Community budget

	1980	1982	1984	1986	1991	1992	1993	1994	1995*	1996**
Net cost of CAP per head (ecu)						94.7	104.6	95.9	101.7	—
EAGGF (%) of total	73.1	63.1	69.9	65.2	64.12	59.5	58.4	53.6	50.0	52.1

Source: European Commission.
* Amended budget.
** Draft.

Table 14.8 FEOGA guarantee section expenditure by sector (%)

	1982	1984	1986	1991	1992	1993	1994	1995	1996
Milk and milk products	26.8	31.6	27.4	17.6	12.7	14.9	12.9	11.6	10.3
Cereals	14.7	10.5	14.4	16.1	17.4	18.3	36.8	39.5*	42.1*
Beef and veal	9.3	11.2	12.0	13.2	13.7	11.7	10.5	13.2	13.4
Sugar	10.0	8.7	7.4	5.6	6.1	5.9	6.3	4.8	4.8
Fruit and vegetables	7.4	7.3	4.2	3.4	3.9	4.9	4.7	5.2	4.2
Olive oil	4.0	4.8	4.6	5.8	5.5	6.5	5.5	2.4	4.4
Tobacco	5.0	4.3	3.5	4.1	3.8	3.9	3.2	3.1	2.7
Wine	4.6	6.0	4.9	3.2	3.4	4.7	3.6	2.8	2.7
Sheep and goat meat	2.0	2.8	2.4	5.5	5.4	6.4	3.9	4.5	3.3
Pig meat	0.9	1.1	1.0	0.8	0.4	0.8	1.3	0.7	0.4
Eggs and poultry meat	0.8	0.7	0.6	0.5	0.6	0.7	0.7	0.5	0.4
Rice	0.4	0.5	0.5	0.4	0.3	0.3	0.1	0.2	0.1
Fishery products	0.3	0.2	0.2	0.1	0.1	0.1	0.1	0.1	0.1
Other (incl. set-aside until 1994)	13.8	10.3	16.9	23.7	20.7	20.9	10.4	11.4	11.0

Source: Compiled from European Commission data.
* Now arable crops including set-aside.

A further problem with the budget is the inequality of contributions between member states. Germany and the UK have borne the majority of the burden of contributions, while France, Italy and Eire have been the main beneficiaries. This has been eased by the refund which the UK received in 1983 and the refocusing of support since the 1992 reforms.

Another major problem is that too much support has been given to fund commodities in structural surplus. Price support reductions in 1984, 1985 and 1986 were attempts to resolve this situation, which did not have a significant effect, and major reforms were undertaken in 1992 by MacSharry (see later in this chapter).

Table 14.9 Guarantee section expenditure by activity (%)

	1982	1984	1986	1992	1993	1994	1995
Set-aside and income aid	—	—	—	—	—	34.6	41.6
Price guarantees/intervention	43.0	43.1	38.4	44.2	47.4	30.8	32.0
Export refunds	38.5	34.6	37.9	29.5	29.2	24.4	19.6
Storage	13.9	18.0	18.3	18.9	15.4	3.2	3.3
Others (including mcas)	4.6	4.3	5.4	7.4	8.0	7.2	3.5

The UK budget contribution

The organisation of revenue collection and expenditure within the EU means that some countries will be net financial beneficiaries while others will be net financial contributors. In general, contributions to the EU budget depend on a country's gross imports, particularly from outside the EU. Receipts from the budget depend heavily on the level of agricultural production within the country and in particular on the degree of self-sufficiency in agricultural products. Agricultural exporters receive more from the budget than do net importers of agricultural products.

This transfer of funds from importers of agricultural products to countries exporting agricultural products within the EU is caused by the heavy emphasis of expenditure from the EU on agriculture. This situation has meant that Britain, only about 65% self-sufficient in all agricultural products, was a large net contributor to the EU budget (prior to 1983); in fact the second largest net contributor after Germany. In 1983 Britain was in seventh position in the EU league table of GDP per head but was one of only two net contributors, the other being Germany.

Britain's net contribution in the three years 1981 to 1983 would have been £3225 million, but a rebate of £2038 million was negotiated by Mrs Thatcher, leaving a contribution of £1187 million or an average of £396 million per year over the period. By 1993 the total budget had grown to about 68 billion ecu. Ranked 8 out of 12 measured by GDP, the UK was still the second largest contributor behind Germany (17.1%), but providing only 4.5% (approximately £2550 million) of the total. Italy, France and the Netherlands were the other net contributors.

Having presented the budget case it must be stated that this 'inequality-of-contribution' situation did not arise unexpectedly. At the time of UK accession to the EU, it was understood that the dominance of agriculture, in terms of EU expenditure, would mean that UK would not receive as much direct financial benefit as other countries where agriculture was a more important part of the economy. Indeed the purpose of joining was to ensure that the UK would not be isolated in Europe, financially, commercially or strategically and that she would gain access to both supplies and markets in continental Europe, unburdened by the trade barriers of any kind.

Table 14.10 shows a ranking of beneficiaries and contributors to the budget compared to the relative GDP levels.

Table 14.10 Ranking gross domestic product (GDP) and net contributors to the EU budget over a ten-year period

	GDP ranking		Net contributions		
	1983 (EU 10)	1993 (EU 12) (ecu/head)	1983 Ranking (£/head)	1993 (%)	1993 Ranking
Belgium	5	5	7	–0.1	6
Denmark	3	2	6	–0.7	8
Germany	2	3	(–1)	(17.1)	1
Greece	9	11	9	–6.7	12
Spain	—	10	—	–4.8	11
France	4	4	3	1.4	4
Ireland	10	9	8	–3.8	10
Italy	8	7	10	2.9	3
Luxembourg	1	1	4	–0.4	7
The Netherlands	6	6	6	1.9	5
Portugal	—	12	—	–3.7	9
UK	7	8	(–2)	(4.5)	2

( ) Indicates a net contribution to the budget in 1983 and 1993.

The Common Agricultural Policy (CAP)

The CAP (detailed originally in Article 39 of the Treaty of Rome, 1957) can be considered the cornerstone in the construction of the EU.

Objectives

(1) To increase agricultural productivity
(2) To ensure a fair standard of living for the agricultural community
(3) To stabilise agricultural markets
(4) To ensure regular supplies of food
(5) To maintain reasonable prices for consumers

Principal methods adopted to achieve the objectives

(1) *A unified (single) market with:*
  (a) Common prices, with the minimum fixed above world market prices
  (b) No customs duties
  (c) Free movement of goods
  (d) Stable currency parities
  (e) Harmonisation of administration, health and veterinary legislation

(2) *Community preference* carried out using levies which prevent 'cheaper food and products' coming from outside the EU
(3) *A common farm fund* – European Agricultural Guidance and Guarantee Fund (FEOGA or EAGGF). This is where the money is 'stored' and paid out to ensure security of food supply to 372 million consumers while protecting (to some extent) the incomes of farmers

There are two parts to the CAP:

(1) Guarantee – stabilises prices and markets
(2) Guidance – aimed at altering the structure of agriculture

Guarantee

The guarantee sector accounts for most of the money spent on the CAP (see Tables 14.6–14.8 for a breakdown). The support schemes themselves were originally based upon the principle of a guaranteed minimum price underpinned by intervention buying. Since the MacSharry reforms of 1992, however, there has been a fundamental change in the way that agricultural and rural incomes are supported. Previously based on market support they are now much more focused on direct payments to those producers deemed to be most vulnerable. The payments are related to the intensity of production for both arable and livestock in an attempt to control and reduce structural surpluses.

Method of operation

The CAP system operates by use of:

(1) Guaranteed minimum prices supported by intervention, and quality control
(2) Direct aid payments for qualifying producers
(3) Import controls on third country trade
(4) Export controls on third country trade

The effect of the CAP on the UK

The UK system of government creates a very transparent method of policy implementation and we therefore have a high level of applying EU directives. This ought not to be over-exaggerated simply because implementation is not so obvious in the other states.

Prior to joining the EU the UK used a system of deficiency payments. If a product did not achieve a guaranteed minimum price in the marketplace then the producer was paid a subsidy to make up the deficiency direct from the Treasury via the Ministry of Agriculture. This ensured that food was relatively cheap to the consumer but taxpayers often accused farmers of being 'feather-bedded' and not operating in a real market. Post 1973 farmers in the UK obtained their returns more directly from the market (via higher consumer prices) and the high guaranteed prices, unrelated to market demand, caused expensive surpluses to build up in many products, as producers received their price signals from the guaranteed system (which was much higher than the equilibrium price created by the unencumbered operation of the price mechanism), rather than from the market place.

The support schemes

Peas and beans and oilseeds

The pea and bean scheme was introduced to encourage the production of proteinaceous crops within the EU as supplies were well below the self-sufficiency level. The result has been an increase in the area of peas and beans grown for animal feed. The scheme provided indirect support for the producer who benefited by receiving at least the minimum purchase price from the compounder. The feed compounder had to pay the minimum purchase price to the producer to qualify for the subsidy available. The subsidy was calculated as 45% of the difference between the activating price, fixed by the Commission annually, and the world market price of soyameal. The subsidy varied widely over the period of the scheme but usually lay somewhere between £40 and £65 per tonne. From 1993/1994 the aid to processors and minimum prices were replaced by the system of aid requiring set-aside by producers taking part in the general scheme. Prices have dropped with the integration of these protein crops into the schemes introduced after the 1992 reforms and, along with oilseeds, they now trade at 'world' prices in European markets. The products covered are dried peas (excluding chick peas), dried beans and sweet lupins. In the oilseeds scheme support is offered for rape, sunflower and soya. There is also a support scheme for non-fibre flax seed which follows the general principles for arable crops under the area aid schemes with a compensatory payment made for set-aside.

Imports are free of levies and customs duties, with the exception of a low rate on peas, and exports are not subsidised.

Cereals

EU cereals production accounts for about 12% of world production. The main products are wheat, barley and maize, each of which has increased in production since the EU began. The Commission determines the intervention price annually for bread-making wheat, maize, barley, sorghum, rye and durum wheat. The intervention price offered to farmers increases each month and is affected by grain quality (see Table 14.11). The grain must be sound, fair and of marketable quality, and will be regarded as such if it is of typical colour, free from abnormal smell and live pests. The grain must also be within pesticide residue limits. Grain with a temperature of more than 18°C in November, and 15°C in December and following months, will not be accepted.

Table 14.11 Quality standards for feed barley and feed wheat

	Barley intervention	Barley futures	Wheat intervention	Wheat futures
Minimum tonnage	100		500	
Moisture (maximum)	14.5	15	14.5	15
Specific weight (kg/hl)	62	62.5	72	72.5
Total impurities	12		5	2
Broken grain	5		5	2
Dirt content		1		1
Sprouted grain	6	8	5	8
Pest damage	5			
Heat damage	3	3	0.5	3
Grain impurities	12	2	5	2
Misc. impurities	3		3	
(of which ergot and garlic)				0.001
(of which ergot)	0.05		0.05	
Hagberg falling number	n.a.		220	
Zeleny index*			20	

* If the Zeleny Index is between 20 and 30, wheat must pass the Dough Machinability test.
Source: GAFTA and Intervention Board.

The prices offered for intervention (see Table 14.13) are not what UK producers receive; this is because of the costs involved in selling to intervention. Table 14.14 gives an example of the real value to the farmer (or merchant) after the costs of selling into intervention are taken into account. The intervention offered price is worth considerably less in cash terms to the sellers than the published figures imply but they do, however, create a 'bottom' to the market so that farmers who could not, for example, obtain at least £100.30 per tonne for feed wheat in November 1996 may have considered intervention, with all its associated administration, the best outlet.

When the local market price falls below the intervention price farmers (or other grain sellers) can offer grain to intervention. The scheme, however, does not begin to operate as a market outlet until November each year and ends in March.

Importing

The relatively high-cost EU producers are protected from cheap imported grain (which is traded at 'world' prices) by an import duty imposed upon it (see Table 14.12). The import duty is the intervention price plus 55% (the reference price), less the costs of insurance and freight import price taken during the previous two weeks. The duty will therefore vary with trends in world market prices. For wheat there are three representative prices reflecting high, medium and other quality, but for protein and maize there is only one. Cereals not supported in the CAP system have a fixed rate of duty imposed which is set annually.

Table 14.12 Import reference prices and duty on cereals 1996–1997 (ecu/tonne)

	Duty paid import reference price		Fixed rate of duty	
	Common wheat, durum wheat and barley	Maize and sorghum		
1996				
July	184.74	198.85	Oats	122.3
August	184.74	198.85	Buckwheat	51.0
September	184.74	198.85	Millet	76.7
October	184.74	184.74	Triticale	127.7
November	186.45	186.45	Wheat and durum wheat flour	236.0
December	188.15	188.15		
1997			Rye flour	231.3
			Brans and sharps	60
January	189.86	189.86	Roasted malt	208.4
February	191.56	191.56		
March	193.27	193.27		
April	194.97	194.97		
May	196.68	196.68		
June	196.68	196.68		

Table 14.13 Intervention prices 1985–1986 and 1995–1996

	1985–1986 (£/t) (July)	1995–1996 (£/t) (November)
Wheat	111.01	97.43*
Barley	111.01	97.43*
Oilseed rape	260.76	
White sugar	335.19	
Butter	1 937.63	2 658.14
SMP	1 076.71	1 664.54

* Increment is 1.1 ecu/tonne/month
Source: Intervention Board.

Exporting

Exporting grain from the EU is most important if surpluses continue. Since EU grain is relatively expensive, to sell it abroad is impossible unless its price is reduced to competitive 'world' levels. To achieve this, refunds (or export restitutions) are offered (at certain times of the year for amounts of grain determined by the Commission) to exporters who wish to move grain out of the EU to designated zones in the rest of the world. This system of export restitutions is a source of irritation to USA, Canadian (and other Cairns group members) exporters who see it as unfair trading, since the EU excludes them from EU markets and then hands out

Table 14.14 Estimated value of intervention prices after costs are deducted; 1986–1987 and 1994–1995

	1986–1987 (£/t) (March 1987)	1994–1995 (£/t) (March 1996)*
Cereals	105.08	106.08
Costs to intervention:		
Haulage	4.00	5.00
Payment delay	3.50	2.00
Co-responsibility	3.57	—
2% weight loss and discounts	2.25	2.50
Overheads	1.00	1.25
Total	14.12	10.75

* Increment per month 1.10 ecu/tonne

large subsidies for EU exporters to compete in their markets. It is also a very expensive system to operate, with refunds amounting to over £60/tonne at times.

Milk products

The EU is the largest milk producer in the world, accounting for about 16.6 % of total agricultural value in 1993. Almost half of the milk is produced in Germany and France and in December 1994 the EU herd was estimated at 21 million with milk production 111.1 million tonnes. Milk production has been reduced since 1983 by the introduction of quotas and levies, with cow numbers down about 27% and milk production about 11%. Although there is a target price for liquid milk set by the Commission, producers do not necessarily receive this. In fact the price is determined for the individual producer by his milk quality and bacterial and mastitis cell counts. The UK receives a price for milk from the various buyers which reflects the value of milk sold for liquid consumption and for processing. The milk market was deregulated in November 1994 after 60 years of the statutory monopoly held by the Milk Marketing Board (MMB). As a result milk prices rose from 21.5 ppl in 1994 to approximately 24 ppl in 1996, as milk processors bid the price up in order to ensure supplies. Since then prices have fallen and some forecasts for 1999 have been as low as 19 ppl.

The EU support for the milk sector is given indirectly by providing intervention buying for skim milk powder (SMP) and butter. When the market price of SMP or butter falls below the intervention price offered, these products can be sold into intervention. The intention is that they will be taken off the market in times of surplus and stored until a shortage occurs when both butter and SMP will be released onto the market. The support scheme has run into difficulties, however, as there have been consistent surpluses of butter and SMP which have now built into the 'mountains' so emphasised by the media (see Table 14.15). this is despite many attempts to dispose of 'cheap' butter inside and outside the EU and efforts to

Table 14.15 EU stocks of products (31 December) (1000 tonnes)

	1983	1990	1991	1994	1995
Durum wheat	660	1 443	3 376	1 407	342
Common wheat	585	6 681	1 991	1 994	2 530
Barley	550	4 074	4 632	6 121	3 175
Rye	214	1 821	3 583	2 263	2 255
Oil seed rape	—	13	13	—	—
Olive oil	120	74	18	187	10
Skim milk powder	1 048	333	416	35	65
Butter	594	251	266	177	59
Beef	323	537	1 010	380	40
Alcohol (000 hl)	—	2 438	2 399		
Maize	—	67	1	2 259	22

Source: European Commission.

compulsorily incorporate SMP in animal feed. The cost of storing these products is prohibitive and the Commission attempted to reduce the surplus (by controlling production with quotas introduced on the 1 April 1984) and therefore the storage costs.

Imports and exports

Export refunds are provided for most milk products, the rate being fixed every four weeks by the Commission. The import minimum price, fixed annually, is related to the target price set for milk and protection is afforded by applying import duties to imported products to bring them in line with EU prices. Quota arrangements can be applied to certain imported products to provide further safeguards if it is deemed necessary.

The cost to the CAP of operating support schemes across a wide range of agricultural products is enormous, not least of which has been the cost of storage for butter, beef, SMP, wine, cereals and olive oil, which increases with structural surpluses (see Table 14.15).

Pressure for reform

Various attempts to reform the support system occurred during the 1980s with (amongst other measures) milk quotas introduced in 1984 (and still with us!) and 'stabilisers' operating between 1988 and 1992, which fixed the maximum quantity on which support could be paid and provided for reductions in support prices in the years when surpluses were created. Milk quotas have effectively controlled milk production and created a 'licence' to produce. Great demand by milk producers (see Table 14.16) raised the value of quota as profitability has been very good, especially in the UK, with deregulation of the industry following the dissolution of the Milk Marketing Board. However, the market has fallen in 1998.

Table 14.16 Value of milk quotas

	Purchased (ppl)	Leased (ppl)
1988–1989	34	4.62
1989–1990	34	5.40
1990–1991	38	5.93
1991–1992	35	7.00
1992–1993	21	4.23
1993–1994	46	5.54
1994–1995	60–65	13.63
1995–1996	66–68	12.80
1996–1997	50–60	10.00

Despite these measures the CAP reached a crisis in the early 1990s because it had been too successful in increasing food productivity. From being a net importer the EU became a net exporter of most agricultural products. This was caused by the adoption of technical improvements and an improvement of farm structure across the EU. The high guaranteed prices (which stimulated the surpluses) were not reduced as supply exceeded demand, and huge stocks, expensive to maintain, were built up (see Table 14.15)

The excess stocks were eventually dumped on the world market, using a cash refund (subsidy) to make them competitive. As a result the cost of the agricultural budget rose 43% between 1987 and 1993, despite continuing focus on 'budgetary discipline' which merely reduced the rate of increase in spending.

Thus high prices were paid internally for products which could not be consumed. These went straight into intervention stores and were maintained there at high cost. Eventually they were subsidised into the world markets, disrupting trade there in cereals, oilseeds, beef, butter and sugar. This system was criticised both by consumer groups in the EU and by the USA, Argentina, Australia and New Zealand who saw their markets shrinking as a result of 'unfair' EU competition. The response was an increase in farm support systems outside the EU and the danger of more international protectionism. Agricultural support therefore became part of the Uruguay round of the GATT (General Agreement on Tariffs and Trade) talks in 1986 for the first time, thus imposing more pressure on the EU to alter its support system. Other strands to the reform were imposed by the fact that in a market-based support system large farms and more intensive farming, tended to receive much more of the subsidy. This added strength to the arguments of increasingly vociferous environmental and animal welfare pressure groups in certain parts of the EU who were determined to see an agricultural policy which encouraged and supported less-intensive production systems.

One dilemma which faced the reformers was rural depopulation. For many years the CAP had been used as an indirect method of maintaining rural incomes in EU areas in which commercial farming could not be justified. In a way it was a disguised social policy aimed at keeping people in rural areas, especially in continental Europe. Any reform of the CAP had therefore to attempt a long-term reduction of cost, by controlling and reducing production and therefore surpluses; ensure that production became more environmentally and animal friendly; while encouraging people to remain in rural areas. The solution accepted was MacSharry's reforms, published in July 1991 and adopted in April 1992.

The MacSharry CAP reforms

The MacSharry reforms produced a fundamental shift away from market support (financed mainly by the consumer), to direct aid (financed mainly by the taxpayer). The previous market support systems are still functioning but have less influence and lower support prices.

Farm income support has been switched to direct aid payments (which are not production related and therefore termed decoupled), which are meant to discourage output, and away from market support. The payments are also related to reductions in crop areas grown and the intensity of stocking in livestock systems, thus providing environmental benefits. The focus of the aid is on small farms, especially in less favoured areas, and aimed to reduce surpluses while improving the economic efficiency of the EU by reducing market prices. Prior to the MacSharry reforms some estimates put EU prices at 22% above the equilibrium.

Cereal sector reform

Arable area payments were introduced in the 1993 harvest to compensate for reducing support prices, these are shown in Table 14.16. The payments are given for cereals, oilseeds, linseed and protein crops which are grown on eligible land, i.e. land in arable, set-aside or temporary grass in the 5 years prior to 31 December 1991. Two schemes are in operation: the simplified scheme and the general scheme.

The simplified scheme

This is for farms producing less than a maximum area of supported crops: approximately 15.6 ha in England, 17.6 ha in Scottish LFAs, 16.2 ha in Scottish non-LFAs and 17.8 ha in Welsh non-LFAs (92 tonnes, as shown in Table 14.17). No set-aside is required of these producers and they receive cereal rate compensation for all eligible crops. In 1994 regional yields were based on the average between 1986–1987 and 1990–1991, excluding the highest and lowest years. 1995 was 60% of actual regional average plus 40% of UK actual average between 1986–1987 and 1990–1991, excluding high and low years.

General scheme

The base on-going rate of set-aside is 17.5%; this is adjusted to an annual figure and causes great difficulty in planning cropping if the rate has not been agreed by June each year.

A basic percentage of the claimed arable area must be set-aside in order to qualify for the area compensation. For rotational set-aside this was 15% in 1993 and 1994, 12% in 1995, 10% in 1996 and 5% in 1997. For non-rotational set-aside it was 18% in

Table 14.17 Cereal guarantee prices, 1993–1996

Prices	1992–1993		1993–1994		1994–1995		1995–1996		Percentage fall 1995–1996 compared to 1992–93	After switch-over adjustment at original reform
	ecu	£	ecu	£	ecu	£	ecu	£		
Target	226	208.14	130	121.22	120	111.9	131.11	109.32	51.32	41.98
Effective intervention	155	142.75	127	118.42	108	100.7	119.19	99.38	35.5	23.10
Expected market/t	155	142.75	130	121.22	120	111.9	131.11	109.32	29	15.41
Compensatory payment	0	0	25	23.31	35	32.6	54.34	45.30	—	—
Threshold (minimum price/t)	222	204.45	175	163.18	165	153.8	—	—	30*	—
Community preference/t	62	57.10	45	41.96	45	41.9	—	—	27.4*	—

1 February 1993: ecu values were increased by 20.7% to account for abolition of the switch-over mechanism.
1992–1993: 1 ecu = £0.920969
1993–1994: 1 ecu = £0.932453
1994–1995: 1 ecu = £0.840997
1995–1996: 1 ecu = £0.833821
* Threshold discontinued after GATT agreement: % based on original figures expected in 1992.

1994, 15% in 1995, 10% in 1996 and 5% in 1997. From 1997 there will be one obligatory set-aside rate, with no rotational restrictions. The rate for harvest 1998 is 5%. The amount received varies between crops (see Table 14.18) and depends on the green rate prevailing on 1 July in the year of harvest. Payments are made between October and December, except for oilseeds which is paid in September and March.

Restrictions

There are regional limits on arable areas claimed (see Table 14.19). If the area claimed exceeds the base area for a region, payments are reduced and an equivalent area of uncompensated set-aside is incurred in the following year. This applied to non-LFA producers in Scotland in 1995, although only 20% of the overshoot was implemented following an agreement made in 1993. Pre 1992-there was a 0.5% price cut for each 1% overshoot; post 1992 it was agreed to be a 1% price fall in the next two years, for each 1% overshoot. For 1997 and 1998 harvests, base area overshoots do not attract penalty set-aside in the following year but base area payments are still cut. The oilseed payments can also be adjusted (up or down) by the maximum guarantee area (MGA) penalty and if the projected reference price differs by more than 8% from the observed reference price for EU oilseeds in January each year. For the 1997 harvest it is likely that there will be an OSR payment cut of 7%, given an EU reference price in January of 21.3 ecu/t and a projected 1997 average of £158/t. The increased planting may attract a 6% area-related cut in payments for 1997. The MGA for food oilseeds was 4.4 million ha in 1995–1996 after 15% set-aside, plus 300 000 ha of non-food oilseeds. Non-food or industrial crops of oilseeds can be grown on set-aside land but will not receive the area payment, they will receive the 45 ecu/t multiplied by regional average cereal yields, for land set-aside in 1993–1994, rising to 57 ecu/t in 1994–1995, irrespective of whether or not a crop is grown for industrial purposes.

For EU industrial oilseed production, no more than 1 m/t of soyameal equivalent is allowed to be subsidised.

Maize

The base area for maize in England was exceeded by 23623 ha, in 1996 (after deducting the under-shoot in other crops). Growers claiming area payment in maize received 58.49% of the set-aside payment. The 1997 area, however, is not subject to uncompensated set-aside.

Set-aside

Set-aside land must be on the eligible arable area which has had a crop in the previous year (on which harvesting was attempted) or has been in set-aside.

There are several choices:

(1) Six year rotational set-aside – land may only be set-aside one year in six.
(2) Flexible set-aside – the minimum rate is 5% and the maximum 50%. Several options exist which include any combination of rotational or non-rotational set-aside. Within this option the rotational set-aside will not have to have a one-in-six rate and the non-rotational set-aside does not need to have a five

Table 14.18 Area payments

£/ha	Cereals			Pulses			Oilseed			Linseed			Set Aside		
	1994/5	1995/6	1996/7	1994/5	1995/6	1996/7	1994/5	1995/6	1996/7	1994/5	1995/6	1996/7	1994/5	1995/6	1996/7
England	193.53	269.17	266.87	359.41	388.79	385.48	366.01	475.79	448.15	481.06	520.61	516.17	315.17	340.94	338.04
Scotland	180.51	254.41	246.08	335.24	367.48	355.44	407.14	502.34	480.83	484.59	585.09	595.92	293.98	322.25	311.70
Scotland LFA	156.97	238.10	236.06	291.53	343.91	340.97	337.50	421.17	413.23	418.47	547.57	456.58	255.65	301.59	299.00
Wales	168.72	236.26	234.25	313.35	341.27	338.36	373.14	465.66	456.88	419.40	456.97	453.07	274.78	299.26	296.72
Wales LFA	108.35	230.78	228.82	201.22	333.34	330.50	373.14	465.66	456.88	269.32	446.37	442.56	176.45	299.32	289.83
N. Ireland	153.71	238.55	236.51	285.47	344.57	341.63	347.00	433.03	424.87	382.09	461.39	457.45	250.33	302.16	299.58
N. Ireland LFA	153.71	229.86	227.91	285.47	332.02	329.19	347.00	433.03	424.87	382.09	444.59	440.80	250.33	291.16	288.68

Source: HGCA.
* LFA = less favoured area

year commitment. There is a guaranteed option, in which land committed to set-aside will receive payments for five years at a minimum of the ecu level prevailing when the option was taken up. Farms coming out of the old five-year set-aside scheme, with more than 50% set-aside, are eligible for additional voluntary set-aside, although the payment for this is lower (£229/ha in 1997). Transfer of set-aside rights is only possible provided that the farm receiving the set-aside is within 20 km of the giver (rotational or flexible), unless the receiver is in an environmental scheme (flexible only), when distance is not relevant. There is a 1% extra set-aside percentage for the transferred set-aside.

Table 14.19 Regional average yields (t/ha)* + base area**

Region	Cereals 1994	Cereals 1995 and 1996	Oilseed 1994/1995/1996	Base areas ('000 ha)	
England	5.93	5.89	3.08	Cereals	3 761 366
				Maize	33 226
Wales LFA	3.32	5.05	3.14		66 000
Wales non-LFA	5.17	5.17	3.14		60 000
Scotland LFA	4.81	5.21	2.84		60 000
Scotland non-LFA	5.65	5.67	3.49		395 000
N. Ireland non-LFA		5.22	2.92		53 000
N. Ireland LFA		5.04	2.92		53 000

*Calculated on historic values between 1986–1987 and 1990–1991 excluding the highest and lowest years.

** Regional base areas are the average numbers of hectares of eligible cereals, oilseeds, protein crops or set-aside during the period 1989–1991

Beef sector

Beef Special Premium (BSP)

The beef headage payments (see Table 14.20) compensate for the removal of variable premiums previously paid in the market support scheme. The current BSP is an extension of the scheme introduced in April 1989. The BSP is claimed twice in the life of any male animal at 10 months and 22 months but animals must have been on the farm for at least two months prior to the claim. The payment is limited to 90 animals per age group, per year and by stocking density limits which were 3.5 livestock units/ha in 1993; 3.0 livestock units/ha in 1994; 2.5 livestock units/ha in 1995 and 2.0 livestock units/ha in 1996. This calculation involves all beef animals and sheep on which premium is being claimed. There is an extensification premium (of 36.23 ecu/head) which can be claimed if stocking density is less than 1.4 livestock units/forage hectare.

Suckler Cow Premium (SCP)

SCP is payable (see Table 14.20 for the rates) on cows of beef breeds whose milk is

Table 14.20 Beef headage payments

	1992–1993		1994		1995		1996	
	ecu	£	ecu	£	ecu	£	ecu	£
Beef Special Premium	60.00	48.00	75.00	69.07	108.68	103.63	108.68	93.09
Suckler Cow Premium	82.50	65.73	95.00	87.49	144.7	138.17	144.90	124.12
Extension Premium for livestock unit/ha < 1.4	30.00	23.90	30.00	27.63	36.22	34.54	36.22	31.03
Stocking rate limit (hd/ha)	3.0		3.0		2.5		2.0	

1993 Green rate at 1 ecu = £0.796666 1994 Green rate at 1 ecu = £0.920969
1995 Green rate at 1 ecu = £0.953575 1996 Green rate at 1 ecu = £0.856563

Table 14.21 Hill Livestock Compensatory Allowances (HLCAs) 1992–1996

	Severely disadvantaged					Disadvantaged area				
	1992	1993	1994	1995	1996	1992	1993	1994	1995	1996
Breeding cows	63.30	49.50	47.50	47.50	47.50	31.65	23.75	23.75	23.75	23.75
Hardy breed ewes	6.50	5.75	5.75	5.75	5.75	—	—	—	—	—
Other ewes	3.60	3.00	3.00	3.00	3.00	2.85	2.44	2.44	2.44	2.65
Stocking rate limit (LU/ha)	1.4	1.4	1.4	1.4	1.4	1.4	1.4	1.4	1.4	1.4

not marketed. Claims must be matched by sufficient headage rights or quota. The quota is based on claims made in 1992. Suckler cow quota can be transferred when the same rules apply as for sheep quota transfer. Cattle must be on the farm for six months prior to the claim. Two payments are made: 60% after 1 November in the year of the claim and 40% paid after the following 1 April (and is subject to national quota ceilings). Extensification is paid with final BSP or SCP payments. BSP and extensification payments are subject to regional claim ceilings. If the limit is exceeded in a region in a year then the number of eligible animals per producer that can qualify for payment is reduced. In 1995, for example, producers received 97.77% of claims in England and Wales and 83.22% in Scotland. Following the BSE crisis the scaleback in 1997 may be higher.

Sheep sector

The Sheep Annual Premium (SAP)

To qualify for headage payments, ewes have to be 12 months old or lamb during the year. Payment compensates for the removal of the previously offered market support of variable premiums. The amount paid is linked to average lamb prices in Europe, varies each year and is received in two instalments: an advance of 60% in October and the balance in April/May. The payment in 1994–1995 was £21.26/ewe in total, in non-LFAs it was £26.94. The supplementary Less Favoured Area (LFA) payment (of 6.64 ecu per ewe) is paid 90% in October and the balance in May. To claim the SAP, farmers must hold the rights to quota based on the claims they made in 1991. Headage limits were applied 1993 and 1994 (but not in 1995 and 1996) at 500 per farming partner in non-LFA and 1000 in LFA: above this number of ewes, the premium is halved. Each year 70% of the quota must be used or leased. If leased, 70% must be used for at least two consecutive years out of five. Quota sale is permitted but 15% is siphoned into the national regime unless it is sold with the farm. In 1996 there was one retention and one application period: application – December 1995 to 4 February 1996; retention–February 1996 to 14 May 1996.

Hill Livestock Compensatory Allowances (HLCAs)

The hills are specially selected for payments due to increased hardship. These are shown in Table 14.21 and set by the UK each year. HLCAs are limited to nine ewes per hectare in the disadvantaged area and subject to upper limits of six ewes per hectare in severely disadvantaged areas.

Milk

The milk sector was left largely unchanged by the 1992 reforms, despite original intentions to cut both support prices, on milk and milk products, and quota levels. It was agreed to review prices and quota annually, and take account of the current market situation. The butter intervention price was cut by 3% in 1993/1994 and in 1994/1995 but there has been no reduction in quota. Indeed Spain, Italy and Greece have all been given quota increases, which adjusted quota levels upwards to meet production levels. Milk quotas will remain until at least 2006, but over-production incurs a penalty of 115% of the target price. In 1995–1996 this was 31.43 ppl, a total value exceeding £49 million for the industry in the UK. At present the EU milk quota is about 114 million tonnes, approximately 13% higher than domestic consumption, although a lot of milk products are exported, this is with the help of subsidies and more than 35% of butter consumed in the EU is subsidised.

Guidance

This part of the CAP is dedicated to improving the structure of agriculture within the EU. It has its origins in the Mansholt Plan put forward in 1968 which attempted to create a long-term solution to the surplus problem. Mansholt recognised that if support prices were increased, to maintain incomes in rural areas, an unendurable surplus occurred. If support prices were decreased then unendurable social and (therefore) political stress occurred in some member states. He proposed to resolve the problem by accelerating the withdrawal of resources from the agricultural sector so that five million people and five million hectare would go by 1980. It was intended that aid should be concentrated on those farms which could become profitable. Directives on agriculture, which became operative in January 1994, resulted in grants to assist in the modernisation of agricultural holdings which were administered through the Farm and Horticultural Development Scheme, which was replaced in 1980 by the Agricultural and Horticultural Development Scheme (AHDS). The AHDS was superseded by the Agricultural Improvement Scheme European Community (AIS(EC)), partially funded from the FEOGA fund and the Agricultural Improvement Scheme (national) (AIS(N)). The schemes placed a greater emphasis on investment to improve the environment, energy saving, and tourism and craft.

Current aid is provided from EU structural funds. Objective 1 aims to assist the structural development and adjustment of regions whose development is lagging behind. Three regions of the UK – the Highlands and Islands of Scotland, Northern Ireland and Merseyside – come into this category. In Scotland the Highlands and Islands Agricultural Programme will provide about £23 million to farmers and crofters for agricultural business improvement, marketing and developments of crofting townships in the 1994 to 1999 period. In Northern Ireland over £45 million was spent in 1995 on such activities as marketing and processing in agrifood and fisheries, human resource development, research and development, improving competitiveness, environmental safeguards, farm diversification, rural development and assistance to the fishing industry.

Objective 5b aims to promote rural development (in areas not designated as Objective 1 areas) by supporting activities such as the provision of business advice, infrastructure development, marketing, diversification and environmental measures. Eleven areas (four in Scotland) are designated for the period 1994 to 1999, based on low Gross Domestic Product, high dependency on agricultural employment, low level of agricultural incomes and low population density. These are:

South West	Cornwall, parts of Devon and Somerset
English Marches	Most of Shropshire and parts of Hereford and Worcester
Northern English Uplands	Parts of Lancashire, North Yorkshire, Cumbria, Northumberland and Durham
Lincolnshire	Excluding Lincoln and other urban areas
East Anglia	The Fens, rural Norfolk and the Waveney valley
Peak District	Staffordshire and Derbyshire
Wales (rural areas)	Dyfed, Gwynedd, Powys and Clywd
Dumfries and Galloway	
Borders region	
Stirling and Upper Tayside	Rural areas
North and West Grampian	

Regional Development Programmes have been drawn up by central and local government authorities and submitted to the European Commission for approval. It is expected that funds from the guidance part of the FEOGA fund will be around £100 million for these programmes.

Less favoured areas

These were designated in 1975, being areas of low productivity coinciding with previously designated hill areas, with some additions. The policy adopted for these areas was intended to alleviate the consequences of permanent handicap, such as high altitude, steep slopes, long winters or lack of rainfall. By 1983 31% of the utilised agricultural area (UAA) in the EU was designated as less favoured; this was increased to 48% in 1986 and details are shown in Table 14.22.

Processing and marketing grant aid was available in England until 31 March 1996 and remains in Scotland, Wales and Northern Ireland, under EU Regulation 866/90. This aid is jointly funded by the UK Government and FEOGA and is aimed at improving or rationalising the processing and treatment of agricultural produce.

People employed in member states

There have been dramatic changes in the numbers employed in agriculture, forestry and fishing since the EU was organised in 1958. Some commentators have suggested that the exodus was equivalent to one person leaving agriculture every minute between 1960 and 1980, and numbers falling from around 20 million in 1960 to around 7.8 million in 1992, in the EU 12, details are shown in Table 14.23. Further changes seem inevitable as there are still a considerable number of people aged 55 years and over, whose main occupation is farming.

Grants are available at 30% (40% in Merseyside) for projects costing at least £70 000 with a maximum of £1.2 million.

Table 14.22 The less favoured areas in the European Union 1986 ('000 ha)

Country	Less favoured areas		Total UAA in less favoured areas	Less favoured UAA as a proportion of total UAA (%)
	Mountain	Other		
Germany	351.5	5 859.1	6 210.6	50.9
France (including overseas depts.)	4 341.9	7 617.0	11 958.9	38.5
Italy	5 164.2	3 289.8	8 454.0	51.1
Netherlands	—	18.9	18.9	0.9
Belgium	—	314.4	314.4	21.9
Luxembourg	—	133.1	133.1	100.0
UK	—	9 859.0	9 859.0	52.5
Eire	—	3 878.7	3 878.7	67.9
Denmark	—	—	—	—
Greece	4 978.8	2 260.1	7 238.9	78.2
Total	14 836.4	33 230.1	48 066.5	48.0

Source: Compiled from European Commission information.

Table 14.23 People employed in agriculture in the EU as a proportion of each member's civilian employment (%)

	1960	1970	1983	1992
Austria	—	—	9.9	7.1
Western Germany	13.8	8.9	5.7	3.5
Italy	32.6	18.6	12.0	7.9
Netherlands	9.8	7.1	5.6	3.9
Belgium	8.7	4.4	3.4	2.9
Luxembourg	16.6	11.1	4.7	3.1
UK	4.8	2.8	2.5	2.2
Eire	37.3	26.8	17.5	13.4
Denmark	18.2	9.5	7.4	5.2
Greece	57.1	40.8	30.0	21.9
Spain	42.3	29.5	18.8	10.1
Portugal	43.9	30.0	26.7	11.5
Sweden	—	—	5.6	3.3
Finland		—	13.1	8.7

Source: Eurostat.

Future reform in the EU system

The agrimonetary system

It seems sensible to abandon the present complex and confusing system and have all CAP payments made in ecu (or Euro). This suggestion may be criticised for exposing producers to unwarranted foreign exchange risks, instead of protecting them from risk via the present (expensive) system. In the absence of EMU embraced by all EU members, this exchange risk cannot be avoided, although it may be ameliorated by hedging on the futures market. This process would become easier by hedging against the ecu rather than the green rate, or by receiving support payments in ecus and paying for inputs in ecus.

Hedging

After the pound sterling left the ERM and 'floated' against other European currencies, it initially weakened against the deutschmark and the French franc. Support payments are calculated in ecus and then converted into the currency of the country in which the farmer operates. In the UK ecus are converted to 'green pounds', the conversion rate being fixed on specific dates for some products, for example with area payments (cereals) the date is the 1 July annually and with sheep and beef premiums the 1 January.

Because support prices are initially calculated in ecus, farmers within the EU are technically dealing in a foreign currency. The effect on support payments of movement in the ecu/£ exchange rate is to either increase their value if the pound weakens or reduce their value if the pound strengthens and is illustrated in Table 14.24.

Table 14.24 The impact of changing ecu/£ exchange rate on arable area payment

Strength of the £	Ecu/£ exchange rate	Arable area payment (ecu/ha)	Sterling value arable area payment (per ha)
Weak	1.25	170	136.0
	1.30	170	130.8
	1.35	170	125.9
Strong	1.40	170	121.4

In Table 14.24 a movement from 1.25 ecu/£ to 1.40 ecu/£ (12%) reduces the area payment per hectare by £14.60. Thus on a 500 ha cereal unit the loss of income would be £7300. If in October 1997 a farmer felt that the pound was likely to strengthen against the ecu, so that on 1 July 1998, when the arable area payment was due to be converted from ecus to pounds, he would get fewer pounds for the ecu payment, then the farmer might consider it sensible to try and protect his income, he could do this by taking out an ecu hedge.

The farmer could approach his bank to sell them ecus for delivery on 1 July 1998. The farmer would be quoted an exchange rate for this transaction, which would be the spot rate on the day of the deal, adjusted for the interest rate difference between the currencies for the period between the deal date and the date on which the ecus were to be delivered: 1 July 1998. The farmer has no ecus to sell to the bank when he makes the agreement! When 1 July arrives and the bank wants the ecus the farmer has sold to it then the farmer takes out an opposite deal which allows him to close out the original one! He asks the bank to sell him the same number of ecu's he sold them. The exchange rate, however, will be different, it will be the spot rate on 1 July 1998.

The result of these transactions and their effect on 'fixing' the value of the area payment is shown in Example 14.1.

Example 14.1 Hedging ecu payments against the pound sterling

James Token farms 750 ha, of which 600 ha is arable for which there is an area payment due in 1998 of 102 000 ecu. On 17 October 1997 the economists are forecasting that the pound will strengthen significantly between now and 1 July 1998 when the ecu payment due will be converted into sterling. The prediction is that the exchange rate will be 1.380 ecu/£, which James believes is likely. If so the value of area payments would be reduced. He decides to hedge against this situation and for once the predictions are right! When James asks the bank to buy him ecus on 1 July 1998, to cover the ecus he sold forward to them in October 1997 it costs him:

17 October 1997 ecu/£ spot exchange rate	1.260
Adjustment for interest differential 17 October 1997 to 1 July 1998	0.001
Exchange rate at which the bank will buy ecus for delivery on 1 July 1998	1.261

Because he took out the hedge, the farmer will receive from the bank:

$$\frac{102\,000 \text{ ecu}}{1.261} = £80\,888.18$$

but because he has no ecus to sell to the bank, he buys them on 1 July 1998 at the spot ratio:

$$\frac{102\,000 \text{ ecu}}{1.380} = £73\,913.04$$

He therefore makes a 'currency profit' of £6975.14. The sterling he receives, however, for his area payments will be at the conversion rate on 1 July 1998, that is, 1.380 ecu/£. He has therefore 'lost' value for the actual area payment receipts compared to the area payment he would have received if the exchange rate had been the one prevailing on 17th October 1997. The loss is £6975.14. Thus he has received the amount of sterling that he expected in October of 1997 by using hedging. This demonstrates that by taking out a hedge of this nature a farmer would fix in advance the value of his area payment income. Whilst he ensures he does not 'lose' money overall, equally he eliminates any possible gain which might be made if rates moved in his favour.

Clearly this hedge could work the opposite way so that the pound weakened, resulting in a currency loss, but there would then be an 'actual gain' in the value of the area payment.

Example 14.1 assumes a 'perfect hedge' and no credit implications for the farmer in obtaining the bank's agreement to the hedging transaction. In practice the following points should be remembered.

(1) Because the green pound does not follow the ecu rate precisely but operates within a banding arrangement the 'hedge' will not be perfect.

(2) There is a risk associated with a hedging contract. If the pound weakened then there would be a currency loss which the farm business would have to be able to absorb. As a result the bank would consider what the risks were and, for a hedge with a maturity date inside 12 months, would expect the business to be able to absorb a loss of 10%, i.e. on a £100k hedge it would be £10k. For hedges over 12 months the 'marginal risk' is taken as 25%.

(3) Although the hedging deal closes on 1 July 1998 the cash for the area payment is not received until October 1998, which would mean that if there was a currency loss, it may require additional overdraft facilities to cover the position between July and the date the area payment was received.

The green money system has been maintained since 1969 in one form or another, because it protects farmers in strong currency countries from price cuts which should be associated with a revaluing currency. Removing this protective system is politically unacceptable in several EU countries and it has been a feature of the CAP that farmers in the EU should be protected against unjustified income losses caused purely by exchange rate alterations. This was the basis for freezing green rates on 23 June 1995 and allowing compensation payments (again complicating the system) for producers in member states who suffer income loss due to appreciable currency revaluations. These payments should not be linked to production and are part funded by the EU and part by national governments.

Economic and Monetary Union (EMU)

The start of EMU will mean that CAP transactions will be done in Euro (the new single currency), and there will be no need for agrimonetary arrangements with EMU member states. For non-EMU countries there will need to be a green rate to convert from Euro to national currencies, unless all payments are made to individual producers in Euro, without converting to national currencies. Since this seems unlikely, it is anticipated that an agrimonetary system along present lines will be perpetuated.

The future reform of the CAP

The cost of the CAP system continues to be a major issue. The reforms of 1992 have been expensive because of the commitment to compensate producers for lower market prices, while continuing to operate a market support system. The compensatory payments (arable aid) costs around 18 billion ecu (1996), although the framework for controlling production exists and cost reductions can more easily be achieved by decreasing direct aid (if the politicians are minded to do so), as the Integrated Agricultural Control System (IACS) is now well established and enforceable. Some financial savings have been made, however, by the avoidance of export refunds for surplus grain, with world grain prices being higher in 1995/1996 than for many years. Grain prices in 1997/1998 were, however, lower.

Apart from political pressure, agricultural spending is governed by rules agreed in 1988 which limit the rate of increase per year to a maximum of 74% of EU GDP. The question remains, can CAP meet the General Agreement on Tariffs and Trade (GATT) commitments without further production controls or serious price reductions? In 1995 and 1996 it has been possible for EU producers to stay within the GATT limits but only because of poor harvests in Europe coinciding with poor harvests in other parts of the world. The agreement has not really been tested in a year when there have been average or good harvests and the success of the MacSharry reforms continues to be measured within the framework of GATT, as well as financial and environmental limits. Future reform is suggested in the Agenda 2000 proposals (see p. 301).

The General Agreement on Tariffs and Trade (GATT) now the World Trade Organisation (WTO)

Signed in October 1947 in Geneva by the USA and 22 other countries, GATT now has 119 signatories (and more than 20 countries applying the rules) who control over 80% of world trade.

There have been eight rounds of GATT talks with the latest, starting in 1986, launched at Punta del Este (Uruguay round) being the first to include agricultural products. The framework for agreement proved difficult to work out but the Accord, based on the Dunkel Plan of December 1991, modified by the Blair House Agreement in November 1992 and the USA–EU negotiations in November and December 1993 was signed in Geneva on 15 December 1993 and ratified in Marrakesh on 12–15 April, 1994. The Agreement applied from 1995 and will be renegotiated in 1999/2000. GATT continued to operate (after the 1993 deal) for an interim period while its main functions have been carried out by the new World Trade Organisation (WTO) from 1 January 1995. The guiding principle of GATT is to liberate trade by negotiating multinational tariff reductions and the elimination of quotas and non-tariff barriers.

Basic principles

There are three basic principles:

(1) *Non-discrimination* or the most favoured nation treatment (MFN) for everyone, means that any concession agreed between any *two* signatories *must* be extended to everyone

(2) *Reciprocity* means that any nation gaining from a tariff reduction must reciprocate by making concessions
(3) *Transparency*, or tariffication, means that *all* non-tariff barriers to trade should be replaced by tariffs so their impact can be measured more easily, and reduced. It encapsulates four elements which are:
   (a) import exclusion (levies and other barriers)
   (b) export subsidies
   (c) internal product support
   (d) sanitary and phyto sanitary, non-tariff barriers

The main players in 1986–1994

The USA insisted on excluding agriculture from GATT negotiations in the 1950s but changed its stance by the 1960s, when the relatively 'young' European Union (EU) resisted attempts to reduce its system of protectionism. The EU's variable import levels and export subsidies on agricultural products could not have been maintained if agriculture had been subject to the same GATT rules as the manufacturing sector. The EU therefore refused to discuss the Common Agricultural Policy (CAP) in the Dillon (1960–1962), Kennedy (1964–1967) or Tokyo (1973–1979) rounds of GATT.

Against this background there were four main 'players' in the Uruguay round.

(1) *The USA*, whose initial position was for free trade but which faced internal pressures for maintaining import restrictions on farm products, farm subsidies and agricultural support.

(2) *The Cairns Group* (including Australia, New Zealand, Canada, Argentina, Brazil, Chile, Colombia, Malaysia, Indonesia, the Philippines, Thailand and Uruguay) favoured free trade because of the structure of their agriculture, the farming methods adopted, geographical isolation and current agricultural policies. They would gain from liberalised trade.

(3) *The European Union (EU)* remained (and remains) committed to supporting agriculture in order to ensure food security and rural population, but was under pressure from many sources because of the cost of the CAP, the poor focus of support, surpluses and international trade disruption.

(4) *Japan*, who favoured protection to ensure food supply, rural incomes and environmental policy.

The Uruguay round set out to prohibit the safeguards (Voluntary Export Restrictions (VER) and Orderly Market Access Agreements (OMAs)); prevent dumping; define and reduce subsidies on exports; extend the MFN principle to services; protect intellectual property rights; phase out quotas on textiles and prohibit local content and export performance rights imposed on inward investment to some countries.

Some quotas have been allowed to remain for reasons of national security, correction of balance of payments (where foreign reserves are endangered) to protect infant industries in developing countries and on agricultural products if justified by national government policies. The talks resulted in agreement on Trade in Services (GAS) and on Trade Related Aspects of Intellectual Property Rights (TRIPS) as well as those made in Agriculture which are discussed here.

The Agreement

Domestic support

A 20% reduction of the aggregate measure of support (AMS), based on 1986–1988 levels. Decoupled (direct aid) payments are in the 'green box' (being production neutral) and do not have to be reduced.

Aid payments linked to limiting production are in the 'blue box' and are exempted under agreed conditions but those which maintain market prices have an effect on trade, are in the 'yellow box' and are targets for reduction under the GATT agreement. Maximum allowed annual AMS levels are shown in Table 14.25. Some exceptions are made in developing countries.

Table 14.25 Maximum EU AMS for agricultural products

Base total AMS (1986–1987 to 1988–1989)	1995–1996	1996–1997	1997–1998	1998–1999	1999–2000	2000–2001
80 975	78 672	76 369	74 067	71 765	69 463	67 159

Because tariffs were high in the EU, even a 20% reduction over six years still leaves a high level of protection. The quality standards required under the minimum access commitments will also prevent significant quantities being imported. The import duty system (for those products already supported under EU schemes) also affords good protection for the producers, being 55% above intervention. From time to time, however, large quantities do come in because, for example, the barley price is used to set the duty for sorghum and there may be a disparity between prices for these commodities. The current AMS is well below the GATT commitment, because the CAP reform put most of the support payments in the 'green' and 'blue' boxes. There was also a fall in subsidised EU exports of cereals in 1995–1996, due to lower EU output and higher world prices caused by poor harvests, and more cereals were used in the EU in animal feed compounds. Intervention stocks thus fell from 33 million tonnes in 1993–1994 to 1.1 million tonnes by November 1996. It is possible that the 'undershoot' on subsidised exports, in 1995–1996, of about 17.5 million tonnes could be carried forward into 1996–1997, and future years except for 2000–2001. It may, however, be opposed by some EU members as contrary to the spirit of the agreement. It is likely that 1996–1997 subsidised exports will stay within GATT limits (at around 30 million tonnes) but for future years, given average harvests, set-aside may have to be increased or prices lowered to world levels, in order to prevent large stock accumulation.

Market access

The tariffication of support measures and a 36% average reduction in tariffs (based on 1986–1988 levels) with at least a 15% reduction on each product. Access for products where there was little or no trade before the agreement rose by 3% in 1995 and should reach 5% by the year 2000 (see Table 14.26).

Exports

A 36% reduction in export subsidy (achieved by reducing volume or value) and a 21% reduction in volume subsidised (see Table 14.27), of the average 1986–1990 levels (except where the average quantity exported in 1991–1992 exceeded that in 1986–1990 such as beef). This aimed at reducing trade distortions caused by 'dumping' of surplus products onto the world markets. The maximum allowed on EU cereal exports is shown in Tables 14.28 and 14.29.

Table 14.26 Market access commitments (tonnes)

	Current access	Minimum access
Wheat	—	300 000
Maize and sorghum	2 300 000	500 000
Brans	475 000	—
Sugar (cane or beet)	1 565 000	—
Mushrooms	62 660	—
Sweet potatoes and manioc	6 857 390	—
Bananas	2 000 000	—
Citrus fruit	45 000	—
Cheese	15 250	104 000
Butter	76 667	10 000
Skim milk powder	—	69 000
Live cattle (head)	194 000	—
Beef and veal	151 050	20 000
Pig meat	—	75 600
Live sheep and goats, sheep meat and goat meat	319 875	—
Poultry	—	29 000
Eggs	—	208 000

Table 14.27 Reductions in subsidised exports

'000t or '000 hl	Subsidised exports			Reduction %
	1991–1992	1995	2000	1991–1992/2000
Wheat and flour	20 255	19 118	13 436	–34
Feed grains	12 199	12 183	9 973	–18
Rice	173	177	145	–16
Rapeseed	—	97	79	—
Olive oil	112	143	117	—
Sugar	1 299	1 560	1 277	–2
Raw tobacco	205	190	128	–38
Fresh fruit and vegetables	1 039	1 108	907	–13
Processed fruit and vegetables	190	194	166	–13
Wine	2 954	2 980	2 433	–18
Alcohol	1 185	1 407	1 147	–3
Skimmed milk powder	254	297	243	–4
Butter and butteroil	273	447	366	—
Cheese	427	407	305	–29
Other milk products	1 208	1 161	938	–22
Beef and veal	1 324	1 119	817	–38
Pig meat	490	491	402	–18
Poultry meat	470	440	291	–38
Eggs	112	107	83	–26

Source: European Commission.

The Peace Clause

For the EU this meant that the CAP, as it existed in 1994, had been accepted by GATT members and should not be challenged. The CAP, therefore is (relatively) safe from external pressure until 2004 when the Peace Clause lapses and when the USA is likely to renew its pressure for CAP changes to reduce internal support and export subsidy even further. The implication is that the CAP is accepted by WTO members at present, but by 2005 it may not be. A new round of negotiations should have been completed by then, however, and it is certain that reducing agricultural support will be high on the agenda.

Impact on CAP

In 1993 when the GATT agreement was signed there was severe consternation in the EU that production, under the MacSharry reformed CAP of 1993, could not be contained within the agreed GATT limits (in the 1993–1996 period). The concern was mainly focused on exportable surpluses (especially cereals) of which only an agreed tonnage could be subsidised, the rest having to be sold out of the EU at world market prices. There were fears that this may necessitate increases in the annual set-aside percentage for cereals to ensure that production was kept within acceptable limits. The export situation is complicated by the interval between the base period (1986–1990) used for calculations, and the implementation period (1995–2000) in

Table 14.28 Maximum expenditure on EU cereal export subsidies

Million ecu	Base total AMS (1991–1992 to 1992–1993)	1995–1996	1996–1997	1997–1998	1998–1999	1999–2000	2000–2001
Wheat	2015.1	2309.0	2105.0	1901.4	1697.7	1493.2	1289.7
Coarse grain	1635.8	1605.7	1493.9	1382.2	1270.4	1158.6	1046.9

Table 14.29 Maximum volume of subsidised EU cereals

'000 Tonne	Base volume 1986–1987 to 1990–1991	Start volume 1991–1992	1995–1996	1996–1997	1997–1998	1998–1999	1999–2000	2000–2001
Wheat*	18 276.0	21 602.1	20 408.1	19 212.7	18 020.7	16 825.4	15 630.0	14 438.0
Coarse grain**	13 725.6	14 259.6	13 690.2	13 120.6	12 551.5	11 981.9	11 412.3	10 843.2

* Including wheat flour, durum and first-stage processed products.
** Including maize, barley, sorghum, rye and processed products.
Source: Euro Commission.

which production and trade patterns have altered. For butter and skim milk powder (SMP) the EU can increase its level of subsidised exports, due to large export levels occurring in the base period, but for cheese the reverse is true. To ease the problem of large immediate cuts (the front loading problem), six equal cut backs have to be implemented on the 1991–1992 figures while the year 2000 targets, of 64% of 1986–1990 expenditure and 79% of quantity, still have to be met.

Conditions in the world cereal markets have prevented the need for increased set-aside. Indeed shortages have led to prices being much higher than anticipated after the MacSharry reforms of the CAP, and after the GATT Agreement to reduce dumping of surpluses. Set-aside in the EU has been reduced and for the present it appears that EU producers can remain within the GATT limits but will probably have (given average conditions) to allow intervention stocks to increase, and be forced to cut quotas for sugar and milk toward the end of the millenium.

The need for CAP to reduce domestic support was easily met by MacSharry's 1992 reform package of reduced market support prices and because direct payments are deemed to be decoupled, in the green box and therefore allowed. The AMS in the base year (1986–1990) includes payments made under CAP to cereal and livestock producers. The AMS calculated for 1993–1996 does not include the area aid or headage payments for livestock.

Further switching of support from market prices to decoupled direct aid payments is allowable under GATT, provided that total payments do not exceed 1992 levels.

The agreement to allow greater access to the EU for imports will probably not appear to have much short-term impact, because of various derogations allowed, some safeguard provisions, and very high initial tariff levels. It is after the next WTO round, if a substantial tariff cut is agreed, that more impact will be felt by EU producers, users and consumers. Eventually, the variable import levels will be abolished and a fixed levy introduced which will be reduced over a given period.

There was also the problem of existing EU intervention stocks, which were substantial in 1993. The EU was allowed to dispose of these onto world markets (see Table 14.30). Surpluses are no longer given preferential treatment and have to be sold at world prices.

Early fears over the CAP – GATT incompatibility have not materialised because of poor cereal harvests in Europe, and the rest of the world generally, due to dry growing conditions. European Commission forecasts based on the expected growth of cereal production (less than 1% per annum) and increased EU usage (because of relatively cheaper cereals with lower market support prices) have not proven to be as wildly

Table 14.30 Additional subsidised exports from the EU (tonnes)

Wheat and wheat flour	8 116 000
Cheese	102 000
Other milk products (excl butter and SMP)	44 000
Beef and veal	362 000
Poultry	253 000
Eggs	16 000
Tobacco	156 000

Source: European Commission.

inaccurate as many commentators were predicting. The future, however, remains uncertain. The framework and formula for liberalising trade in the agricultural sector has been established and it is likely that there will be severe pressure for much greater steps to be taken in the next WTO round. The enlargement of the EU is also an unknown factor. How will the next round of talks include the new entrants?

Enlarging the EU

The last enlargement (1995), involving Austria, Finland and Sweden, went relatively smoothly since they are all rich, industrialised economies. Ten CEECs have European Agreements already signed or pending: Poland, Hungary, the Czech Republic and Slovakia (the Visegrad Four); Romania and Bulgaria, Latvia, Lithuania and Estonia, plus Slovenia. In theory they could all join within 15–20 years, but 2002 is often mentioned as the likely date when the next enlargement will occur. Turkey, Cyprus and Malta are also pursuing membership.

For agriculture these countries will pose a significant integration problem. Their production potential is formidable, although at present only Hungary, Bulgaria, and Estonia are net exporters. They produced an amount equivalent to 20% of the agricultural output of the EU last year and, given reasonable improvements in technology, could easily achieve 30%. It is clear therefore that the CAP will have to be adjusted before the CEECs enter, if a budgetary and production crisis is to be avoided; this reinforces the view that lower support will be offered to producers in future which will enable an easier (and cheaper) integration. The EU will probably concentrate on structural and technical assistance to these new members prior to accession, and on improving the access of current members to the new markets.

The cost of enlargement is almost impossible to estimate. It is difficult to determine the net contribution which new members might make, especially since it is likely that there will have been significant CAP reform by then, and another WTO round. The Commission suggests that the cost might be 10–15 billion ecu per year, compared with a predicted EU budget for 2000 of about 42 billion ecu. This would mean a 35% increase in the CAP budget, although this extra cost would be balanced by an extended market for current EU members, plus whatever budget contributions the new members make. It may not, therefore, be impossible to accommodate.

Likely outcome of future CAP reform: Agenda 2000

Whenever further serious reform occurs (probably before the next WTO round in 1999) it is most likely to be an attempt to pre-empt the WTO decision of reduced market price support and smaller compensatory (direct) payments. The Commission have already signalled their intentions by publishing the Agenda 2000 proposals which continue the trend towards detaching agriculture market support from maintenance of rural income while directing support to less favoured areas and small producers, especially in the southernmost states, to prevent rural degradation. (A further focus on the principle of modulation which aims support where it is most needed.) Environmentally acceptable practices will also be encouraged. More emphasis is being placed on a social market economy, which will allow greater freedom to farm in the long term, while providing a safety net of support for those who are not able to withstand free market competition. The continuing challenge is how to exploit modern agricultural technology in the most environmentally acceptable way (with cross compliance needed in order for producers to qualify for subsidies), without exacerbating rural depopulation by disadvantaging small land-owners in less favoured areas, or creating market instability.

An ideal situation to reach would be to eliminate intervention and export subsidies prior to the 1999/2000 WTO round. This seems an unlikely achievement in the present climate, especially with the uncertainty surrounding BSE, EU enlargement and the single currency. Agenda 2000 aims to address these issues. Its main objectives are:

- Increased competitiveness internally and externally (WTO compatible)
- Food safety and quality; obligations to consumer
- Fair standard of living and stability of rural incomes
- Integration of environmental goals into CAP
- Simplification of legislation
- Promotion of sustainable agriculture
- Alternative income opportunities for rural communities

Agenda 2000 sector proposals

Cereals

(1) Intervention price to be cut by 20% in 2000 from 119.19 ecu/t to 93.5 ecu/t
(2) A non-crop-specific area compensatory payment of 66 ecu/t (currently 54.34 ecu/t) to be paid on all cereals, oilseeds, linseed and voluntary set-aside (and potato starch)
(3) Durum wheat to receive an additional payment of 344.5 ecu/ha
(4) Set-aside reference rate to be 0% – small producers (less than 92 t) remain exempt even if set-aside is needed in the future. Voluntary set-aside remains, but extraordinary set-aside to be abolished
(5) Quality standards for intervention to be set annually as required by market conditions
(6) Monthly increments on intervention prices to be abolished
(7) Area aid can be reduced if market prices are higher than expected
(8) Silage maize to be excluded from support
(9) Reference yields for maize to be the same as for cereals
(10) Export levies may be used to ensure internal supplies but may not be explicitly linked to world prices
(11) Payment dates altered from 1 January to 31 March from 16 October to 31 December as at present. Sowings to be notified by 31 May
(12) Ceiling on payments which may be 80 000 ecu per farm (i.e. modulation) although it is possible that aid limits for big farms may be adjusted as follows:

100 000–200 000 ecu reduced by 20%
200 000 ecu+ reduced by 25%

Oilseeds

(1) Oilseeds area compensation to be reduced to 66 ecu/t from 186 ecu/t

Protein

(1) To receive a supplement on the cereal support of 6.5 ecu/t to keep them competitive with cereals

Dairy

(1) Quotas to be extended to 2006
(2) A 2% increase in quota with priority given to young farmers and farms in mountain areas
(3) Intervention prices for butter and SMP to be reduced by 15% from 2000–2001 to 2003–2004
(4) A Dairy Cow Premium (DCP) of 145 ecu/cow to be paid in compensation for a milk (support) price reduction and 70 ecu per cow in compensation for the beef price reduction
(5) The DCP to be paid by 100 ecu/cow direct from the EU, with 45 ecu/cow put into a 'pot' which is to be nationally distributed (this is expected to amount in total to 228 million ecu in 2000, rising up to 911 million ecu by 2003) per head or per ha of permanent pasture. The national (non-rotational) base area to be set for grassland which has been in grass for more than five years

Beef

(1) Intervention price to be cut by 30% between 2000 and 2002
(2) Suckler Cow Premium (SCP) to be increased up to 215 ecu/cow from 145 ecu/cow
(3) Beef Special Premium (BSP) for steers to be increased up to 232 ecu from 109 ecu (in two payments); and for bulls up to 368 ecu from 135 ecu (only one payment). A so called 'national envelope' which may work as follows: BSP for steers 170 ecu from 109 ecu (two payments) and for bulls up to 220 ecu from 135 ecu (in one payment); these to be paid direct from the EU. The rest of the payments will be determined by national government (with 1962 billion ecu available from 2002 in the EU; 266 million ecu in the UK)
(4) SCP rights to be reduced to 1995 or 1996 levels plus 3%
(5) BSP national ceiling remains but the limit of 90 payments per unit to be abolished
(6) Calf processing premium to be discontinued from 2002
(7) Extensification premium to be increased from 36 ecu to 100 ecu for farms having less than 1.4 livestock units (LSU)/ha

Agenda 2000 impact

These proposals, if adopted, are expected to:

- Eliminate cereal export subsidies
- Expand cereal exports
- Reduce cereal profits (slightly)
- Reduce oilseed profits (greatly)
- Increase the wheat area grown
- Raise beef intervention stocks

Criticism of these proposals are:

- They do not decouple support enough and there may be WTO problems
- CAP cost will rise possibly to 50 billion ecu per year by 2001
- They do not meet the challenge of new members
- They are deficient on environmental measures

The decision on the adoption of these proposals will be taken by the Council of Ministers (possibly in late 1998). For the present, discussions continue throughout the EU.

Renationalisation of CAP using the 'national envelope' also seems a dangerous policy since cohesion could be abandoned with the gap between rich and poor member states, in terms of agricultural and rural development, probably coming a long way behind short-term expediency and crisis policies in the political arena.

Key demands of future support will continue to be based on the following in an attempt to increase the competitiveness of EU farm products:

(1) The reduction of market support until EU prices are as close to world prices as possible
(2) Direct subsidies to compensate producers for reducing prices
(3) Further decoupling of support from production
(4) The redirecting of funds away from producers who are technically intensive and/or economically efficient, and towards those who are most in need: the Modulation policy
(5) Greater focus on support being limited to environmentally acceptable methods of production, referred to as cross-compliance

This will enable the EU to negotiate the next WTO round with almost all its agricultural support measures in the Green box (i.e. production neutral) and therefore not subject to reduction by international agreement. The EU will then have only to reduce the level of transitional, direct aid compensatory payments for producers forced to reduce output and 'suffer' lower market support prices, in order to ensure that CAP stays within budget limits, especially when the EU is enlarged by the Central and Eastern European Countries (CEECs), and also ensure that the EU can take advantage of export opportunities in a free trade environment.

It is clear, in summary, that the reforms will be driven by the need for cost control in the long term; environmental acceptance (cross-compliance) and, to a lesser extent, animal welfare issues; rural depopulation; further decoupling of support from production and the quest for so called modulation and cohesion within the member states.

15 Agricultural Marketing Organisation

Having looked briefly at the EU policy which affects UK agricultural marketing it is necessary to consider the ways in which it is possible to improve the marketing of farm products. The marketing chain, which has already been identified, for food between producer and consumer is complex. The possibility exists for the producer to become more involved personally in the activities of the chain, or to have an organisation become more involved on his behalf, so that better links are established with the customer and consumer, and profit potential is enhanced. The process of 'shortening' the chain and organising it better is referred to as integration. Integration is deemed to be vertical when farms and firms at different stages of the chain combine, and horizontal when farms or firms at the same stage combine. In some cases, where farms and firms combine, there are elements of both vertical and horizontal integration: a potato production and marketing group, for example, where farmers have joined together to grow potatoes and have purchased their own grading and packing plant to prepare the potatoes for sale through an established wholesale or retain outlet owned by another company.

The achievement of integration can be considered in four areas:

- Ownership
- Statute
- Co-operation
- Contract

Ownership

One firm or farm can own all the processes necessary in the chain to produce, process and distribute the food to the consumer, this would be an example of a fully integrated firm.

Pick-your-own enterprises come under this heading; they have increased in recent years for field scale fruit and vegetables and, while being important sources of cash income for individual farmers, are not very significant for the agricultural sector as a whole.

The problems with pick-your-own are:

(1) The area to grow: the size of the market has to be estimated and, even when this can be done with any degree of accuracy, bad weather when the crop is

ready can mean that none of the public arrive to pick so the farmer has to arrange alternative means of harvesting and disposal. Obviously there is a limit to the amount of flexibility possible so this may ultimately limit the size of the operation

(2) Parking facilities: providing a car park can be difficult for some farmers, especially near the picking area
(3) Providing cash for changing money
(4) Providing toilets and a play area for children
(5) Advertising
(6) Long hours and pilfering
(7) Pets, hygiene and the possible fire hazards

The advantages are:

(1) A good margin on cash sales
(2) Direct sales to the consumer
(3) Diversification of the business

There are also some regulations which have to be taken into account, these are weights and measures, public liability, the Workplace (Health, Safety and Welfare) Regulations and of course VAT.

Statute

The statutory marketing schemes which various UK governments have set up and imposed upon the farming community were an attempt to organise and integrate the marketing chain, so that improved efficiency would benefit all concerned.

Legislation and marketing organisations

The depression of the 1920–1930 period focused agricultural attention on marketing. The high costs of distribution gave rise to the idea that 'middlemen' were making unwarranted profits and that producers were not getting a fair deal because of the poor bargaining position in which they found themselves. In 1922 the Linlithgow Committee was set up to enquire into selling and distribution costs in agricultural marketing. The committee report in 1925 concluded that the marketing margin (the cost added to the product after it leaves the farm gate) was too wide, and channels of distribution were too complex, thus leading to high costs. They also thought that there were too many small mills and markets, and they recommended producer co-operation.

In 1927, as a means of improving efficiency in marketing, two Acts, on horticultural produce sales and auction markets, tried to improve the practices involved in distributing and selling. The realisation that there was a distinct lack of information concerning marketing and its environment led to the publication of the Ministry Orange Books which presented details on marketing and how to improve it, using examples from other countries.

The Agricultural Produce (Grading and Marketing) Act in 1928 gave government the power to inspect produce offered for sale and to standardise trades. None of these measures was, however, very significant in solving the problems and so a further major move was made in 1931 with an Agricultural Marketing Act which enabled the setting up of marketing schemes proposed by producers themselves. This enabled the statutory organisation of marketing if two-thirds of producers were in favour. The legislation was deemed necessary as attempts to improve marketing by voluntary co-operative measures had not succeeded, because farmers not participating could grain from the co-operative effort without taking part or making any commitment to the venture.

The powers of the 1931 Act meant that no outsiders could operate in the schemes without the approval of the scheme organisers, hence total control went with the schemes. The schemes were an attempt to strengthen the position of producers, in relation to the large buyers, and to stabilise prices in order that the Cobweb Cycle did not continue to make market conditions chaotic as they had been during the Great Depression of 1922–1933; for example, the price index of agricultural products 1927–1929 = 100, 1933 = 77.

The 1931 Act, despite its powers, failed to produce any action on import controls. The UK became an unwilling dumping ground for excessive production resulting from other countries and it was not until November 1931 that import duty was imposed on some horticultural products. In March and November 1932, respectively, the Import Duties Act and the Ottawa Agreements Act resulted in regulation of imported quantities of food and food products and the 1932 Wheat Act, which included a subsidy scheme, extended the policy of protecting home production.

The 1933 Agriculture Act, based upon reactions of producers and the recommendations of re-organisation commissions preparing the milk and pig schemes, provided powers to promote and conduct producer cooperation. Certainly greater powers to control quantities sold by each producer, based upon previous sales, and to restrict imports were provided.

By 1938 there were schemes for hops, milk, potatoes, pigs and bacon. Hops, milk and potatoes arose out of producer schemes, while pigs and bacon were developed as a result of proposals from a re-organisation commission, since the problem faced here was to control imports rather than simply stabilise prices and productivity. As well as the boards there were commissions for wheat, sugar and livestock. These were set up by the Government where no schemes were proposed by producers. The commissions, comprised of independent members, administered subsidies and determined other aids necessary for the Government.

During the war only the milk and hops boards continued while all production and utilisation was controlled by the Ministry of Food.

In 1946 the Lucas Committee was set up to report on the working of Agricultural Marketing Acts. Reporting in 1947 the committee observed that producers had made little use of the Acts to increase efficiency. It was recommended that a two-tier organisation be instituted having commodity commissions, with full trading and regulatory powers, and producer marketing boards, with no power to directly interrupt distribution and marketing. The 1947 Agriculture Act, in its endeavour to provide '... proper remuneration and living conditions for farmers and workers in agriculture and an adequate return on capital invested in the industry...', laid down the annual review of agricultural commodities: wheat, barley, oats, rye, potatoes, sugarbeet, fatstock, milk and eggs, plus wool in 1950. It also provided deficiency payments which guaranteed prices for producers in these commodities.

Further policy measures led to some of the pre-war boards reappearing (potatoes 1955), while new boards were formed for wool, tomatoes and cucumbers (1950), MMB Northern Ireland (1955), sugar (1956 – this took over from the abolished Sugar Commission), eggs (1957), and the Pig Industry Development Authority (PIDA, 1957), which disposed of the assets of the Pigs and Bacon Marketing Board and the Bacon Development Board. In 1965 the Home Grown Cereals Authority (HGCA) was established to improve the marketing of cereals grown in the UK.

Some of these organisations are no longer in existence. In 1967 the Meat and Livestock Commission (MLC) was established and took over the functions of the Pig Industry Development Authority. In 1971 the Egg Marketing Board ceased to operate and the Eggs Authority was set up, this ceased to trade in 1986. Also disbanded in 1971 was the AMDEC (Agricultural Marketing Development Executive Committee), and in 1982 the CCAHC (Central Council for Agricultural and Horticultural Cooperation) was absorbed into the newly formed Food from Britain organisation, while the HMB is now English Hops Ltd. In 1994 the Milk Marketing Board was disbanded in favour of several producer cooperatives, the largest of these being Milk Marque, and the Potato Marketing Scheme ends in 1997.

Marketing organisations

1932 Hops Marketing Board (HMB) (now English Hops Ltd)
1933 Wheat Commission
1933 Pigs and Bacon Marketing Board
1933 Milk Marketing Board (MMB) (disbanded 1994)
1935 Sugar Commission
1935 Bacon Development Board
1936 British Sugar Corporation (BSC) (1)
1937 Livestock Commission
1950 Wool Marketing Board (WMB)
1950 Tomato and Cucumber Marketing Board
1954 Fatstock Marketing Corporation (FMC) (2)
1955 Potato Marketing Board (PMB) (disbanded 1997)
1955 MMB (Northern Ireland) (disbanded 1994)
1956 Sugar Board
1957 Egg Marketing Board (EMB)
1957 Pig Industry Development Authority (PIDA) (3)
1960 Horticultural Marketing Council
1965 Home Grown Cereals Authority (HGCA)
1967 Meat and Livestock Commission (MLC)

1971 Eggs Authority (Ceased trading in 1986)
1982 Food from Britain

(1) Taken over by Beresfords 1983
(2) Set up by Farmers Unions as an ordinary limited company because a scheme to set up a meat board was rejected by the Government
(3) Took over from operations of the Pigs and Bacon Marketing Board and the Bacon Development Board

The powers of the boards

The boards themselves were given varying powers, as can be seen in Table 15.1. By far the most powerful was the MMB, which totally controlled milk supply in the UK, either directly or by licence. The other boards existed to regulate supplies, improve quality of products, advise the Government of the needs of farmers, allocate receipts to producers, and act as agents for the Intervention Board of the Ministry of Agriculture (MAFF) in administering various EU or UK-based support or disease eradication or control schemes.

Though termed marketing boards these organisations really concentrated their efforts mainly on supply regulation (and development) rather than on expanding sales through exporting or creating new products. Increased pressure of over-supply made the need to export more imperative and, during the late 1980s, greater effort, especially by the MMB, was expended in this direction.

The commissions and authorities are not trading organisations like the marketing boards and they do not buy or handle goods, though they could be given powers to

Table 15.1 The main powers of the marketing boards

	Milk	Potatoes	Hops	Wool
Area	UK (five regions)	GB	England	UK
Monopolistic supply control	yes; and price discrimination	no; supply regulated by riddle and quota	yes; by quota	none
Advisory committees	powerful	consultative	consultative	consultative
Grading/quality control	yes	yes	no	yes
Processing and distribution	yes	yes	no processing	yes
Allocation of total receipts	yes	no; only for a part of the crop in time of surplus	no	yes

do so under extreme market conditions. These organisations are mainly concerned with improving marketing, by carrying out research and disseminating information, but they do act as agents for the Intervention Board for Agricultural Products and the Ministry of Agriculture (MAFF), in administering the schemes where necessary.

Co-operation

The increased interest in co-operative production and marketing since the early 1970s reflects the recognition by farmers that they needed to improve performance in both areas. Many farmers were either unable or unwilling to do more as individuals to improve marketing, but they recognised that co-operation offered a means of resolving some difficulties which they were facing. The grant aid provided by governments has also given much practical encouragement for co-operative ventures.

Development of the agricultural co-operative movement

1844 The Rochdale Pioneers, a group of weavers and industrial workers, decided to set up shops and stores to provide household goods, because wages were poor and the existing shops tended to be owned by the mill owners who charged high prices or provided goods in lieu of wages. This first co-operative (although there is evidence to suggest that corn mills were jointly owned in England up to a century before the pioneers started) was therefore a consumer co-operative.

1860 The co-operatives combined to form a united body, the Cooperative Wholesale Society (CWS). The aim of this society was to supply goods at reasonable prices. Profits were not retained but given back to members in proportion to value of goods purchased, 'dividend'. (All sales were made virtually at cost price.)

1867 The Agricultural and Horticultural Association was founded in Manchester by Edward Greening. It was quite successful and was registered as a company because the new Industrial and Provident Society Act had no provision for this type of organisation. It ensured good quality and kept prices down but was wound up during World War I.

1869 Cumberland farmers tired of uncertain fertiliser supplies founded a society: Aspatria Agricultural Cooperative Society, the oldest in the UK. Several others arose spontaneously among farmers, mainly in cheese and dairy production, under Danish influence.

1894 Plunkett, a Member of Parliament in Ireland, decided that the owners of small farms would benefit from co-operation and he managed to found the Irish Agricultural Organisation Society (IAOS). In 1919 The Plunkett Foundation for Cooperative Education was set up.

1902 The Agricultural Organisation Society (AOS) began in England; this was a central association of co-operatives supported by well-known farmers. A few years later the Scottish one (SAOS) was formed. There were 19 co-operatives with 4814 members at the time.

1902–1910 Development of many co-operatives on individual initiative, e.g. dairies, egg packers, wool, fruit. These organisations recognised the need for central buying to compete with large suppliers.

1912 Farmers Central Trading Board (FCTB) began; this was a buying agency which nursed co-operatives along.

1913 The CWS opened its own Agricultural Department.

1918 The FCTB was transformed into the Agricultural Wholesale Society as a central buying organisation for co-operatives. This organisation had high hopes but the management was poor. The war boom was followed by the depression and, caught with high stocks, it was forced into liquidation in 1924.

1924 Because of the collapse of the FCTB the prestige of the AOS suffered and it was wound up on the understanding that the NFU (set up just pre-war) would look after co-operatives, but they never really did this and there was no real leadership in co-operatives.

The situation in 1920: 1558 agricultural co-operatives, with 200 000 members and £17 million turnover.

The situation in 1929: 230 agricultural co-operatives, with 67 000 members and £9.8 million turnover.

In the industrial depression the cooperatives tried to survive and maintain prices by collectively refusing to sell at poor prices; this failed. Because of this depressed situation, Government action resulted in the Agricultural Marketing Acts of 1931 and 1933.

The 1931 Act provided legislation for the setting up of compulsory Marketing Boards, provided that producers were in favour.

The 1933 Act authorised the control of imports for any commodity under a Marketing Board.

When the MMB was set up there was no need for co-operatives in the milk sector. During this time (1930–1931) the need for a central organisation again became very apparent but there was confusion over who should have power in this organisation:

(1) The NFU?
(2) A co-operative union?
(3) Should it be independent?

1936 The Agricultural Cooperative Managers' Association was started. Its members exchanged views and information designed to represent the business side of the movement and has remained a practical and powerful influence in the affairs of agricultural co-operation in England.

1938 A Commonwealth Conference was held on agricultural co-operation. The members met frequently until 1949 but A.W. Ashby was the only economist interested.

The war years stimulated general cooperation within the population and in the agricultural sector

1945 Membership of co-operatives was 90 000 with an annual turnover of £20.5 million.

1946 The Agricultural Cooperative Association (ACA) was formed. The NFU were reluctant to approve it but did so with the proviso that it must not be political.

1954 The Farmers' Central Organisation (FCO) was set up by the NFU which duplicated the work of ACA. Soon afterwards the NFU Development Company set up a subsidiary, the Farmers' Central Trading Company Ltd, this was supposed to be a national wholesale purchasing agency of agricultural co-operatives. Neither of these received much support from the existing co-operative movement.

1956 Sanity! ACA and FCO were amalgamated to form the Agricultural Central Cooperative Association (ACCA). This was closely linked to the NFU (who had members on the working parties). This organisation presented a much better image to the public, although the problems with the NFU were not quite settled. The larger farmers found that by 'grouping', buying in bulk and for cash, they could get better terms than co-operatives. The result was the formation of farmers' groups, usually closely associated with the NFU.

1961 Agricultural Central Trading Ltd was formed to bring the groups together. Groups had to buy from societies and had to help one another; this was not successful and after a few months co-operatives went their own way. ACT was left to service the groups.

1962 Agricultural Marketing Development Executive Committee (AMDEC) was set up to look into projects to improve agricultural marketing.

1964 ACCA asked Dr Joseph Knapp (USA) to prepare a report on its problems. He recommended that ACCA should be fully independent with increased financial support from its members, and that the government should help with the financing of research and education.

1966 June: the first part of the recommendations was implemented and the ACCA once more became ACA, and a working party of ACA and NFU began looking into the other problems.

1967 The 1967 Agricultural Act provided for the establishment of a Central Council for Agricultural and Horticultural Cooperation.

1971 AMDEC disbanded and grants for production groups abolished.

1972 ACA was converted into ACMS (Agricultural Cooperation and Marketing Services) which included NFU representatives.

1975 Grants were re-instated for 'approved' forage groups.

1980 The NFU withdrew its interests from ACMS and formed its own Marketing section. Also a Central Coordinating Council 'Super-council' was proposed by Detta O'Cathain to correlate all UK co-operative marketing activities. No immediate agreement was reached on funding or the constitution of this body.

1982 The Food from Britain campaign was started and £14 million from Government funds was provided to promote British exports of food.

Principles of co-operation

A co-operative in its purest form can be defined as 'a business organisation, organised for and funded and controlled by its member patrons to provide services and/or goods for its member patrons on an at-cost basis'. The principles upon which the co-operative movement based its organisations originally were:

(1) Profits. Any money surplus generated during a trading year may be returned to the members in proportion to the amount of business they have done with the co-operative in that year, or could be retained and added to general-reserves, or retained and allocated to members
(2) One man, one vote
(3) Open membership
(4) Do not seek to force competition
(5) Financing may be in the form of equity investment (shares) from members or as revolving loans from members usually in proportion to usage

These general principles are still the basis of co-operation, although the application of the principles depends upon the constitution of the organisation.

There are three basic constitutions which a co-operative organisation can choose to adopt.

(1) *Partnership*. Must have less than 20 members and be registered under the Partnerships Act 1908. These tend to be small and close-knit groups of farmers.
    (a) Taxation is the same as that of a normal trading partnership, in that the individuals who are partners are charged income tax on the preceding year's taxable income
    (b) Normal relief is given against the profits for capital allowances
    (c) The taxable income is added to the individuals' income and charged at the relevant marginal rate

(2) *Co-operative society (industrial and provident society).* Must be registered under the Industrial and Provident Societies Act, and be based upon the following principles:
   (a) Any trading surpluses made during the year are returned to members on a patronage basis, retained in general reserves, or in allocated reserves
   (b) One man, one vote, not capital contribution shareholding, under most circumstances
   (c) Open membership
   (d) Twenty thousand pound limit on share capital
   (e) Limited return of dividend on loan and capital share linked to 'safe' rates
   (f) Ninety percent of the members must be involved in agricultural businesses but it is permissible to trade with non-members up to 10% of turnover without losing the benefits of mutuality, this to be agreed with the Inland Revenue
   (g) A minimum of 2 members

(3) *Co-operative company.* Must be registered under the Companies Act, and if the articles meet the requirements mentioned above, the company is, to all intents and purposes, the same as an industrial and provident society. The difference is that there is no limit on share capital or dividend

For both an industrial and provident society and a cooperative company

(1) Members are only liable for debts up to the nominal value of the issued shares (limited liability)
(2) Controlled by a board of directors (some elected, some appointed) and can enter into contracts, borrow money (though the amount is controlled by the articles of association and directors' borrowing powers) and own assets
(3) Taxable profits are calculated after deduction of capital allowances and rebates, interest and dividends. Thus only retained profits are taxed

Production groups or marketing groups

(1) Can be either an industrial and provident society or a co-operative company; that is, they are incorporated
(2) Can be a partnership with less than 20 members; that is, they are unincorporated
(3) Can use the principle of mutual trading where funds contributed to the organisation in excess of costs continue to belong to the members as contributors and not as investors or shareholders. In this case the group is not taxed and the surplus is only taxed when it is returned to the members

The financing of co-operative organisations

(1) Organisations and share capital: companies pay more of the operating costs in proportion to their patronage. In some cases (grain storage groups) there is a revolving fund for making capital repayments, which operates over five or ten years. A reserve fund is usually retained within the organisation for 'insurance' or expansion purposes
(2) Loan capital: industrial and provident society or company
(3) Grants
(4) Other commercial sources: in the past some co-operatives have been able to borrow from banks at 1% below normal commercial rates.

Machinery syndicates

These have been organised by small groups of farmers in order to share the capital and running costs of large specialised machinery. The most widely used form of machinery sharing is now machinery rings, many of whom have memberships greater than 250.

(1) Each member is responsible for a share of plant and machinery and maintains that responsibility
(2) Each member can claim part of the available capital allowances for each machine

There is no constraint on who joins a cooperative. For example, a current trading partnership or limited company can do so provided they agree to be bound by the rules.

Types of co-operative

(1) *Production.* Less in evidence now since the emphasis on grant aid moved to marketing in the mid 1970s. Essentially producers worked together to produce silage, potatoes, or dairy products, by sharing men, machinery, and knowledge. In this way the timing of operations could be (but was not always) speeded up, provided the whole operation was organised properly from the outset.
(2) *Buying groups.* These are usually formed (and may only be loose, informal organisations) to purchase large quantities of fertiliser or feed in order to obtain big discounts off the retail price.
(3) *Marketing (and storage).* These two quite often go together since, for example, the ability to store grain allows flexibility in when to sell so that the best deal can be achieved. There are many such groups throughout the country. Most will allow members to choose when to sell their own grain by operating a pool system. If the member selects the harvest pool for barley, he will receive the average price obtained for barley sold in the August period. If he chooses the

'long pool' he will receive the average price for barley sold between harvest and June and, since this may cause cash flow problems for some farmers, will be given some money 'on account' through the year with a balancing payment in July, when the final value is known.

There are many examples of livestock marketing groups which operate successfully on behalf of farmer members by having a manager who carefully selects stock to suit the particular needs of different abattoirs, in order to achieve highest prices for stock.

Marketing groups are the most abundant because:

(a) Farmers are traditionally less skilled at marketing and more prepared to allow someone else to do this job for them.
(b) Grant aid has been more generous and long lasting for marketing groups, especially joint ventures between the food processing industry and farmers, because most improvements were possible in this area of farm business management.

(4) *Requisites provision*. These are usually displayed in a shop or warehouse which members and the general public can buy from. (e.g. West Midland Farmers).
(5) *Service*. These exist to provide a service to members; for example, rabbit or pigeon clearance.

The benefits of co-operation

(1) Provides improved links with the buyers and therefore it is possible to determine clearly what quantity, quality, time and place will best serve the needs of both parties.

(2) Creates improved bargaining power. This can be achieved by producers 'bulking-up' the crop and livestock products so that they can offer larger tonnages of grain or potatoes. Thus in local areas farmers can negotiate a better deal since buyers only have to go to one source for all their supplies and are prepared to pay a premium for that.

(3) The fixed costs associated with the capital investment required to improve production and marketing are shared between the members of the group. Membership of a grain storage group, for example, will mean that storage can be reserved, this will cost about £35 per tonne, as a capital contribution, plus an annual payment of £6–8 per tonne. Compared with the cost of erecting on-farm storage shown in Table 16.10, the co-operative is very cheap. Another area of importance is in machinery syndication where the capital cost of mobile pea viners or forage harvesting equipment is almost prohibitive for the individual, but groups can afford to buy the benefits of speed and efficiency which these 'Titans' bring.

(4) Co-ordination of production is much easier within a group. Members can be informed of particular market needs regarding quality in, for example, a variety or breed which will achieve premium prices. It is also possible to ensure that not too much of one particular variety is grown, and that a sequence of

produce reaches the market throughout the season, creating orderly marketing while providing the producer members with the best advice available.

(5) Co-operatives can afford to pay for the services of the best managers. This means that they are being advised by specialists who are in constant touch with 'the markets' and any changing conditions. The manager can increase the security of an outlet for the group members by determining specific market requirements demanded by each (a task which most farmers are unable to accomplish) and then encouraging his members to produce to specification.

(6) The profit potential will be increased, although this is not necessarily the *raison d'être* of co-operatives, and the profits previously taken by 'middlemen' will in part be enjoyed even if extra costs associated with an increased involvement in processing are incurred.

(7) Financial assistance used to be offered in the form of grant aid and banks have in the past provided loans at preferential rates. Some grant aid is now offered through Objective 5(b) of the European Social Fund's scheme for group ventures.

To receive grant aid, projects must be eligible and approved under the current guidelines.

(8) Members grouped together often find that there is a 'spin-off' benefit from discussing improved techniques as a byproduct of other formal or informal co-operative meetings.

In summary it should be emphasised that requisite groups are convenient and cause price stabilisation; production groups allow access to the benefits of large equipment, through the investment of relatively small capital sums by individuals and the pooling of labour resources. Marketing groups improve market negotiating power while allowing the members access to specialist managers which they would not otherwise afford, and the benefits of pool pricing can be gained. The importance of co-operatives in EU countries is shown in Table 15.2.

Some problem areas in co-operation

(1) *Finance.* Many co-operatives, especially in the early 1960s, suffered from poor financial control and hence got into difficulties
(2) *Management.* The wrong staff were selected to manage the co-operatives, many of them not having much experience of corporate business
(3) *Commitment.* Initially co-operatives did not demand 100% commitment from their members in terms of trading. Thus it was difficult to estimate how much trade a co-operative would do in a year.
(4) *Poor produce.* Because members were not expected to market all their products through the co-operative, there arose a tendency to off-load poor-quality products through the co-operative while disposing of the best to the

Table 15.2 Agricultural products sold through co-operatives in 1992 (%)

	UK	Belgium	Denmark	Germany	Greece	Spain	France	Ireland	Italy	Luxembourg	The Netherlands
Pig meat	20.0	15	95	23	3	5	80	55	10	35	25
Beef and veal	5.1	1	59	25	2	6	30	9	13	25	16
Poultry meat	0.2	—	—	—	20	8	30	20	—	—	21
Eggs	18.0	—	59	—	3	18	25	—	8	—	17
Milk	4.0	65	92	56	20	16	50	98	33	81	82
Sugar beet	0.4	—	—	—	—	20	16	—	—	—	63
Cereals	19.0	28	48	—	49	17	70	26	20	79	65
All fruit	38.0	62	90	30	51	32	45	14	40	10	78
All vegetables	19.0	72	90	60	12	15	35	8	13	—	70

Source: European Commission Directorate General for Agriculture.

Table 15.3 Agricultural products sold under contracts concluded in advance in 1992

	UK	Belgium	Denmark	Germany	Greece	Spain	France	Ireland	Italy	Luxembourg	The Netherlands	Portugal
Pig meat	70	55	—	—	5	—	30	—	—	15	35	—
Beef and veal	1	90	—	—	3	—	35	—	—	—	85	—
Poultry meat	95	90	—	—	15	—	50	90	—	—	90	—
Eggs	70	70	—	—	10	—	20	30	—	—	50	—
Milk	98	—	—	99	30	—	1	10	—	—	90	—
Sugar beet	100	100	100	100	100	100	100	100	100	—	100	—
Cereals	30	22	65	—	3	—	10	10	—	—	50	—
All fruit	60	98	100	90	85	—	90	100	—	—	85	95
All vegetables	—	—	—	—	100	—	—	—	100	—	—	100

Source: European Commission Directorate General for Agriculture.

highest bidder. Co-ops thus got a bad reputation and were placed in a weak bargaining position.

Contracts: as a means of integration

In 1967 the Barker Committee reported on contracts in agriculture. There are many different contractual agreements made in agricultural production and marketing and only the main types are outlined here.

(1) *Market specifying* (e.g. bacon contracts, tomatoes)
   (a) The producer controls production but the market outlet is controlled by the buyer
   (b) The price is not usually specified
   (c) The quantity and quality of the produce are not usually specified. If quantity and quality are specified the producer must meet these requirements and the buyer's risk is diminished. Also the quality standards accepted are varied from year to year
   (d) The risk of crop failure remains with the farmer and there is no chance of speculative price gains
(2) *Production management* (e.g. peas, heavy hogs and sugar beet)
   (a) There is a partial transfer of management where the buyer provides some inputs and advice
   (b) The buyer can plan ahead and control the harvesting time and quality of the produce
   (c) There is little of value for farmers in this type of contract except good advice and some credit on inputs
(3) *Resource provision (wage or integration)* (e.g. broilers, horticultural crops and pigs). The buyer provides a market outlet, helps with production decisions and supplies resources, the danger to the producer is the borrowed money element with the collateral being the promise to supply.
   (a) Wage: the buyer retains ownership of the resources and pays the farmer for his labour input (e.g. heifer rearing)
   (b) Integration: this varies from the provision of say seed to a day-to-day involvement in the management of the farm
   (c) The processor lends working capital to the farmer who finds the fixed capital. The danger is that the contract may not be renewed for long enough to cover the long-term costs of the fixed capital

Table 15.3 shows the importance of contractual arrangements for selected products in the EU.

16 Grain Marketing

Why bother with grain marketing?

Farmers grow cereals to make a profit. It is vital to get the crop husbandry right in order to achieve technical efficiency but of equal importance is getting the best price to achieve economic efficiency. So much attention to detail is paid in getting the husbandry right that it is essential not to waste or devalue this effort by placing so little emphasis on disposing of the grain efficiently and effectively. Once the 'lid' is lifted off the UK grain market situation, however, it is soon apparent that to get the marketing right requires as much expertise, organisation and 'know-how' as does growing the crop. Many farmers, realising this, have opted out of 'marketing' their grain, preferring to sell by their established method, which may or may not be the best. With cost pressure increasing in the last few years, however, and market prices falling in 1997, farmers are finding it difficult to maintain their profit margins. This cost pressure has caused a 'rethink' of marketing performance by many farmers with the result that much more attention is being focused on improved marketing organisation and pre-sale planning for grain disposals.

UK/EU/world situation

Grain markets have been greatly affected by the break-up of the USSR and the ensuing chaos surrounding land ownership and therefore production and distribution of grain for national use. Exports from the USSR and the Central and Eastern European countries (CEECs), former Soviet satellites, virtually ceased. Imports also fell, mainly due to lack of funding caused by currency weaknesses fuelling inflation in the CEECs.

The UK market for cereals is very much influenced by supply and demand on the world markets and it is necessary to take a brief look at production, consumption, stocks and trade in grain throughout the world.

Table 16.1 shows that yields rose steadily up to 1986, while the area grown fell. Production doubled from 846 million tonnes in 1961 to 1662 million tonnes in 1991, with consumption going from 832 million tonnes in 1961 to 1595 million tonnes in 1986 and falling back to an estimated 1373 million tonnes in 1996. World stocks of grain fell fairly steadily until 1975–1976, rose again until 1985–1986 and fell again between 1971–1972 and 1975–1976. Since then stocks have fluctuated dramatically to 1996. During this period consumption exceeded production in 1965–1966, 1970–1971, 1980–1981, 1991–1992 and 1993–1994 to 1996, and at their lowest point stocks still amounted to 142 million tonnes! This has of course created a downward pres-

Table 16.1 World wheat and coarse grain: production, utilisation, stocks and trade

Year	Production			Use	Trade	Stocks	Stocks as a percentage of use
	Area (m/ha)	Yield (t/ha)	Tonnes (m/t)				
1960/1	646	1.31	846	832	72	199	24.0
1965/6	659	1.40	921	955	110	142	14.9
1970/1	671	1.64	1 102	1 143	109	165	14.4
1975/6	717	1.74	1 246	1 236	150	148	12.0
1980/1	723	2.00	1 447	1 463	215	190	13.0
1985/6	710	2.31	1 662	1 595	180	308	19.3
1990/1	546	2.58	1 417	1 387	173	273	19.6
1991/2	541	2.50	1 352	1 376	199	246	17.8
1992/3	542	2.63	1 421	1 378	192	285	20.6
1993/4	532	2.53	1 353	1 385	176	251	18.1
1994/5	527	2.63	1 390	1 405	182	248	17.7
1996	—	—	1 329	1 373	188	175	12.7

sure on prices and in fact world prices in 1979 were higher than in 1989.

Stocks have been fairly stable in the last five years, falling slightly in 1995 and more dramatically in 1996 following another poor world harvest. Consequently, prices reached high levels on world markets, although in the UK prices have been very low in the 1997–1998 marketing year due to oversupply and reduced market support.

The influence which the UK, and indeed even the EU, has on the world market is put into context when it is noted that of the 1.6 billion tonnes of grain produced world wide only about 12%, 175 million tonnes, is produced in the EU, and between 20 and 23 million tonnes, about 1.5%, in the UK.

The UK and EU become more influential, however, when trade is considered, as the EU tends to have a grain surplus each year. The UK usually has an exportable surplus each year ranging from 6 to 9 million tonnes. Greater emphasis is therefore focused on grain quality, the GATT export limits and the impact of set-aside.

Perusal of the export and import figures for grain shown in Tables 16.2 and 16.3 reveals that the majority of grain which is traded in the world moves from the West, mainly Canada, the USA, the EU and Argentina, to the East, especially the USSR, China, Egypt and Japan. Thus the movement of grain is a very important factor in the world grain market and has a considerable influence on prices, especially when there are hold-ups in shipping grain because of bad weather or transport strikes.

The farmer's decision on the movement of spot prices, and indeed forward and future prices, will be determined by their knowledge of world markets and the factors which influence UK grain prices.

Factors which influence grain market prices

Many factors have an influence on grain market prices in the UK but they do not all have a significant influence every year. UK grain prices are related to world prices.

Table 16.2 World wheat and coarse grain exports ('000 t)

	1985/6	1988/9	1992/3	1993/4	1994/5	1995/6*
USA	60 790	101 500	90 100	74 800	147 200	158 800
Canada	22 560	17 850	24 600	22 900	29 400	25 100
Argentina	15 832	6 430	13 500	12 100	19 300	16 400
Australia	21 045	13 050	12 100	15 000	28 700	18 800
EU-12	23 361	33 300	30 100	25 400	31 700	24 800
E. Europe	4 882	4 685	—	—	—	—
USSR	1 000	500	—	—	—	—
China	6 750	5 000	21 400	23 300	7 600	2 200
S. Africa	1 300	2 000	4 700	3 400	6 800	1 800
Thailand	4 000	1 950	—	—	—	—
Others	3 482	7 800	82 600	75 100	1 900	28 500
World total	168 314	195 785	279 100	252 000	272 600	274 600

* Forecast.

Those factors which affect world prices will therefore influence UK prices and must be considered. As climatic, economic and political conditions alter, then different factors will become relatively more important in influencing prices, causing them to be forced up or down as buyers and sellers interpret the changing market conditions which they face.

Climatic factors

These can have a very marked effect on prices in a world, UK or even regional context. When the weather is bad at drilling, heading or harvesting, yields will be decreased and scarcity of grain will push prices up as buyers bid amongst themselves for a smaller amount of grain (or vice versa if weather conditions are good).

Bad winter weather (in the USA, the UK and the EU) can also prevent grain transportation both nationally and internationally, which will cause short-term rises in grain prices.

The weather conditions will also have an influence on grain quality with, for example, a damp, wet harvest causing an increase in sprouted grain, which will reduce the quality and value of milling samples; while dry conditions will cause pinched and shrivelled grains. Dry conditions in many parts of the world in 1994–1995 and 1995–1996 caused poor harvests and saw grain stocks fall to the lowest level for 15 years.

Weather abnormalities within a year can also cause market price fluctuations, as occurred in 1978 when the UK experienced a very cold spell of weather in March and April. The result was that livestock farmers could not turn all their animals out to grass, as usual, and a much higher demand for feed concentrates by farmers led to the feed compounders having to go back into the market and buy more grain in

Table 16.3 World wheat and coarse grain imports ('000 t)

	1985/6	1988/9	1991/2	1992/3	1993/4	1994/5	1995/6*
EU-12	7 850	6 200	4 500	3 600	3 700	9 100	10 900
E. Europe	8 865	7 842	2 400	6 800	3 900	4 500	2 700
Japan	26 940	27 350	26 700	27 000	26 700	45 600	45 000
China	7 300	16 500	16 900	7 400	6 900	18 300	23 100
USSR (former)	28 700	25 700	39 800	27 700	19 100	6 700	6 800
Egypt	8 500	7 400	7 200	7 700	7 500	12 000	11 300
Algeria	3 490	6 200	3 500	4 000	4 200	8 300	6 400
Morocco	2 179	1 525	1 600	2 900	3 600	1 200	2 800
Nigeria	1 100	250	—	800	1 200	—	—
Tunisia	1 087	1 815	600	700	700	1 800	1 700
Libya	400	600	1 500	1 000	1 000	1 700	1 500
Brazil	4 501	1 265	6 200	6 800	6 600	9 600	8 400
Israel	1 700	1 800	1 600	1 900	1 900	3 600	3 300
Saudi Arabia	7 875	4 800	6 600	5 400	5 000	9 200	10 200
Venezuela	1 775	2 500	1 700	2 200	2 400	3 200	3 000
Mexico	1 860	2 100	7 000	5 600	5 400	12 700	12 300
Malaysia	1 785	2 150	1 100	1 000	1 200	2 800	3 200
Korean Republic	7 125	9 600	10 300	9 800	10 200	21 000	20 500
Pakistan	1 562	2 200	2 300	2 800	1 700	2 300	2 000
Iraq	2 300	4 100	1 900	400	1 000	1 000	1 000
Iran	3 900	4 070	3 500	4 200	4 200	4 700	6 100
Others and unspecified	37 520	62 068	52 200	149 400	133 900	93 000	92 400
World total	168 314	195 785	199 100	279 100	252 000	273 000	274 600

* Estimated.

order to compound their feeds. This raised cereal prices quite dramatically during that period, while in previous months of the same year prices had been low due to the large harvest.

Economic factors

Supply

Physical yields and quality will be influenced by the weather but the area which farmers grow tends to be influenced by profitability (unless land retirement is forced upon them), that is, as a crop becomes more profitable more is grown. Profitability is influenced by price/cost ratios and technical efficiency. Technological advances in controlling diseases have led to an increase in cereal crop yields, while high guaranteed prices have raised profitability. The area of cereals grown in the UK therefore rose until set-aside was imposed in the MacSharry reforms of the CAP.

Supplies increased dramatically during the 1970s and 1980s but the recent attempts at surplus control have caused a reduction of cereal area in the UK and EU.

Demand

The demand for cereals mainly from developing countries has risen but has not kept pace with the rate of increase in supplies at current prices. Thus it has not been possible to dispose of surpluses produced in the UK, the EU Canada and, especially, in the USA even at 'world' prices in countries where direct and indirect demand for grain is increasing.

Forecasters predict an increased demand for cereals, as population grows, and attempts to control production in the EU and North America will be adjusted to match demand under the GATT/WTO deliberations. Even so the picture by 2020 may be bleak with a shortfall of 150 m/t predicted, assuming no increase in cereal area. Shortfalls are forecast in Sub-Saharan Africa, the Middle East, Southern Asia, the Far East and South and Central America, while Europe, North America and Oceania have surpluses.

Grain prices: are farmers always right?

Traders in any market can influence prices by acting upon their expectations of future trends. Farmers are no exception to this and when enough farmers feel that grain prices will fall next month they tend to sell grain in order not to suffer from lower prices. The increased market supplies tend to result in lower prices (assuming demand does not increase). When they believe prices are going to rise next month they hold onto grain, creating a short-term 'scarcity' which may cause prices to rise. It is often felt, therefore, that farmers are able to predict market trends correctly! There are many other factors which influence grain prices besides the supply and demand situation from UK farms.

Guaranteed prices

The prices of grain in the EU are fixed by the agreement of the Council of Ministers in consultation with the Commission and since the mid 1970s the basic intervention price was good enough (until the 1992 reforms), in continental Europe, to encourage excess production, especially of feed barley and wheat. The UK green pound overvaluation situation, until June 1980, resulted in UK cereal intervention prices being unattractive to farmers. Intervention therefore played little part in the UK marketing scene until the 1980 and 1981 harvest years. The EU surpluses of grain are normally sold into Eastern Europe and North Africa (see Home Grown Cereals Authority, Marketing Note), while UK exports go mainly to Poland, Spain, Algeria, Italy, Belgium/Luxembourg and the USSR. EU exports to third countries are subsidised by export restitutions, which means that the European Commission pays exporters a subsidy from the FEOGA fund to sell grain outside the EU. These subsidies are necessary in order that relatively 'high priced' grain from the EU can compete on the world market with 'cheaper' grain in and from other countries, and to ensure that transport costs will also be covered. Intra-EU trade receives no export restitutions but, due to the vagaries of the green currency situation which created MCAs, it was sometimes worthwhile for countries in the EU to trade with a partner when a hefty MCA subsidy, now abolished, was available (see Table 13.4).

The EU regime for cereals influences the market price in that it fixes a 'bottom' to the market. Although market prices may go some way below the offered intervention level guaranteed it still acts as a basic price from which to work. The effect on the market price is that price peaks which occurred fairly regularly in the past have been smoothed out since the instigation of the EU regime. However, in some years price peaks have been appearing towards the end of the season as grain sellers anticipate the export market 'opening-up', with EU restitutions being paid to encourage exports. Thus grain is held in store in order to benefit from the anticipated export drive, which will in itself tend to raise market prices (as less grain will be available within the EU and exporters will be prepared to pay higher prices to obtain grain for resale abroad).

Transport costs and transportation

The trends in haulage costs are shown in Table 16.4 from the HGCA survey.

The average for all journeys was 53 miles in 1985 and 62 miles in 1996. Excess charges for delays in loading or unloading, of more than two hours, averaged £17.80/hour in 1996.

The cost of transporting grain has a major influence on the price of grain in the UK. As UK and world transport costs rise, then profit margins (on grain imported from the USA and Canada, and on the grain exported from UK) begin to fall. The result is that unless prices rise, less grain moves about the world.

There are also physical problems with the movement of grain.

In the USA the locks on the Mississippi tend to slow down the passage of grain to the East for export. In the UK facilities at many of our major grain exporting ports are not modern enough to allow large ships in, and only coasters of up to 20 000 tonnes are able to work between many of these ports and the continent of Europe. If the ship size could be increased, then costs/tonne would decrease, UK grain would be more competitive abroad and our exporters would be able to dispose of more grain each year.

Table 16.4 GB haulage costs (excluding VAT)

Distance (miles)	1980 (£/t)	1985 (£/t)	1990 (£/t)	1995 (£/t)	1996 (£/t)
10	2.71	3.18	3.21	3.53	3.51
20	3.04	3.50	3.50	3.83	3.86
50	4.10	4.45	4.39	4.73	4.90
100	5.51	6.04	5.86	6.24	6.65
125	6.05	6.84	6.59	7.00	7.52
150	6.59	7.63	7.33	7.75	8.39

Political factors

The influence of world affairs is felt in the grain markets. President Carter's embargo on grain sales to the USSR in autumn 1980 caused grain prices to fall in anticipation of the fact that there would be more grain available in the West. The much publicised intention of President Reagan to re-open the grain trade with the USSR raised market prices, but the possibility still existed that the USSR might interfere in Poland, in which case grain from the USA would once more be prevented from going to the USSR. In fact, on 29 December 1981 President Reagan suspended talks on a long-term grain supply agreement between the USA and the USSR because martial law was imposed in Poland. Events such as these in present times rapidly alter prices. Other political events, such as strikes, can also have an effect on grain prices. The lorry drivers' strike in the UK in 1979–1980 had some peculiar effects. Grain could not be moved (except by private lorry owners) in the UK for the period of the strike. Grain exports from the UK were therefore prevented and foreign buyers bought grain elsewhere, leaving an increased grain supply in the UK, which caused depressed prices after the strike. During the strike, because grain was difficult to obtain, feed compounders were willing to pay increased prices for any grain which was moving, so that a short-term price rise occurred for those sellers who were able to deliver grain.

The agreements which are made between governments to provide aid for certain foreign countries also have an influence on grain prices. UK aid to Poland, for example, provided Poland with the means (in foreign exchange) to purchase large quantities of grain from the UK. This meant that prices remained stronger as grain supplies were shorter in Western Europe.

Psychological factors

The farmer is always right in his assessment of grain market prices! If farmers collectively think that grain prices will rise after Christmas they tend to store grain, so that very little moves onto the market, supplies are scarce and prices rise. After Christmas when prices are high the farmers congratulate themselves on being correct. If enough farmers feel that prices will be low after Christmas so that storage will be uneconomical, then they tend to sell grain before Christmas; grain supplies onto the market are high and prices are forced downward by the buyers. Thus once more farmers are correct. (Perhaps it is advisable to do exactly the opposite to the action of most farmers!) There are several other instances of buyer or seller expectations actually having an influence on market prices and it is well known that panic buying or selling can soon set in amongst traders of any commodity, often based upon no more than a well directed rumour – or indeed real information. The prospect of an invasion of Poland is one example; the United States Department of Agriculture (USDA) forecast of a tremendous harvest failure in Russia or China will increase grain prices; the expectation of export markets opening up can also have the same effect.

Several factors which affect grain prices have been identified but it is not always possible to determine which ones will be influential in any one year. The farmer who can anticipate these factors, or who allows a specialist to market for him, will usually remain ahead of the rest of the farming community, who continue to market their own grain without specialised knowledge.

The next section explains the choices available to grain producers and shows in detail how they can decide which one will best suit their purpose in achieving that extra £1 or £2 per tonne.

Marketing choices and decision making

There is no blueprint or formula which can be produced to ensure that farmers will always make the correct marketing decisions and achieve the highest prices possible for their crops and livestock. It is essential, however, for farmers to be aware of the marketing methods and opportunities available to them. There are four methods which a farmer can use to dispose of his grain:

(1) Spot market sales
(2) Forward contracts
(3) Futures contracts
(4) Intervention sales

The spot market

Spot market prices for grain are determined by negotiation between buyers and sellers at the time of sale. These prices will be a reflection of grain quality and the market supply and demand situation. As a result of these influences on the market, spot prices fluctuate as the market supply and demand alters, or is expected to alter in the near future.

Conditions of a spot contract

When a farmer has sold grain on a spot basis, it means that he has entered into a contract with a buyer to deliver grain to that buyer. The price of the grain will be the price offered and accepted on the day upon which the agreement takes place, and will be related to the quality of the grain, probably on a formula basis. There are no statutory quality standards set down for grain sold on a spot basis. There are, however, guidelines for wheat decided by United Kingdom Agricultural Supply Trade Association (UKASTA) and the National Farmers' Union. Increased emphasis on intervention quality standards has meant that more grain is traded with these standards in mind (see Table 14.11). Other factors, such as specific gravity (bushel) weight), weed and miscellaneous impurities, sprouting, heat damaged or shrivelled grains, will all be assessed by the buyer before he finally pays for the grain. The spot contract will state the time and place of collection of grain, the quantity of grain and the quality standards expected.

Many spot sales are agreed verbally but the same details are agreed and a contract still exists between buyer and seller. Normally grain will be collected within 7–28 days of the contract being agreed. Payment for the grain will usually reach the seller before the end of the month following the month of collection. Payment may occur more quickly than this, however, and it will depend upon the relationship between buyer and seller. For example, if the buyer wants more grain from the same source later, he may pay very quickly for the first lot. In some cases where farmers sell grain to a feed compounder and then buy concentrate feeds back, no money will change hands, except a small figure to balance the transaction, reflecting the difference in value between grain sold and feed purchased (this is referred to as 'contra trading'). All spot contract prices are agreed on the basis that the buyer pays for transport costs and they are said to be quoted on an 'ex-farm' basis.

Who makes spot contracts?

Spot contracts can be made between two farmers, between farmer and miller, or farmer and compounder. Indeed anyone involved in trading grain can be party to a spot contract.

Advantages and disadvantages of spot market selling

The spot market is susceptible to sudden price fluctuations. Those farmers who are interested in marketing may feel that this enables them to choose their times to sell, and provides an opportunity for obtaining windfall profits as a result of sudden shortages. Any farmer who grows or stores grain to sell by this method, anticipating that spot prices will rise at some future date, bears the risk of prices falling. For farmers who need to sell grain at harvest time, because of lack of storage space or cash flow problems, this method leaves little room for manoeuvre and they may be forced to accept a low market price on the day when they have to sell.

The frequent variations in supply and demand which give rise to the price fluctuations in the spot market created the need for a method of selling where future grain supplies are agreed at guaranteed prices. Thus the forward contract emerged as a method of disposing of grain.

Forward contracts

Forward contracts are agreements between traders to buy and sell grain at a fixed price on a specified date in the future. The prices quoted for forward contracts relate to grain delivered on a specific date between one and six (or more) months into the future; an element of forward planning is therefore brought into the marketing of grain.

Conditions of a forward contract

Prices for grain are quoted for several months ahead. In November, prices for wheat and barley are quoted for December, January, February and possibly March. This means that the farmer selling grain can be certain of receiving a set price for the grain he is holding, once he has signed a forward contract to sell. In December 1996 the spot price for feed wheat was £90/tonne and the forward prices for wheat delivered in January, March or May 1997 were £91.30, £93.50 and £95.10 per tonne, respectively. This means that the farmer could have had £90 per tonne in December 1996 (spot price) or £95.10 per tonne in May 1997 on a forward contract. The forward contract will specify price per tonne for the grain when it is delivered, the quantity of grain, the required quality of grain, and the time of collection. All the precise details will be determined by negotiation between buyer and seller.

Payment will usually reach the seller before the end of the month following delivery and may in practice be much quicker than this.

Transport costs are paid by the buyer so that forward prices are quoted on an 'ex-farm' basis.

Who makes a forward contract?

Forward contracts may be agreed between two farmers, a farmer and a miller or compounder, or by any party involved in trading grain.

Advantages and disadvantages of forward contracts

A forward contract allows more stability to be brought into marketing grain, because the farmer knows how much money will be obtained for his grain, several months before it is actually delivered. The problem is that the agreed quantity of grain must be delivered at the specified quality and there are frequent disagreements between the buyer and the seller over quality. The farmer usually has to rely on the quality testing of the buyer, as it is unlikely that farmers will have testing equipment which is acceptable to the buyer, or else he has to have an independent test carried out. The latter situation is to be recommended but it is naturally a time-consuming process and any grain actually tested in this way may not necessarily prove to be representative of the whole lot of grain which will be tested when it arrives at the buyer's premises, so disagreements may still occur. There is no fool-proof method of overcoming this invidious situation. The farmer is in a weak bargaining position and should pay more attention to sampling and testing procedures. With regard to quality therefore (and to a lesser extent quantity) forward contracts create an inflexible position for the seller and cancellation of a forward contract is difficult without incurring a penalty.

The possibility of default on a forward contract and the desirability of being able to cancel a contract without incurring any penalty led to the development of a more flexible contract, called the futures contract.

Futures market contracts

In the era of colonial trading, the unpredictable nature of buying and selling and the problem of organising shipping led to the development of meeting places or coffee houses where merchants met to transact business and exchange information. As trading became centralised in these places they developed into commodity exchanges and clearing houses which organised and guaranteed standardised grades and contracts for the various commodities. These contracts became known as futures contracts. Futures contracts and futures markets exist for many commodities from grain to precious metals, but the principles of trading remain the same in each case. Grain in particular lends itself to standardisation and the rapid fluctuations in market prices, which can be stabilised by futures trading, made grain futures a worthwhile proposition. Many grain futures markets have existed since the 1870s in the UK, culminating in the present London market (There was also a smaller futures market in Liverpool, now no longer operating.) The agricultural commodities, wheat, barley and potatoes, are traded on the LIFFE exchange and control is exerted through the Recognised Investment Exchange (RII) and the London Clearing House (LCH). The operation is overseen by the Securities and Investment Board (SIB), so trading is regulated by independent bodies.

The purpose of the futures market

(1) It brings together buyers and sellers to a central market
(2) It allows buyers and sellers to follow market fluctuations as they occur and the exact value of grain is known at any time
(3) It enables the value of grain to be protected over a period of time on behalf of farmers, merchants and compounders, all of whom are in the process of holding grain before use or sale. Thus a degree of stability is brought to grain market prices
(4) A standard contract is created which specifies quality, quantity and time of delivery, again imposing some control on the market

The aim of the futures market is to offer an alternative for a trader to market his grain in place of, or in addition to, the physical market.

Some terms used in futures trading

Clearing house: The organisation which registers, monitors and guarantees dealings on the futures markets and carries out the financial settlements

Position: To have contracts bought or sold on the futures market

Bull: Trader who expects prices to rise

Bear: Trader who expects prices to fall

Long: A person who has *bought* futures contracts and has not yet closed out

Short: A person who has *sold* futures contracts and has not yet closed out

Closing-out (liquidation): Buying or selling to offset an existing contract

Deposit (initial margin): Money paid through the broker to the clearing house as a guarantee of fulfilment of a futures contract (brokers receive interest on this money)

Margin money (margin calls): The demand through your broker for additional *cash* when the price moves against you

Basis: the difference between the futures market price and the local market price

Stop-loss (stops): A price set by the trader at which his broker will close out his contract. It is a means by which losses can be limited in adverse market conditions

Tender: Delivery to fulfil a futures contract

Hedging: The sale (or purchase) of futures contracts in order to gain protection against price decreases (or rises) in the future. The 'hedger' has grain or potatoes, which he will want to physically sell (or buy) later in the year

Delivery points: Stores designated by the futures exchanges where the physical product (grain or potatoes) can be delivered to fulfil a futures contract

Carrying charge: Refers to storage charges, insurance, interest charges and any other costs.

GTC (good till cancelled): Refers to an instruction by the trader, to his broker, to sell or buy at a predetermined price. The instruction stands until the trader changes it

Cover:

(a) Purchase of futures to close out a short position
(b) Security provided to cover price fluctuations on established contracts

Last trading day: The last day in a delivery month when trading can take place, contracts outstanding after this time must be settled by delivery

Overbought/oversold: A view that the price has risen/fallen too much in relation to the factors which influence prices

Liquid market: A market where buying and selling is easy because there are many traders willing to participate with large quantities at small price differentials

Spread: The simultaneous purchase and sale of two different contracts in anticipation of a beneficial change in their relative prices

Straddle: The purchase of product in one market and its sale in another market at the same time

Switching: Buying a product for delivery in one month while simultaneously selling the same product in another delivery month on the same exchange

Premium: The cost of purchasing an option

Strike price: The price at which a commodity can be bought or sold when exercising an option

Use of the futures market

The futures market can be used for:

- Hedging
- Speculation
- Procurement

Hedging

The farmer is really only interested in hedging, which means that he is intent on guaranteeing a known price for the grain which is already in store or growing, for the month when he expects to sell the grain physically.

The person engaged in hedging is *not gambling* in the market, as the speculator does, because any change in futures prices will be offset by a parallel change in physical grain prices.

Speculation

The speculator is not interested in using the physical grain; he enters the market to buy grain futures at a price which he considers to be relatively low, in anticipation of the grain futures price rising (in the few days or weeks), when he will sell the futures contracts and make a profit. The speculator therefore bears the risk of losing money if he makes the wrong decision and has to sell cheaply (after his initial buying operation), but accepts this risk in order to place himself in a position to gain any profits that are to be made if the price of his futures contracts rise. Traders who engage in speculation on the futures market are essential because, without them to accept any losses when prices fall or take profits when prices rise, the price stabilising functions of the market could not be maintained.

Procurement

Some traders use the futures market to assure themselves of physical supplies of grain at a known price on a specified date. Between 5 and 10% of grain traded on the futures market is physically moved between buyer and seller.

Futures contract conditions

Futures contracts are a form of security. They are traded like any other security and may be bought and sold many times before the agreed delivery date.

(1) A futures sale contract consists of a promise to deliver a specified quantity and quality of grain at a specified calendar date in the future
(2) The types of grain traded on the futures market are feed barley and feed wheat of EU origin; no milling wheat or malting barley is concerned

(3) The months in which futures contracts mature (that is grain is physically delivered or the contract closed-out) are September, November, January, March and May for wheat and barley, with July being an extra trading month for wheat
(4) The grain is traded in 100 tonne lots
(5) Prices are quoted in pounds and pence with a minimum movement accepted of five pence per tonne. (So the smallest increase or decrease in price which can be made at a time is 5 p per tonne.)
(6) Grades and quality standards for the grain are determined by LIFFE (see Table 14.11)

Example 16.1 How does the futures contract work?

In October 1995 I Will Grabbit (farmer) decided to get involved in a futures contract to hedge against grain prices being very low in September 1996.

Step 1

October 1995 – Grabbit sells 100 tonnes feed wheat (through his broker) on a futures contract for delivery in September 1996 at £99/tonne. This price is crucial in that the farmer must decide what price he needs to make the grain profitable to grow. In this case he has decided that £99 per tonne is adequate. He instructs his broker to sell the grain at £99/tonne or more. If the buyers in the market (brokers working on behalf of buyers) are not willing to offer the required price the farmer will either have to accept a lower price or not engage in futures trading.

Step 2

Assuming that a buyer has paid £99/tonne the farmer now has two options open to him, one of which he must take in (or before) early September 1996.

Option 1: Physical delivery of grain to fulfil the contract

Farmer Grabbit delivers 100 tonnes of feed wheat to an authorised grain futures store (there are about 130 stores authorised on the UK mainland) of his own choice, that is, the nearest one to his farm.

Problems associated with this option

(1) The farmer has to pay transport costs
(2) The farmer has to get the storekeeper to agree to accept the grain
(3) The quality standards for futures delivery of grain must be met or no Warrant of Entitlement will be obtained (see next section for details on a Warrant of Entitlement)
(4) The farmer will be breaking his established trading links with local buyers which may be detrimental to him. Provided that the grain quality is correct, the grain will be accepted by the agreed store. Payment for the grain will be received by the farmer within seven days of delivery once the Warrant of Entitlement is sent to his broker
(5) The deliverer pays the first 14 days rent (approximately £3.50/lot/day)

Procedure for delivery of grain against futures sales contracts

(1) The deliverer must first find a store registered for futures delivery, usually the one nearest to his premises.

(2) Get the storekeeper to agree to take the grain and sign the Warrant of Entitlement.
(3) This 'Warrant' ensures that he will maintain the grain at the required standard and that the tonnage mentioned has been received into store. The Warrant is thus 'title' to the goods and as such is negotiable and transferable.
(4) In order to 'tender' against a futures contract the grain must be in the store by the 'notice day'.
(5) First notice day for each futures quoted position is eight days before the beginning of the delivery month, i.e. to delivery against November futures, first 'notice day' is 24 October or the previous business day if this is not a working day. In January, it is the second business day after Christmas day. To deliver against the November contract the grain must be in store and the Warrant signed by the storekeeper between 24 October and 23 November (provided that this is a business day).
(6) Although the first notice day is eight days prior to the tender month, any working day after that until the 23rd of the delivery month constitutes a tender day.
(7) Last notice day is 23rd of the tender month, except in July, when it is the 14th.
(8) The deliverer of the grain must then inform the futures broker of his intentions to deliver. The broker completes and presents a 'tender form' to the clearing house by 1100 hours on the day prior to the tender day.
(9) Cash payment is due seven days from date of tender.
(10) The Warrant of Entitlement which has been signed by the storekeeper and returned to the deliverer of the grain is sent to the futures broker
(11) The futures member is unable to obtain payment for the grain until the Warrant has been lodged for 24 hours.
(12) Within the terms of the contract the delivered grain is entitled to remain in store for 14 days rent free to the buyer. This means that the deliverer gives (i.e. pays for) 14 days rent after tendering. This rent is not more than 3.5 p per tonne per day at present (£3.50 per day per 100 tonne load) but is subject to periodic review.

The costs and risks, however, involved in this option are rarely worthwhile, so most traders go for Option 2.

Option 2: 'Closing-out' the contract

The farmer instructs his broker to buy 100 tonnes feed wheat on the futures market for September 1996 delivery. *This transaction must occur at the same time that the farmer physically sells his grain* or the farmer is leaving himself open to any price fluctuations which may occur in either physical grain prices or futures contract prices (i.e. he becomes a speculator). The buying transaction involves the same amount of grain (100 tonnes) as the farmer originally sold but this time trading is in the opposite direction. This transaction means that Farmer Grabbit can fulfil his original futures contract when he agreed to deliver 100 tonnes of feed wheat in September 1996, while giving himself the opportunity to sell wheat locally. This course of action is the only one open to farmers involved in futures trading whose grain does not meet futures quality standards or who live a long distance away from the nearest futures store (in excess of 20 miles). By selling locally of course, on a spot or forward basis, the farmer will incur no transport costs.

The precise timing of the 'closing-out' operation (that is, buying a contract after initially selling as Farmer Grabbit is doing) will depend upon the farmer's situation *but* at the latest it must be done in the first week of the month in which the initial contract matures (in Farmer Grabbit's case the first week in September 1996), otherwise it may not be possible to buy a futures contract for that month as quotations may not be offered by market traders when the maturity date of a contract is very close.

Since Farmer Grabbit is short of storage space and has cash flow problems he wishes to sell wheat in early September. On 5 September he physically sells 100 tonnes of wheat to a local merchant for delivery in seven days at £95/tonne. At the same time he must ask his broker to *buy* 100 tonnes wheat for September 1996 delivery at £95/tonne (or as close to £95/tonne as possible). He makes a profit of £3.70/t on the futures transactions (sold at £99/t, bought at £95/t) having paid £30 brokerage. Added to the cash price of £95/t this means a net income of £98.70/t.

Why does the farmer bother with a futures contract?

(1) By using this method the farmer is protected from market price fluctuations. In Example 16.1, where the September market price falls, the farmer gains by £3.70/tonne because he would have received only £95/tonne for the wheat on physical sale on 5 September 1996 without his futures transaction.

(2) In Example 16.2, when the September market price rises, the farmer could have gained up to £4.00/tonne by having no futures transaction, waiting until 5 September and then selling wheat on a spot basis locally. It is, however, always easier to determine the best course of action with hindsight, when it is too late to do anything about it, and in October 1995 he could not know precisely that spot prices would rise. He is therefore unable to plan his cash flow with any certainty unless he uses futures trading.

(3) Having obtained the initial price of £99 on a futures sales, in October 1995, he is guaranteed to receive a figure very close to that price in September 1996. So 11 months before physically selling the grain it is possible for him to plan his cash flow.

Other crucial factors that may change

The assumption that the local spot market prices and the futures market price will be the same at the time he arrives at his 'closing-out' operation (in September 1996) is not entirely accurate. There is usually a difference between the futures price and the local spot market price (the futures is usually higher) of between £2 and £4. This reflects the difference in the quality of the grain and allows for an element of transport costs which must be borne by the futures seller who delivers grain. Differences may also be caused by the vagaries of demand and supply in local areas. Nevertheless, prices in the two markets do tend to move in unison and are reasonably close when 'closing-out' becomes necessary.

The use of the futures market therefore allows forward planning, on a price basis, once the farmer has obtained the initial price which he felt was necessary to ensure profitability (£99 in the example).

Example 16.2 How a farmer becomes involved in a futures contract, and how the contract works

(1) In October 1995 the farmer decides upon the price he thinks he will need in September/October 1996 for grain to make the enterprise profitable.

(2) Assuming that he decides on £100/tonne for feed barley, he instructs a broker to act on his

behalf in the futures market to sell a 100 tonne lot of barley for £100/tonne or more. If no buyer is willing to pay £100/tonne for barley he will either have to accept a lower price or not become involved in a futures contract at all.

(3) Assuming there is a buyer willing to pay £100/tonne, the broker makes the sale of 100 tonnes of barley for delivery in, say, September 1996. At this time of sale, the broker will call for a cash deposit of around 10% of the value of the contract (possibly less), which is deposited with the clearing house. This deposit is required and additional 'margin' money may be called for to maintain the solvency of the market when the price moves against a person holding a futures contract. If margin money was not requested from contract holders incurring losses the clearing house may not be able to cover profits made by other traders. The margin money is calculated on a daily basis at 12.30 pm. Calls must be met within 24 hours or contracts may be cancelled or closed-out by the broker.

The farmer in the example, having sold 100 tonnes of barley in October 1995, for September 1996 delivery at £100/tonne, will deposit £500. If the value of September 1996 barley rises to £101/tonne in November 1995, he will have to deposit another £100 with the clearing house via his broker.

This money is held and will be paid to the contract broker as a 'profit' when the contract is closed. So 'losses' occur for a contract holder when the price of a sold tonnage of grain rises.

If the value of September 1996 barley remains stable then no additional margins will be called for, and if the value falls then the farmer will receive a profit at the time the contract is closed, along with his initial deposit.

The figure of £98.60/tonne (see Fig. 16.1) is close to that of £100.00/tonne which the farmer initially wanted in October 1995.

In this example the futures price rose between October 1995 and September 1996; this reflected an increase of similar proportions in the spot (physical) market, and a futures trading loss of £400 was incurred plus interest costs and brokerage charged (see Fig. 16.2). The farmer would have been better off not having a futures contract, in this example, but this fact only becomes apparent with hindsight. The crucial point to note is that he received a price for the barley of £98.60/tonne, which is very close to the amount of £100/tonne which he budgeted for in October 1995.

Margin calls can be minimised using 'stop losses' or avoided altogether using 'options'.

Options in futures trading

The farmer can purchase an option when trading futures in order to prevent himself from having to pay cash margin money should the market price move against him.

If he sells 100 tonne (1 lot) of feed barley for September delivery at an agreed price, by purchasing this option the farmer avoids the payment of margin calls. He would only exercise this option if the market price for September rises; otherwise he would not use it.

The cost of an option of this type is negotiated between the two parties concerned, usually the broker and his client, and could be of the order of £2/tonne for grain or £15 to £25/tonne for potatoes.

Fig. 16.1 Expenses incurred by the farmer who sells 100 tonnes of feed barley for September 1996 delivery

Futures transaction	£	Physical transaction	£
October 1995: sells 100 t September 1996: barley at £100/tonne	10 000		
September 1996: buys 100 t September 1996: barley at £104/tonne	−10 400	Sells 100 t barley locally at £104/tonne	10 400
Loss on futures (represents total margin calls)	−400		

	£
Income from physical sale	10 400
Loss on futures trading in this example	−400
Costs of trading:	
Brokers and registration fee	30
Interest (from cash flow in Fig. 16.2)	56.36
	9 913.64 or £99.14/tonne

The contract unit for wheat and barley options is 100 tonnes (20 tonnes for potatoes) and the minimum price movement 5 p/tonne (10 p/tonne for potatoes). Strike price increments are £1/tonne (£5/tonne for potatoes) and trading months are the same as for the underlying futures (April and November for potatoes).

The option expires at close of business on the second Thursday in the month preceding the trading month for wheat and barley (for potatoes it is on the fifteenth business day of the month preceding the trading month). Instructions to exercise or not exercise the option must be given to the clearing house one hour after the option expiry. Options expiring in-the-money are automatically exercised.

The value of options

In-the-Money

The premium 'charged' for an option is set where buyers and sellers agree in a free market.

Call option

If the current price of a futures contract is £110/t then the right to buy (call option) wheat futures at £105/t will be more valuable than the right to buy at £115/t. This option has an intrinsic value of £5 and is in-the-money by £5.

Fig. 16.2 Futures trading cash flow

	Oct	Nov	Dec	Jan	Feb	Mar	Apr	May	June	July	Aug	Sept
Futures price (£/t)	100	101	101	101.50	102	102.25	103	103	103	103.50	104	104
Receipts:												
Sales (physical)												+10 400
Deposits returned												+500
Profits												—
Sub-total												+10 900
Expenses:												
Deposit	500											
Margins		100		50	50	25	75			50	50	
Brokerage	30											
Sub-total	530	100		50	50	25	75			50	50	
Net cash flow (NCF)	530	100		50	50	25	75			50	50	+10 900
Cumulative (NCF)	530	630	630	680	730	755	830	830	830	880	930	+9 970
Interest at 8%		3.53	4.22	4.25	4.61	4.98	5.17	5.71	5.74	5.78	6.16	6.2
Cumulative interest		3.53	7.75	12.00	16.61	21.59	26.76	32.47	38.21	44.00	50.16	56.36
Cumulative balance	530	633.53	637.75	692.0	746.61	776.59	856.76	862.47	868.22	924.00	930.16	+9 913.64

Put option
The right to sell (put option) wheat futures at £105/t when the current price of wheat futures is £110/t is not profitable. Whereas the right to sell at £115/t when the underlying futures price is £110/t is in-the-money £5, and has an intrinsic value of £5.

At-the-money
A call or put option whose strike price is equal to (or approximately equal to) the price of the underlying futures. If the November wheat price is £114.80 then the nearest option strike price (the at-the-money) will be £115/t because option price intervals are £1/t for wheat.

Out-of-the-money
A call or put option that has no intrinsic value because, for a call option, it has a strike price above the underlying futures; or as a put option it has a strike price below the current underlying futures.

If the November wheat is currently trading at £105/t, a put option with a strike price of £101 will be £4 out-of-the-money. It is possible by selling the futures contract, to receive £105/t; and by buying the £101/t put option the right to sell futures at £101/t is purchased.

Clearly this is a likely loss-making transaction which, depending upon the time to expiry of the contract, will be reflected in the premium for the option.

Using options

Example 16.3 illustrates how buyers of grain (or potatoes) can use options. They would purchase an option to sell grain at an agreed price in the future to protect their closing-out operation and, again, remove the threat of margin calls, should the market price fall in their trading period. For £2/tonne (the cost of the option) he has:

(1) Avoided margin calls of £400
(2) Benefited from the rise in price in the physical market

He is now better off by £200 because he bought the option.

How the futures market operates

Buying and selling of contracts is carried out by full floor members of the market. Trading occurs across the 'floor' of the futures ring, on an 'open outcry' basis. This means that a broker with grain to sell will ask 'what is (the price of) September barley'. The reply will come from a prospective buyer '100 (pounds)'. The seller will accept by saying 'agreed', and the number of lots of grain to be traded will then be agreed. This transaction will be recorded by the brokers on a daily business slip, and contracts are passed to their clients each evening. All contracts are registered by the broker, with the clearing house, by 11:00 hours the following day. Purchase and sale contracts, that is the 'closing out' of contracts, may be connected by notifying the

Example 16.3 The effect of options in futures trading

	Financial situation		
	Futures (£)	Option (£)	Physical intervention (£)
February 1996			
Sells 100 tonnes feed barley for September delivery at £102/tonne	10 200		
Purchases an option to buy September barley at £102/tonne which costs £200		–200	
September 1996			
Physically sells 100 tonnes of feed barley to a local compounder for £106 as the market is short of barley and prices are good.			10 600
At the same time wishes to close-out his hedge by buying 100 tonnes September delivery barley. Without the option he would have to pay £106/tonne for this, representing a loss on futures trading of £400 and, in fact, this would have been called from him as margin money as the price rose through the trading period from February to September.			
With the option he can now exercise the right to buy September barley at £102, which he does and because he held the option he will not have been liable for any margin calls, the broker will have to pay these.	–10 200		
Gross trading position	—	–200	10 600
Net income			10 400

clearing house of the details, and the broker's account will be credited or debited accordingly.

Points to note

(1) If the farmer sells more grain on the futures market than he grows (or has in store), then he is speculating on the futures market with the possible result that he could lose, or make, a lot of money.

(2) The farmer must have confirmation of the futures purchase from the broker

before physically selling the grain locally. Otherwise he becomes an unintentional speculator and is open to market price changes which can occur very quickly and may result in losses.

Futures: some basic rules

(1) Get to know your broker so well that you do not have any worries about trusting his judgement.

(2) Start trading with a small amount to begin with; ask your broker about the possibilities of a 'dummy' trade to see how the market will work for you before real trading begins.

(3) Keep in close contact with your broker and with the movements in the market.

(4) Set price objectives before you start trading.

(5) Use *stops* to protect profit and limit losses in any futures trading, or

(6) Ask your broker about *options*, especially for products where the price is likely to fluctuate a lot during the season (potatoes). For a negotiated predetermined amount (e.g. between £15 and £25/tonne on potatoes) these options will protect you from paying any margin money if the price goes against you in the market. (It is rather like protecting your no claims bonus on car insurance by paying a small increase in premium.)

(7) If you are going to trade in and out during the year:
   (a) Don't be greedy!
   (b) Don't try to guess tops and bottoms in the market; far bigger traders are involved in causing price movements and their actions cannot be predicted accurately.
   (c) Be prepared to accept losses especially if you are 'speculating' on price movements.
   (d) Cut losses and run profits.

(8) If in doubt stay out! If in doubt get out!!

(9) Be prepared to pay margin money and prepare your bank manager for possible cash requirements as a result of your trading. This will ensure that you can stay in the market until the time is right to close-out your initial hedging operation. If you cannot meet margin calls within one to two days the broker may be forced to close-out your contract which can result in losses for which you will be liable.

Summary of futures trading

Advantages

(1) Hedging allows:
  (a) Price stability to be guaranteed for up to 12 months ahead.
  (b) Flexibility in choice of physical outlet of grain by the farmer.
  (c) Easier business planning.
(2) Use of the futures market protects the farmer from market price fluctuations *but* not from the possibility of crop failure or poor weather conditions; the farmer still bears these risks.
(3) The existence of the futures market and the fact that it is used by grain traders ensures a greater stability of grain market prices as a whole, whether farmers use futures trading or not (and at present only a few farmers and producer groups do, although this number seems likely to grow as CAP market support falls). Grain traders use the market to 'fix' the price at which they will be able to buy grain in the future so that they are not susceptible to grain market price changes as they used to be.

Intervention

When the market price which farmers receive for grain falls below the guaranteed minimum 'intervention' price (Table 14.13) set by the European Commission (and agreed by the Council of Ministers) each year, then farmers or merchants (or anyone) holding grain can offer it to the Intervention Board for Agricultural Products in the months when buying in is operational. The EU intervention system is organised to deal with feed wheat, milling wheat, durum wheat (used for making spaghetti and pasta), feed barley, maize and rye. There is no intervention system for oats and only feed wheat, feed barley and milling wheat are of real significance in the UK in relation to intervention at present.

Intervention value depends on:

(1) Month of delivery, as price offered varies through the year.
(2) Quality of grain. The standards are shown in Table 14.11. Grain which is better than standard quality achieves a bonus (calculated as a percentage of the July basic price) and grain which is only just above minimum intervention quality will incur a discount.
(3) The Intervention Centre nominated by the seller in relation to the store stipulated by the Intervention Authority, and the location of the grain.

Problems with intervention selling

(1) Meeting quality standards, especially if harvesting conditions are difficult.
(2) The testing procedure for grain, especially milling wheat, has incurred a lot of

criticism in the past. Discussions are taking place to improve and harmonise procedures for a satisfactory, efficient and standard means of testing the milling quality of wheat throughout the EU. In the past opinion within the trade has been that the dough machinability test, the Zeleny test, and measurement of specific gravity are a long way short of satisfactory.

Quality tests for grain

The quality standards for intervention and futures delivery were described earlier. The quality required for malting barley is set out below, along with the main methods of testing for quality.

Protein

Protein is found in all the various components of grain. On average 10% of the grain weight is protein. Proteins are rich in nitrogen, but they are not the nitrogenous compounds found in cereals. Cereal proteins are divided into four types:

(1) Albumins
(2) Globulins
(3) Prolamins
(4) Glutelins

Albumins and globulins are found in the embryo and the aleuron cells. Prolamins and glutelins are in the endosperm, a store of energy for the germination of the grain, which consists of starch and protein.

When measuring the protein content of wheat flour it is the accepted practice to calculate and state the percentage protein at 14% moisture level.

It is vital for the miller to assess the gluten content in wheat since this is the ingredient which creates the elastic and 'bouncy' quality in the dough.

Gluten and protein content are linked and so by measuring the protein percentage the ability of the wheat to make good bread can be determined. Most home-grown wheats have protein contents of between 9 and 12%. Usually premiums are paid for protein contents of 11% and over, provided that the sample is acceptable in all other respects. Several factors affect the protein content of wheat, variety being most important, along with crop husbandry and weather patterns. These factors are well documented, although there are continuing discussions amongst farmers and academics concerning the 'best' practices to adopt, and will not be developed here. The way in which grain is dried (and stored) will also have an effect on its milling quality since it is possible to damage the wheat germ by overheating the grain. This may render it useless for bread making. Table 16.5 shows the combinations of moisture and temperature at which grain is dried that will damage grain in particular time periods.

Table 16.5 The effect of moisture, temperature and time on grain during drying

Moisture (%)	In 1 min (°C)	In 10 min (°C)	In 30 min (°C)
15	72	67	64
17.5	69	64	61
20	67	62	59
22.5	65	60	57
25	63	58	55

The results in Table 16.5 suggest two recommendations:

(1) The maximum 'safe' drying temperature to maintain milling quality is 60°C
(2) If the grain is above 25% moisture or a tray dryer is used the temperature should be reduced

The longer that grain is subject to heat and the higher its moisture content, then the lower should be the temperature, in order to prevent milling quality damage.

The protein standards required for intervention seem hard to achieve but it depends on the percentage dry matter at which protein is quoted. Grain traded at 16% dry matter needing 11% protein will be only 9.5% protein at 14% dry matter.

Alpha-amylase: the Hagberg test

This test is used to determine the amount of alpha-amylase in the grain (see Table 16.6). Alpha-amylase is an enzyme which is naturally produced in the grain when germination begins (although in some varieties it is active even before germination has started). Its function in the grain is to break down starch into sugar to facilitate the germination and growth of the grain. If it is very active the characteristics of the wheat flour are affected in that during baking the dough becomes 'sticky' rather than elastic and does not make very good bread. The activity is expressed as the Hagberg falling Number, which relates to the time it takes to prepare and carry out the test, measured in seconds. It is unusual for millers to accept wheat with a Hagberg of less than 180 and with improved varieties now being grown a figure of 300+ is sought.

Table 16.6 Hagbergs and sprouting in grain

Hagberg	Alpha-amylase and sprouting activity
<150	High amylase, numerous sprouted grains
150–200	Average amylase
201–400	Low amylase, few or no sprouted grains

Dough machinability test

There is only a weak correlation between the machinability test and bread quality. It is not used by millers in the UK. The millers in Germany use this measure because they cannot deal with a sticky dough; in this country we use additives to overcome this difficulty where necessary.

Zeleny indicator test

This is used to test the quality of the protein in wheat. The wheat is milled to flour in specified mills. The flour is mixed with lactic acid and alcohol. The mixture is allowed to settle and the height of the sediment is measured. The higher it is the better is the protein quality. This is a good test of protein quality but is slow to carry out and only one or two laboratories in the UK can do it. It is used on the Continent because they mix varieties much more.

SDS test

This is carried out by the Flour Milling and Baking Research Association for a similar purpose to the Zeleny test (i.e. to test protein quality). It is a fast and cheap test but is not such a good indicator of protein quality.

Specific weight

There is no standard method and no determined way of calibrating instruments used because there is no reference measure to check against. No-one knows, for example, what height of pot to use, what shape it should be or the height from which grain should be poured into the pot. In any event this measure is only 'a general indicator of quality', a measure of bulk density, and is not much good for determining quality. It is used in the UK because it is required by the trade, especially for exporting and for intervention.

The Chopin Alveograph

Developed in 1937 in France, this is used to predict the baking quality of wheat, by determining the type of wheat protein. It is therefore possible to assess whether the wheat will be suitable for bread or biscuit making (see Table 16.7), and the test is used mainly for UK export markets where millers buy unknown varieties or blends. It is important for UK growers and traders to know which varieties have a good score in order to aim at export markets. It is not necessary for UK growers to have the equipment or use the test.

The test is complex, involving five stages, and is done over a three-day period.

(1) One kilogram of wheat is cleaned and dried to a standard moisture content over 24 hours
(2) The wheat is milled in a Chopin mill and 55% of white flour has to be extracted before the sample is left for 24 hours

Table 16.7 Varietal suitability to end use, 1995 and 1996

Variety	Suggested use	Specification required 1995		Actual specification 1995	
		P/L	W	P/L	W
	!00% Bread	<0.8	>200		
Hereward				0.48	234
	Blending	<0.6	—		
Hereward				0.48	234
Hunter				0.6	105
Riband				0.41	79
Soissons				0.62	291
	Biscuits	<0.5	<110		
Riband				0.41	79
	Animal feed	—	—		
Brigadier				1.18	86
Hunter				1.42	83

Source: BCE, HGCA.

(3) A 250 g sample of the flour is placed in the alveograph and sodium chloride solution is added before the dough is mixed for 7 min
(4) The dough is rolled out and five 'discs' cut before resting it for 20 min
(5) A disc of the dough is put onto the alveograph plate and air is pumped into the dough. The pressure (P) is recorded as is the length of time it takes the bubble to burst, in relation to resistance and pressure (L). The dough expands into a bubble which eventually bursts

This process is repeated with the other four 'discs' of dough and the measurements are recorded (see Fig. 16.3).

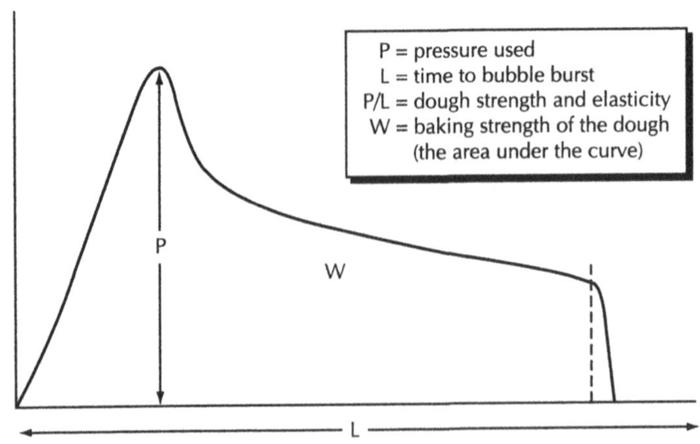

Fig. 16.3 The P, W and L measures in the Chopin Alveograph. Source: BCE, HGCA

Moisture

There are three types of water in grain:

(1) Free water
(2) Absorbed water
(3) Water in the constitution

Moisture tests should drive off (1) and (2) but not (3) or the grain will be burned. There are three methods of testing for moisture:

(1) Drying grain at 130°C for 3 hours
(2) Drying grain at 120°C for 4 hours (Marconi method)
(3) Drying grain at 105°C for 4 hours (Wheat Act method)

Each has varying accuracy when compared to intervention standard and the best method to use is (1). However, it is important to determine which method a buyer is going to use.

Many of these tests are used to determine whether grain will be acceptable for intervention. The test for moisture is one of the most significant for the farmer since most rejections occur because grain is too wet.

When grain is rejected at the store because it is not of the correct quality standard, the seller will incur the transport costs of taking it home.

For these reasons farmers accept lower market prices than are offered by the intervention system. This amount varies between £10 and £15 per tonne below intervention price and is related to the quality of the grain and its location.

For good-quality grain located near an intervention centre, the seller will require a market price as high as intervention offered price, or he will sell the grain into intervention.

For poor-quality grain, located a long distance from a centre, the seller will accept between £12 and £15 per tonne less than intervention, on the open market, in order to avoid transport costs to an intervention store, the costs of upgrading the grain to intervention standards, plus the risk of having grain rejected at the store and the delay in payment.

Nitrogen

A good malting sample of barley should have a nitrogen content below 1.6%, and 1.8% is usually considered to be too high.

In addition, a maltster is looking for a bold grain, free from admixture, heated grains and mould.

The higher the nitrogen content in the barley, the less starch there will be and the lower the malt extraction. Nitrogen contents are determined by variety, amount and timing of nitrogenous fertiliser applications, as well as by growing conditions.

To facilitate the speed of throughput during the extraction process the maltster wishes to know how quickly barley will germinate. Ideally he would like to do the whole extraction job within six days, but there will be obvious difficulties if the barley takes ten days to germinate. The barley is therefore tested for its germinative energy and germinative capicity, which give an indication of its speed of germination.

For export bread wheat would normally be expected to be 14% moisture, 230 Hagberg, 76 kg/hl specific weight, 10% protein (at 14% H_2O) and have between 140 and 180 (W) plus 0.4 to 0.6 (P and L) on the Chopin Alveograph.

Importance of intervention in the UK

The intervention system was little used in the UK until 1980–1981 due to green rate problems which kept intervention prices artificially low between 1973, when the UK joined the EU, and 1980.

When the value of sterling rose in the 1980s, intervention prices became more attractive and more grain was produced (see Table 16.8).

Intervention in the UK has been more important for feed barley because it tends to attract a lower market price than wheat. (There is also a larger surplus of barley in the UK and EU.)

Table 16.8 Intervention purchases of cereals in the UK ('000 tonnes) 1977/78 to 1986/87

	77/8	79/80	80/1	81/2	82/3	83/4	84/5	85/6	86/7
Barley	21	14	730	348	1 201	377	1 157	1 421	87
Wheat:									
milling		2	87	37	517	24	14	0	0
feed			58	31				3 188	23
Rye			0.3	0.1				0.2	—
Oil seed rape		1	6						—

The effect of the intervention system on market prices

Intervention influences market prices in that it puts a bottom on grain prices. Although a great deal of grain is sold below the intervention price, because there may be a £12 to £15 gap between market price and intervention price per tonne, when intervention prices rise, so will market prices. This occurs because feed compounders and merchants will buy grain at the lowest price on the open market which farmers will accept. If the gap grows too large, more than £15 per tonne for example, then intervention, with all its problems, becomes worthwhile. So market prices generally do not fall more than £15 per tonne below the intervention price, as farmers would begin to offer grain to intervention and feed compounders and merchants would not be able to obtain grain, so they must increase the prices which are being offered to farmers.

Decisions on selling grain

The four methods have been outlined but a knowledge of how these methods operate is only the first essential aid to marketing decisions. It is necessary that a farmer selling grain can determine, at any time, in which month he should sell grain and which method will provide him with the best price.

Cereal price trends in the UK

A review of grain prices in the UK reveals a fairly stable pattern from year to year, which is only broken by supply changes caused by abnormal weather.

Figure 16.4 shows price trends throughout each year. Price changes week by week are not regular and depend upon the market influence.

For some cereals there tends to be a relatively high price for the first tonnages of new-crop grain onto the market in August and then a price dip in September/October.

Up until 1984 cereal prices rose each year by £2 to £4/tonne. This was a result of the price support schemes in the EU which increased basic prices for cereals annually.

In 1984 continuing surpluses meant that support prices for cereals were decreased and a threshold system meant that when production exceeded the threshold tonnage, in a particular year, the intervention price was reduced in the following marketing year by 1% for every extra 50 000 tonnes. In the 1986–1987 year a co-responsibility levy effectively reduced the intervention price by £3.37 per tonne,

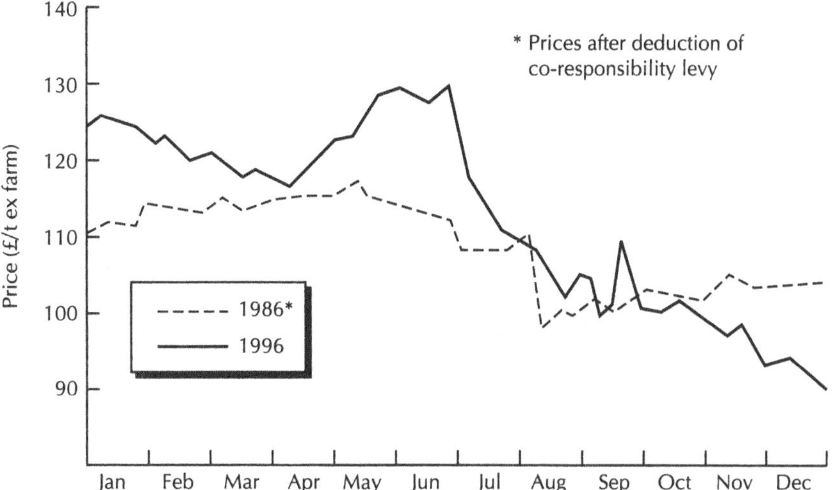

Fig. 16.4 Price trends for feed wheat.

£3.53/tonne in 1987–1988, and intervention quality requirements were changed considerably.

Figures 16.5 and 16.6 show the way in which futures prices and spot prices are closely related, with futures usually a few pounds higher than spot, which reflects the difference in quality standards and makes allowances for transport costs.

The farmer is now faced with deciding which method of sale to use and at what time of year to sell his grain. There are several factors which must be considered before a decision can be made.

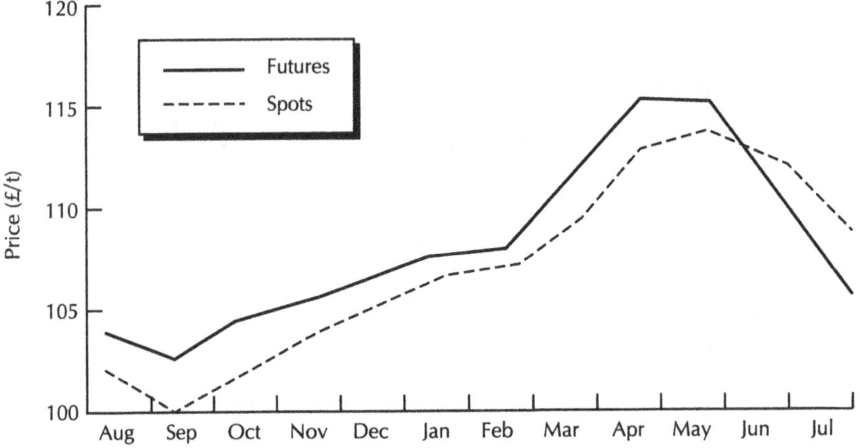

Fig. 16.5 Feed barley spots and futures (nearest position) prices.

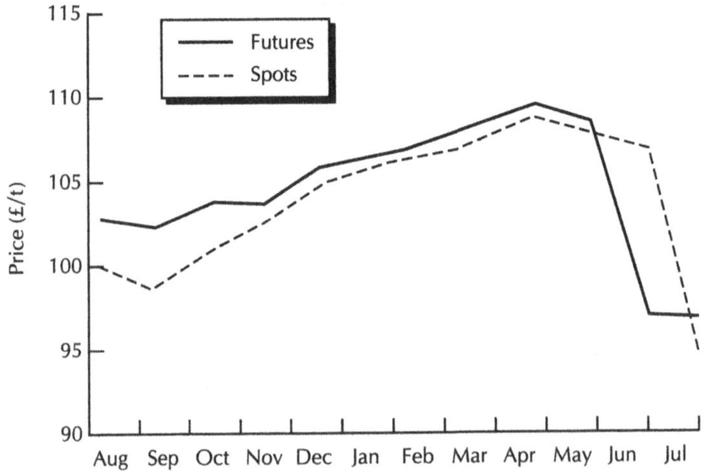

Fig. 16.6 Feed wheat spot and futures (nearest position) prices.

The benefits of storing grain

(1) It is possible to obtain higher prices
(2) Decision-making remains flexible
(3) Buildings are being utilised (this may not be a good reason for storing grain in some circumstances)

Disadvantages of storing grain

(1) Interest charges at bank rate or opportunity cost of money 'tied-up' in grain rather than cash
(2) Losses in store: quality and quantity
(3) Capital and operating costs of UK storage facilities; cost of keeping grain up to standard over time
(4) Risks of price fluctuations

Basically what the farmer has to do is to evaluate, on a daily basis, the prices being quoted to him for selling grain by the various methods. He must be able to decide whether a spot price will provide a better return for grain than will a forward contract price for two or three months ahead, or a futures contract price for up to 12 months ahead. He must also be able to calculate whether the intervention price provides a better return than the spot price. In a calculation to determine which method and which month to choose to sell grain the farmer must compare the gains from storage with losses incurred by storage. One way of determining this is to construct a partial budget.

In Example 16.4 interest is calculated (simple interest) using the spot price of £107/t in February as the base price. So the interest cost of keeping grain to sell on a March forward contract is 1% of £107 = £1.07. Interest for April forward is 2% of £107 = £2.14, and so on.

For intervention the payment delay is taken as four months, so interest for February intervention is 4% of £107 = £4.28, and so on for March.

For futures the deposit of £500 will have to be paid on the day the contract is made (i.e. 31 January 1996) so that money will be gone from the business for two months on a March contract and four months on a May contract on a per tonne basis. This is not a significant amount.

The calculations show that the best deal will be obtained by selling to intervention in March, even though the money will not be received until July. Of course, if the business is under pressure for cash before July, then May futures offer the best alternative because the cash will be received at the end of May.

The same procedure is followed to check prices for feed barley, malting barley, milling wheat and oilseed rape and other costs can be introduced if the farmer is able to estimate them.

It is therefore possible for the farmer to determine the best options open to him. In effect the methods of calculation are simply showing him the present (undiscounted) value of future monies which will come into the business (at different times depending on the options available) if he does not sell grain at the earliest oppor-

Example 16.4 Choosing the best method and time of sale

In January 1996 the farmer has 100 tonnes of feed wheat in store. The prices quoted to him on 31 January 1996 are shown below.

Since it is assumed that the spot money will be received on 28 February, it is possible to compare £107/tonne on 28 February with the other prices quoted.

The interest rate is 12% pa.
Estimated transport costs are £4.50/tonne to intervention.
The futures deposit is £500.

		£/t
Spot	(delivered 14 February)	107.00
Forward	March (money on 28 March)	108.00
	April (money on 28 April)	111.50
	May (money on 28 May)	113.00
Futures	March (money on 28 March)	111.55
	May (money on 28 May)	115.80
Intervention	February	122.59
	March (delivered on 28 March)	124.18

	Losses (£/t)		Gains (£/t)	Net (£/t)
Forward:				
March (money on 28 March)	1.07 (i.e. 1% of £107)		1.00	–00.07
April (money on 28 April)	2.14 (i.e. 2% of £107)		4.50	+2.36
May (money on 28 May)	3.21 (i.e. 3% of £107)		6.00	+2.79
Intervention:				
February (money on 28 June)				
interest	4.28			
transport cost	4.50	8.78	15.59	+6.81
March (money on 28 July)				
interest	5.35			
transport cost	4.50	9.85	17.18	+7.33
Futures:				
March (money on 28 March)				
interest on deposit	0.10 (2% × £500)			
interest on grain	1.07 (1% × £107)			
brokerage	0.30	1.47	4.55	+3.08
May (money on 28 May)				
interest on deposit	0.20 (4% × £500)			
interest on grain	3.21 (3% × £107)			
brokerage	0.3	3.71	8.80	+5.09

tunity. He can thus compare the various prices quoted to him with the current spot price after accounting for interest and other costs.

In this analysis no value has been, nor can be, put on losses in storage or on costs of maintaining grain quality in store. It is not worthwhile to try to estimate these as they will obviously vary between farms and between years. However, the risks of losses will be the same for each lot of grain so this omission will not adversely influence the preceding calculation methods.

The aforementioned methods of analysing grain price options provide a means of helping farmers in their decision making. However, if the farmer is convinced that the spot price for feed wheat is likely to rise, for example by £9/tonne between February and May 1996, then he need not bother to calculate the present value of the forward prices because the spot price in February is obviously better than any of the forward prices. the same argument holds true for the futures prices in this example (but not for intervention). The problem of course is that if the farmer is wrong in his assessment of the way in which spot prices are expected to move, and if, say, on 28 May the spot price has only risen by £3/tonne, the farmer will have made no provision to 'lock-in' a price for his grain and will be forced to sell spot at a lower price than he could have received if he had taken out a forward or futures contract.

Further considerations on futures trading

If the farmer has in store, for example, 250 tonnes of feed barley or 370 tonnes of feed wheat, he cannot trade these precise amounts on the futures market because trading takes place only in lots of 100 tonnes. Given this situation, when the farmer believes that the futures market price provides the best option, then he has two courses of action open to him. First he can cover himself for two lots of feed barley on a futures contract (200 tonnes) and sell the other 50 tonnes using the next best alternative, plus futures contract for 300 tonnes feed wheat and sell the rest of the wheat (70 tonnes) by the next best option. Secondly he could trade more feed wheat and barley than he actually possesses, that is sell 300 tonnes feed barley and 400 tonnes of feed wheat on futures contracts. This would mean, however, that he is indulging in speculation, by selling wheat that he does not possess, with the possible risk of losing money which that involves (see previous section on methods of selling grain).

Other considerations in grain marketing

Losses due to drying and cleaning grain

A great deal of grain coming from the field nowadays needs to be dried down to a level which will ensure good keeping quality and its acceptability for possible sale to intervention or the futures market. In many cases where grain is dried on farm it is also cleaned, usually after drying as cleaners tend to work more efficiently on 'dry' grain, but in some cases before drying as it is expensive to dry rubbish.

It is important that producers are aware of the percentage weight loss arising as a result of drying and cleaning, so that they do not overestimate the size of the saleable crop. As shown in Table 16.9 and Fig. 16.7, by drying 100 tonnes of grain from 20% to 15% moisture the farmer would be left 94 tonnes of grain at 15% moisture. This is calculated using the formula:

$$X = \frac{W_1(M_1 - M_2)}{100 - W_2}$$

where X = weight loss, W_1 = original weight, W_2 = final weight, M_1 = original moisture, M_2 = final moisture.

At 20% moisture in Fig. 16.7 there is 80 tonnes of dry matter and 20 tonnes of water. At 15% after drying there is 80 tonnes of dry matter and 14.12 tonnes of water and therefore a field yield of 100 tonnes at 20% moisture will be reduced to a saleable yield of 94.12 at 15%. When the grain is cleaned as well, a further 2–5% may be lost in the cleaning process, depending on how clean the grain is initially.

The majority of UK grain is stored in bulk, using on-farm floor drying and storage systems. These systems cannot guarantee:

(1) An even moisture content in a bulk sample of grain
(2) That each 20 tonne load will be the same as the sample taken to assess its quality

Add to this the difficulties inherent in an on-floor drying and storage system of separating different varieties and qualities of grain, and the problems faced by the farmer of uneven drying resulting in variations in grain moisture content. It is often

Table 16.9 Weight of water lost when drying grain (kg)

Initial moisture (%)	Final moisture (%)									
	19	18	17	16	15	14	13	12	11	10
25	74	85	96	107	118	128	138	148	157	167
24	62	73	84	95	106	116	126	136	146	156
23	49	61	72	83	94	105	115	125	135	144
22	37	49	60	71	82	93	103	114	124	133
21	25	37	48	60	71	81	92	102	112	122
20	12	24	36	48	59	70	80	91	101	111
19	—	12	24	36	47	58	69	80	90	100
18	—	—	12	24	35	47	57	68	79	89
17	—	—	—	12	24	35	46	57	67	78
16	—	—	—	—	12	23	34	45	56	67
15	—	—	—	—	—	12	23	34	45	56

Source: MAFF.

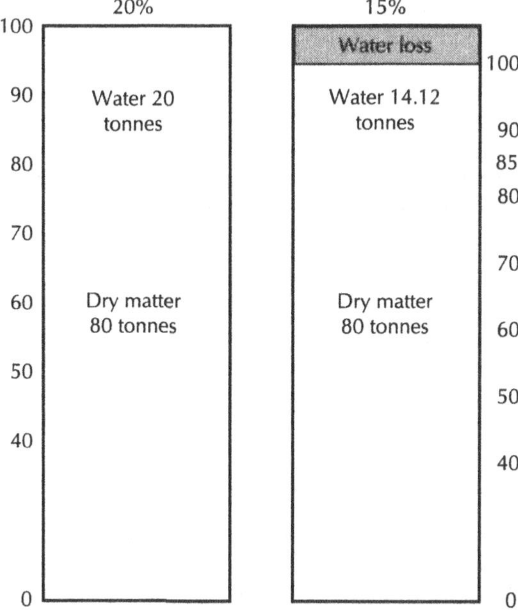

Fig. 16.7 One hundred tonnes of grain dried from 20% to 15% moisture. Source: MAFF.

necessary for the grain near the air inlet of an on-floor dryer to have a moisture content of 11% in order to ensure that the top layer is down to 15%. This is expensive for the farmer in terms of running costs of the dryer and loss of saleable weight in the grain.

At the present time costs of buildings and equipment (see Table 16.10) would prohibit the outlay on new grain storage and drying systems to supersede the on-floor systems, 25% of which are more than 25 years old.

The only viable proposition is to improve intake and drying facilities, i.e. 12 to 20 t/hour drier and then store grain on-floor in bulk stores where only occasional aerations would be necessary. This will enable farmers to put grain into store at known moisture percentages.

Table 16.10 Costs of grain storage buildings

Type of store	1981/2 (£/t)	1985/6 (£/t)	1996/7 (£/t)
Bin storage and continuous dryer in new building	110	150	130
Floor storage and continuous dryer in new building	105	120	120
Floor storage and low temperature on-floor drying	80	100	130
Outdoor free standing bins with drying floors	60	75	70
Bins in existing buildings with existing dryer	30	50	50

Costs quoted may vary + or – £10 per tonne.

Losses in store: pests and diseases

Invertebrate pests

The incidence of these pests is increase on on UK farms. This trend has been recognised since the 1950s as greater quantities of grain have been stored on farms. Many pests are present in the stores before harvest, and as preharvest control is the easiest to carry out fumigation is recommended when grain stores are empty. This process can be greatly facilitated by correct design of buildings.

Grain is a durable commodity and will store for long periods without sophisticated facilities. Provided that it is kept cool (below 12°C) and dry (below 14% moisture) then any insects and mites present will not be active to seriously damage the grain. However, the grain should be inspected regularly and the temperature checked in order to monitor the condition of grain in store.

Pest, mite and micro-organism problems

Some of the earliest reliable information concerning grain storage problems appears in the reports of the War Committee produced during the 1914–1918 war. These reports were largely ignored but show that these problems are not new. They have become more important as quality standards, specified by grain buyers, including intervention, have become stricter, especially since 1973. Before that date the fairly low insect infestations were tolerated by traders because most grain was used for animal feed inside the UK and it was a relatively low value product. As the use of continuous-flow dryers increased and more grain was grown and stored on farm, the infestation by, particularly, saw-toothed grain beetles and probably other pests increased because the grain tended to be warm on entering the store. Hot-spots encourage insect activity and mould growth on a wide scale.

During the 1970s stricter standards were demanded by the EU for intervention purposes, to obtain the phytosanitary certificate for exporting grain, and also to meet the requirements of the UKASTA quality wheat conditions.

Infestation and damage

Mould growth is not a serious problem in UK grain and normally only occurs when grain is stored at more than 16% moisture or when insect activity has raised the grain temperature.

The results of surveys on insect pests in farms (1997), commercial grain stores (1992) and feed mills (1992) are shown in Table 16.11. These surveys were conducted by the Infestation Risk Evaluation Group in association with ADAS.

A survey of 228 farms in 1973–1974 revealed that mites tend to be more widespread on farms. The main ones detected are shown in Table 16.12.

Both insects and mites cause physical damage to the grain by consuming the germ (saw-toothed grain beetle – STGB) or the endosperm (grain weevil and rice weevil larvae) or destroying the embryo (mites). The loss of weight due to these pests is not, however, as serious to the farmer as their mere presence. With higher standards

Table 16.11 The incidence of insect pests on farms, feed mills, central stores (%)

	Central stores	Feed mills	Farms
Saw-toothed grain beetle (*Orizaephilus*)	19.7	46.0	4.8
Flat grain beetle (*Cryptolestes*)	17.2	28.3	4.9
(*Sitophilus*)	22.2	57.2	4.2
Any of the above	39.5	69.0	9.7

Table 16.12 The incidence of mites on a sample of farms (%)

	Farms
Glycyphagus destructor	67.2
Acarus siro	25.3
Tyrophagus longior	19.5
A. farris	13.0
T. palmarum	3.0

being demanded by buyers, grain infested by live pests will be rejected by intervention, feed compounders or millers if unacceptable levels are reached. This is because the nutritive value of the grain is seriously reduced by pests and some mites have a strong smell which taints the grain and renders it unfit for use.

Detection of infestation

Low levels of infestation, less than 1000 insects/tonne, for example, are difficult to detect unless at least 1 kg of grain per tonne is sampled. Insects tend to congregate in the warmest spots, however, and it is important to check the most likely sites – under conveyors, discharge spouts, peaks and ridges and any obviously warmer areas. Temperature is the best indicator of insect and mite activity and this must be recorded, on a weekly basis, for the whole of the grain stored. In the USA a test has been developed which identifies an insect muscle protein called myosin, using the enzyme linked immunosorbent assay (ELISA). Quantitative and qualitative testing can detect the level of insect contamination.

Control of pests

Physical methods

One of the oldest but still very effective methods of protecting grain during storage involves reducing its moisture content. If grain is dried to, and maintained at, a moisture content of 12% or less it can be stored for long periods with only limited risk of infestation. However, many contracts specify a moisture content of 14.5%, which encourages storage at this level. Unfortunately, this offers no protection from insects and mites and is very close to the level at which serious mould growth does occur. Many of the infestation problems experienced by the British grain producer and storer in the past stemmed from the use of 16% as the trading standard for moisture content. Mites are particularly sensitive to moisture content and will not develop to serious levels below 14% but thrive at 16%.

Keeping grain temperatures as low as possible will also help prevent infestation as the saw-toothed grain beetle requires a temperature of more than 18.5°C to develop. Below 12.5°C none of the insect pests can complete their life-cycle, but in practice it is unlikely that grain can be cooled sufficiently to kill insects or to stop mites breeding. However, using a system of low-volume ventilation it is relatively easy to reduce grain temperature to less than 10°C and quite feasible to achieve temperatures of 5°C or less during the winter months. Refrigerated storage seems to offer few advantages under UK conditions.

Storing grain in air-tight bins or in reduced oxygen atmospheres will prevent infestation by insects and mites and will also inhibit the growth of most micro-organisms. At the moment, however, the practical difficulties associated with air-tight storage limit its usefulness. If it is to be used successfully, purpose-built stores or bins are needed. The production of low-oxygen atmospheres is also complicated and often at least as expensive as drying.

Chemical control measures

In the UK practical physical control measures will not eliminate all insects and mites, although they help to reduce their numbers. Also in many cases grain is stored under less than ideal conditions and some form of direct pest control is then needed.

Fumigation with a toxic gas gives quick control of established infestations but can only be carried out by a licensed servicing company and not by farmers. When it is used effectively fumigation gives complete control but offers no protection from reinfestation and costs around £1/tonne.

Some control measures can be used by farmers and will provide a virtual guarantee that grain remains pest-free throughout storage. These chemicals are used in two ways: applied to the structure of empty stores or admixed directly with the grain. Structural treatments are widely used, their object being to kill pests sheltering in the empty store and so prevent freshly harvested grain from becoming infested. These treatments are generally well within the scope of most farmers, and are not expensive. They will control the majority of insects and mites in a clean, structurally sound store but are unlikely to give complete control. If the standard of cleaning is less than perfect or the structure of the store is not sound it is likely that many pests will survive the treatments.

Admixing pesticides with freshly harvested grain is also a relatively simple operation and in some circumstances will ensure that grain remains pest-free for six months or more. Admixture treatments are relatively cheap at around 15 p/tonne. Several pesticides have been cleared through the Ministry's Pesticides Safety Precautions Scheme for admixture with grain. The properties of each chemical are slightly different and the choice between them depends upon several factors, including the pests to be controlled, the period of protection required and the type of store. All pesticides break down rapidly on grain at above 25°C and the advantage of long-term protection is soon lost. Therefore, they should only be applied to cool, dry grain.

Strains, or mutations, of some insects and mites that infest grain have developed resistance to some pesticides. This has not, so far, produced many serious difficulties in controlling grain pests in the UK but if some overseas strains become established real problems could result. Most of the low-cost pesticide treatments would be ineffective and much greater use would have to be made of fumigation and physical control. This would undoubtedly increase both the risk of infestation and the cost of storing grain. It is, therefore, important that every effort is made to use pesticides effectively, as this is the best way of preventing the development of resistance.

Vertebrate pests

The pests primarily concerned are:

- ❑ Rats
- ❑ Mice
- ❑ Feral pigeons
- ❑ Sparrows
- ❑ Collared doves
- ❑ Starlings

The actual losses attributable to these pests are not known. It is easy to overlook the significance of these losses because the pests are not always very obvious. Losses arise due to:

- ❑ Grain consumed (Table 16.13)
- ❑ Grain damaged by rodents (three times that eaten).
- ❑ Grain contaminated by droppings and urine, rodents produce 18 000 droppings and rats 3.5 litres of urine per year.

Average infestations of rodents usually number around 100 individuals, whereas bird infestations are normally lower.

Other problems

(1) Carriage of disease is a risk, apart from grain damage. The most serious human ones are Weil's disease – leptospirosis – and salmonella; for animals they are foot and mouth, swine vesicular disease and ringworm which can be transmitted, causing serious financial losses.

(2) Structural damage. Among items damaged are electrical cables, water pipes,

Table 16.13 Food consumption by pests

	Consumption per day by individuals (g)	Consumption per year by individuals (kg)	Consumption per year by 100 individuals (kg)	(% of 100 t of stored grain)
Norway rat	30	111.00	1 110	1.1
House mouse	5	1.8	180	0.18
Feral pigeon	50	18.25	1 820	1.82
House sparrow	9	3.3	330	0.33
Collared dove	45	16.4	1 640	1.64
Starling	45	16.4	1 640	1.64

foundations, doors, guttering, floors, ceilings and containers, all of which increase costs and make grain storage less profitable.

Prevention of damage

Proofing: to prevent pests coming in.
Hygiene: to make the environment less desirable and control pests.

These methods are best used together.

Proofing and hygiene

Broadly speaking, in grain storage it is the surface area from which feeding can occur that will largely determine the level of availability and hence loss.

Thus every effort should be made to reduce access to the grain by rodents and birds. The subject of proofing is well covered by Jenson (1979).

A good deal of our grain is stored in circumstances far from ideal and here the objective must be to ensure that the environment around the grain supports as low a population as possible.

The main factor is to avoid dead-spaces, i.e. under floors, in walls and in the ceiling, for it is here that spillage is difficult to recover, poisons are difficult to place and rodents can remain undisturbed. Obviously general tidiness and avoidance of vegetation around stores makes the environment less habitable to rodents.

Clearly the situation as regards bird pests is more difficult and efforts are almost exclusively aimed at preventing access to stored grain by proofing.

Spillage in and around the store will act as an attractant to both rodents and birds and should therefore be avoided.

Control

Rodents

In spite of all efforts to exclude rodents, infestation will sometimes occur and direct control will be required. The most efficient rodenticides are the anti-coagulants, with which 100% clearance can be achieved. Warfarin was first introduced in 1950 and has been followed by the second-generation anticoagulents coumatetralyl, brodifacoum and flocoumafen and difethialone, which are used to combat resistance to warfarin. It is essential, however, that the rodents are able to feed for a number of consecutive days on the bait and therefore treatment should ensure that surplus bait is readily available to the rodents over three of four weeks, during which time all the individuals should have a chance to consume a lethal dose. In a situation in which alternative foods are available it will be necessary to make the poisoned bait more attractive by the addition of oil or sugar. This, with the careful placement of bait, should ensure success. Sometimes, however, a more prolonged treatment will be necessary, with critical appraisal of the technique should it prove less than totally effective.

Should difficulties be encountered (failure is more often than not due to inadequate treatment) then alternative measures include the use of the sub-acute rodenticide calciferol or the acute rodenticides, such as zinc phosphide. The acute rodenticides, however, rarely lead to 100% control and can result in poison and bait shyness, which may complicate later treatments.

Alternative approaches include the use of traps and fumigation. The former are labour intensive on a large scale and are unlikely to lead to clearance. The use of fumigation is a specialist technique which requires outside assistance and is not very effective because the rodents live elsewhere.

All other methods of control are fairly ineffectual and this includes the new ultrasonic devices in which emitted ultra-sound is claimed to deter rodents. Published data indicate that these devices have a very limited effect and even then only for a few days. The whole subject is covered in depth by Buckle and Smith (1994).

Birds

The species with which we are mostly closely concerned, i.e. feral pigeon, house sparrow, starling and collared dove are all on Schedule Two of the Protection of Bird Acts and as such may only be controlled by authorised persons. The Wildlife and Countryside Act 1981 prohibits poisoning or stupefying birds. However, stupefying substances can be used by authorised persons, under licence from MAFF, for use against birds on Schedule 2, Part II, provided that the chemical is approved under the Food and Environmental Protection Act 1985. The drugs allowed are alphachloralose alone or in combination with seconal, used mainly against sparrows and pigeons.

The answer to bird problems lies once again in the manipulation of the environment to reduce the amount of food available to the birds, both in the store and in the surrounding area. Work in Manchester has shown that so far as feral pigeons are concerned the flock size is directly proportional to the food available. Thus control of spillage is essential. Therefore:

(1) Proof the grain stores to prevent access by birds
(2) Improve hygiene around the store (spillage of foods) to avoid attracting birds

Summary

Some financial loss occurs through direct consumption and fouling. In addition, there are risks of disease transmission to employees and livestock and potential for structural damage. Thus, expenditure on proofing, hygiene and control to mitigate this loss is desirable.

References

Buckle, A.P. and Smith, R.H. (eds) (1994) *Rodents, Pests and Their Control.* CAB International, Wallingford.

Jenson, A.G. (1979) *Proofing of Buildings Against Rats, Mice and Other Pests.* HMSO, London.

17 Livestock Marketing

The announcement in Parliament (on 20 March 1996) that Bovine Spongiform Encephalopathy (BSE) in cattle was linked to Creutzfeld Jakob Disease (CJD) in humans devastated the UK and EU beef industry. Prior to that beef production in UK was nearly 1 million tonnes with 28% exported (valued at £500 million), the EU beef mountain was virtually gone and, although beef consumption was falling, beef remained the favourite meat in most households.

As a result of the BSE crisis, meat consumption, on a gently rising trend since 1985 (fuelled mainly by an 84% rise in poultry consumption), fell by 2% in 1996 compared with 1995 (Table 17.1).

The 1996 beef market was down 19% in total, after recovering somewhat from an immediate 40% fall in March, with stewing steak and mince particularly badly hit. Recovery was assisted by the Minced Beef Quality Mark (MBQM), launched in June 1996, and the other actions taken to bolster consumer confidence. Sheep meat increased slightly in 1996, as did fresh and frozen pork, and processed pig meat products. Poultry continued its upward trend.

The disarray caused by BSE will probably be felt for 10 years or more. Underlying trends, however, remain, even if consumer confidence is restored, and attention must still be paid to trends away from fat in meat and also a consumer resistance to meat which is suspected of having hormones or antibiotic residues. Animal welfare is also a major issue for some UK consumers and is likely to become more so.

Concern about the level of fat in our diet has led to studies on diet and health promoted by the Government. This has important implications for the meat producer since the Committee on Medical Aspects of Food Policy (COMA) report recommended that the whole population should be advised to:

(1) Reduce the amount of fat in their diet, particularly saturated fat
(2) Replace fat with fibre-rich carbohydrates
(3) Identify those individuals with a high risk of developing heart disease and encourage them to follow a special diet

More recently advice has been given to avoid hydrogenated fats which may be linked to cancer and disease of the cardiovascular system.

Despite the publicity on dietary fat (which could be taken out of context since the most significant factors in causing heart disease are age, sex – males are more susceptible, family history, smoking, and then eating habits) and the desire by consumers to eat leaner meat, beef and sheep producers have not created much change in carcass composition or fat percentage. Pig meat has altered significantly

Table 17.1 UK production, consumption and trade of meat 1990–1997 ('000 tonnes)

		1990	1991	1992	1993	1994	1995	1996	1997
Beef and veal	production	1 001	1 020	960	859	916	974	701	694
	consumption	1 001	1 014	999	903	920	895	733	786
	import	174	192	198	203	148	173	180	162
	export	124	136	143	188	235	274	58	10
Mutton and lamb	production	370	385	355	347	352	364	345	343
	consumption	429	424	378	338	343	360	375	371
	import	145	117	119	120	111	140	150	150
	export	80	80	105	123	126	140	122	122
Pork	production	740	778	781	801	827	782	801	825
	consumption	775	781	778	814	806	763	806	826
	import	85	80	91	115	111	131	147	147
	export	51	77	95	100	133	150	146	142
Bacon	production	179	175	167	181	191	197	192	196
	consumption	434	424	400	403	415	422	421	428
	import	260	254	234	228	229	229	234	237
	export	5	5	5	5	5	5	3	5
Poultry	production	1 218	1 289	1 319	1 399	1 470	1 510	1 564	1 616
	consumption	1 283	1 331	1 422	1 481	1 561	1 581	1 638	1 684
	import	134	137	170	159	186	212	213	207
	export	60	77	79	77	95	138	141	135
Offal	production	165	170	160	151	158	163	136	140
	consumption	219	210	207	187	184	192	164	172
	import	64	54	70	57	53	52	46	44
	export	11	13	20	19	23	31	18	12
Total meat	production	3 674	3 817	3 741	3 737	3 913	3 990	3 739	3 814
	consumption	4 137	4 183	4 184	4 125	4 230	4 213	4 136	4 267
	import	863	834	880	880	838	937	968	947
	export	330	388	448	511	616	737	489	426

since 1975; back fat decreasing more than 20%, lean meat up by more than 5% with fat with fat trim and subcutaneous fat falling.

The BSE story

Transmissible Spongiform Encephalopathies (TSEs) have been known to exist since the 1920s. There is no definitive answer regarding what causes them, and no cure for those who become infected, although TSEs seem to be related to prions (proteinaceous infectious particles). It is not clear whether prions cause TSE or are symptoms of it. The prion proteins seem to redirect refolding of normal proteins simply by contact with them, causing sponge-like hole in the brain. Prions have no DNA or RNA and act differently from known organisms.

TSEs seem capable of crossing the species barrier, birds, rodents and mammals have all suffered with different and fatal forms of the disease; they also cross the blood/brain barrier which is a major defence mechanism in the body which is not normally breached easily. They are immune to heat treatment at 121°C, unlike germs and viruses, and only highly concentrated bleach has any effect on them. They survive ultraviolet and gamma radiation and no immune response has been detected, in animals or people.

There is no reliable diagnostic test on suspected sufferers, and only detailed autopsy clearly identifies disease caused by TSEs.

Cause

BSE, first identified in 1986, can be transmitted through injections, transplants and eating the infective material (i.e. brains, spinal cords, retinas, and offal) from infected animals. Infected animals normally do not show symptoms for 30 months (although there have been some exceptions). The higher the dose of infected material the quicker the symptoms appear. It was thought originally that it took a 100 g dose of infected material to induce BSE, but now it seems that only 1 g can do it, it simply takes longer for the symptoms to appear. Evidence suggests that BSE can be maternally transmitted, although this occurred in a recent 7 year study, in less than 10% of cases.

The BSE outbreak is thought by many to have occurred in British cattle because of a change in the way recycled animal protein was treated prior to incorporation in compound animal feed. Until 1982, animal byproducts were 'cooked' in batches at 130–900°C then boiled in hydrocarbon solvents to extract the fat. Then a continuous system was introduced which operated at lower temperatures (120°C) for a shorter cooking time, and did not use solvents to extract the fat as the market for fat was no longer so lucrative.

An alternative (or possibly associated) cause has been suggested. The use of organophosphate compounds for the eradication of warble fly in cattle is cited by some as a prime suspect. The UK has had a compulsory warble fly treatment programme for cattle for many years and the nature of the compounds used and the lifecycle of the warble may prove to be BSE connected, although there is no scientific proof so far.

In 1986 the first BSE case was diagnosed and by 1993 there was an epidemic. To date over 160,000 cases have been reported in the UK, by far the highest of any country (Switzerland, 214; Ireland, 120; Portugal, 37; France, 16; Germany, 4; and Italy 2; with some of these cases being traced back to UK exported feed or animals).

Transmission

Transmission occurs in three ways:

(1) Inherited
(2) Iatrogenic hormone therapy, corneal grafts, implanting electrodes, injecting brain cells or using badly sterilised instruments)
(3) Sporadic (non-specific mechanisms). The majority of cases, over 95%, are sporadic

Table 17.2 shows the numbers of cases of BSE and CJD in the UK. BSE cases reached a peak in 1992 and 1993 and are now significantly lower. CJD cases have risen from around the mid-twenties during the period 1986–1988, but there is no clear pattern after that, although numbers of cases are on an upward trend, reaching a peak of 59 in 1994, then falling in 1995 to 46. The more recent discovery of a new variant of CJD has also caused great concern and is the focus of much attention, primarily because it appears in younger people. Up to March 1998 23 people (mainly aged 15–30) had died from nv CJD.

Table 17.2 UK BSE and CJD cases

	BSE	CJD	New variant*
1986	0	26	
1987	0	24	
1988	2184	23	
1989	7137	32	
1990	14181	32	
1991	25032	36	
1992	36682	51	
1993	34370	46	
1994	23944	59	
1995	14298	46	3
1996	7876	59	10
1997**	2007	34	8

Source: MLC, Dept of Health.
* Included in total.
** Up to October.

The risk

The risk attached to getting CJD is suggested to be lower than dying in a railway accident, a domestic accident, a road accident, from influenza, natural causes or, the highest risk of all, smoking (Ford, 1996). Nevertheless, the uncertainty surrounding BSE, especially the appearance of a new variant of CJD in 1996, no reliable diagnosis, no clear cause, no established disease process and no cure, has generated tremendous emotion within the consuming public. Confidence and trust in politicians, scientists and civil servants has almost completely disappeared, and although beef consumption has recovered somewhat, after a crash in March 1996, there is still enormous suspicion and mistrust surrounding any official statement regarding food safety.

Action

Several initiatives have been taken to try and restore consumer confidence. In May 1996 the Minced Beef Quality Mark was introduced by the MLC. Mince sales have shown steady improvement since then. The Farm Assured British Beef and Lamb Scheme (FABBL) started in 1992 has attracted 22000 members. An increase in membership fees from £25 per year to £75, by introducing independent inspection, has raised fears of a fall in members. This may jeopardise the whole concept of the MLC's meat industry 'umbrella' scheme – Assured British Meat (ABM) – which embraces assurance for auction markets, hauliers of livestock, processors and some retailers. If the FABBL does not include enough producers to service the major supermarkets (who retail more than 60% of fresh meat) then the processors are likely to require a compulsory licensing scheme supervised by Public Health Authorities. The schemes aim to guarantee to the consumer the quality and traceability of meat and meat products from source to the consumer. Various methods of identification are being tried, including electronic tagging. It seems, ultimately, that the only completely fool-proof method will be DNA fingerprinting, though the cost effectiveness has yet to be calculated.

Since restoring consumer confidence is paramount the industry will be wise to adopt a method which is perceived as error free, if the export ban on British beef and beef products is to be lifted and UK sales increased. In December 1997 sales of beef-on-the-bone were banned in the UK. In March 1998 the ban on beef exports from Northern Ireland was lifted by the EU Council of Ministers due to the system of traceability which proves the origins of animals. This system has been started in the remainder of the UK for animals born after August 1996. The Food Standards Agency is further evidence of government efforts to improve hygiene and raise consumer confidence.

Schemes

By May 1996 the government had three objectives:

(1) To protect consumers from any risk, however remote, that BSE may be transmissible
(2) To eradicate BSE in the UK cattle herd
(3) To prevent transmission to other animal species

This resulted in two directives;

(1) 'To exclude from cattle feed all ruminant material which could convey BSE to cattle'
(2) 'To exclude from the human food chain every bit of those (relatively few) parts of the bovine carcass which laboratory experiments show could convey BSE infectivity, if the animal were infected, and which could convey BSE to man, if it were transmissible to man'

The consequence was:

(1) The Over 30 Month Slaughter Scheme (OTMS). This was introduced after the ban was imposed on all meat from animals over 30 months entering the human and animal food chain. By the end of 1996 over 1.1 million cattle had been slaughtered, 350 000 of them being prime steers and heifers, the rest dairy cows and the dairy herd has fallen by 4%. This scheme provided compensation as detailed in Table 17.3. The backlog of animals eligible for this scheme caused problems, and difficulties existed in getting cull cows through the scheme. Payments are limited to 560 kg live weight.

Table 17.3 Compensation for the OTMS scheme (p/kg live weight)

	May 1996	October 1997
Basic payment	85.6	62.6
Prime cattle supplement	25.0	—
Cows (from 4 August 1997)	—	55.7

(2) Calf Processing Aid Scheme (CPAS). This is an EU scheme (available since 1992) not used in the UK until 1996; calves had to be male, from designated dairy breeds and not more than 20 days old. The CPAS was extended to beef calves from November 1996. Payment of around £100 per animal is made to the processor with the producers receiving between about £80 for dairy and £100 for beef calves. Disposals totalled 415 000 in 1996, to compensate for the loss of live exports, however, more animals were disposed of through this scheme than would have been exported. The scheme was extended to beef calves in an attempt to prevent the build up of future beef mountains in intervention stores.

(3) Additional payments of 850 million ecu were agreed by the Council of Ministers in June 1996. This has been distributed through:
   (a) The Beef Payment Scheme. This provided about £60 per head for adult clean cattle sold for slaughter between 20 March and 30 June 1996.
   (b) The Beef Special Premium and Suckler Cow Premium Top-ups. These payments were made to compensate producers for lower market prices as indicated in Table 17.4.

Table 17.4 Premium top-up payments

	ecu per head	£ per head
BSP	23	19.70
SCP	27	23.13

Payments are related to claims made on animals in 1995.

In Florence on 21–22 June 1996 it was agreed that the UK would meet certain preconditions in order that the ban on UK beef exports could be progressively lifted. These included:

(1) Animal passports
(2) Implementations of OTMS scheme
(3) Meat and bone meal retrieval (11 000 t retrieved to date)
(4) Implementation of Accelerated Slaughter Scheme
(5) Improved removal of Specified Bovine Offal (SBO) from carcasses

Attempts are continuing to get the export ban lifted for England, Scotland and Wales but it is proving to be very difficult. Both the UK and Germany have agreed to implement the European Commission decision to remove specified risk material (SRM) from carcasses of cattle, sheep and goats from 1 January 1998. SRM includes the heads, tonsils and spinal cord of cattle, sheep and goats over 1 year old and the spleen of all sheep and cows. These items will be banned from food product, cosmetics and medicines. In the UK, vertebral columns of all animals are banned from mechanically recovered meat.

Accelerated Slaughter Scheme

The scheme is based on identification and destruction of 'cohorts' of BSE cases. A cohort is defined as an animal born between 1 July 1989 and 30 June 1993. The cull is thought to involve around 147 000 animals and the groups to be targeted are shown in Table 17.5.

Compensation for females is based on 90% of their first lactation replacement value or market value, but for males it is market value only. Additional compensation is available if more than 10% of the herd is lost.

Table 17.5 Slaughter cohorts

Birth	Action
1 July 1989–14 October 1990	Voluntary early slaughter
15 October 1990–30 June 1991	Compulsory slaughter
1 July 1991–30 June 1992	Compulsory slaughter
1 July 1992–30 June 1993	Compulsory slaughter

Cattle passports

These are compulsory on all animals born or imported after 1 July 1996. It is illegal to move animals without a passport, although exceptions exist for those less than 28 days old.

Beef Assurance Scheme

Producers in this scheme will be allowed to sell animals over 30 months but under 42 months old for human consumption provided that the herd:

(1) Is a specialist beef herd
(2) Has been established for at least 4 years
(3) Has no BSE history
(4) Has not been fed meat and bone meal for 7 years
(5) Has been fed concentrates from a ruminant-only feed mill for the past 4 years or no concentrates at all

Other effects of BSE

(1) Beef intervention was reintroduced in April of 1996. 70 000 tonnes of beef went into intervention in the UK. This represents 10% of UK production and 16% of all beef stored in the EU
(2) BSP will be paid only once in the life of young bulls, not twice as previously, with the aim of reducing the weight of carcasses
(3) A new extensification premium will apply in 1997. For a stocking rate of less than one livestock unit per forage hectare the payment will be £42.12 per BSP and SCP claim compared to the standard price of £29.16
(4) By the end of 1997 about 2 million cattle had been slaughtered and their carcasses rendered into meat and bone meal which was destroyed by incineration, including some in power stations

The cost of BSE

Table 17.6 Estimated costs of UK BSE compensation schemes (£m)

1988–1996 average	1996–1997	1997–1998	1998–1999
190	1 030*	760*	700*

* These estimates exclude costs associated with such things as beef subsidies, selective culling, unemployment compensation and substitute beef imports or medical care.

Quality standards

Beef animals are marketed in the trade using the EU scheme (Fig. 17.1). This was introduced in 1982 and uses carcass shape and external fat cover to create classes into which carcasses can be placed. The scheme is an attempt to facilitate the marketing of animals by having a standard means of describing physical characteristics.

The marketing of livestock

In some respects the producer of livestock is in a very difficult position since he is not normally in direct contact with the consumer but usually sells to the wholesaler or processor (often referred to as the customer). This frequently results in one set of quality requirements, especially with regard to fat, coming from the customer, i.e. the wholesaler, while the consumer demands a different quality standard. In essence, the butcher requires more fat in the carcass than the consumer does and the producer for the most part is confused by the contradictory signals he receives.

The producer needs to determine more specifically what his customer wants, and if this approach proves to be unsatisfactory should try to construct some more direct links with the consumer by dealing directly through a farm shop or by marketing direct to the large retail outlets, possibly using a co-operative venture of some sort.

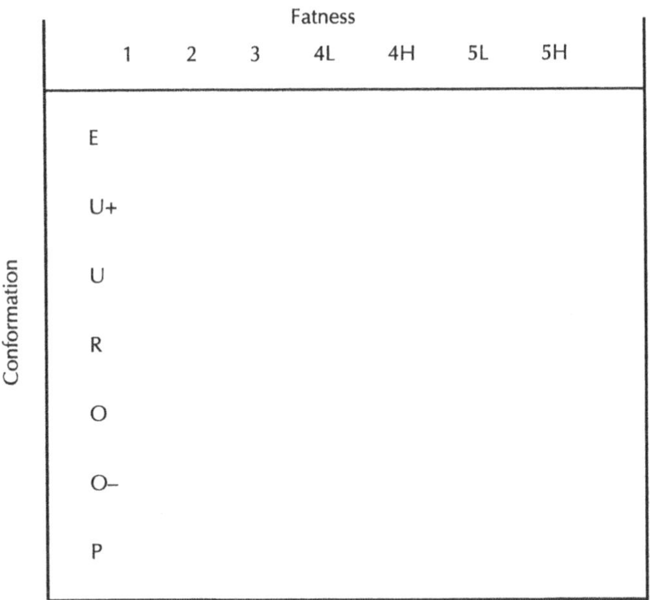

Fig. 17.1 Beef carcass classification. Source: MLC.

Methods of selling livestock

The choice is between selling liveweight (hoof), or deadweight (hook). At present, the majority of cattle and sheep are sold liveweight through the auction market, whilst the majority of pigs are sold deadweight direct to the abattoir. There are several factors to consider before deciding which method to select.

Liveweight: advantages and disadvantages

(1) Price is fixed in a freely competitive market.

(2) Supply and demand is reflected in the return achieved.

(3) The auctioneer has a vested interest in getting the highest price possible for the seller; so the markets continue to get an adequate supply of livestock and his throughput remains healthy.

(4) Cash flow is excellent and usually sellers are paid on the day that the animals go under the hammer.

(5) The money at live auction is secure.

(6) Advice on which stock to send to market can be provided by the auctioneers. In the past, this was a weakness of the system but the most enlightened

auctioneers are now prepared to visit the farm and assist in selecting stock for market. This has the added advantage that the producer discusses market requirements with the auctioneer and should lead to mutual benefits for all concerned in livestock marketing and meat processing.

(7) Some farmers feel it necessary to accompany their animals. The time spent travelling to market and seeing the animals sold is unnecessary and could be spent more productively at home. However, many producers combine the visit with other business or social activities which enable them to keep in touch with the farming scene, while others do not attend the market at all, preferring to 'trust' the system.

(8) The live auction is always subject to price fluctuations in response to market supply and demand. While the price received may be 'fair' for the day there is not much comfort in knowing that a price reflects market supply and demand if it is below the budgeted price required to maintain profitability. There are many factors which can create a short-term fall in demand (for example, bad weather or illness preventing buyers from arriving) or increase in supply. Certainly large sheep producers have to take care that they do not flood small local markets with finished lambs or prices can be severely depressed. In the final analysis the producer can withdraw the animal from sale if it does not reach the price he wants. However, if it is already at 'finished' quality he cannot keep it much longer or it will not be at the correct grade on return to the market, and he will have incurred the extra transport costs and the risk of transmitting disease from the market to his herd or flock.

Deadweight: advantages and disadvantages

(1) Generally, the price will be quoted for stock before they go into the abbattoir.

(2) The buyers will also select stock at the farm, especially if the producer is on contract with them. By this means, there is a good link between buyer and seller which allows a greater knowledge of market conditions and requirements.

(3) The quality of each animal is assessed after slaughter based on seen carcass. The killing out percentage (the amount of saleable meat to the total weight of the animal) is then known, being around 55% for cattle, 50% for lambs, and 70% for pigs, and the conformation can be accurately calculated. This compares with the live auction where the animals are graded by touch and feel, a largely subjective method.

(4) Premium payments are usually made for good quality and level delivery, particularly for pigs and especially on contract selling of all animals. Often it is worthwhile having a contract as this can pay up to two pence per kilogram more for pigs, with slightly less for cattle and sheep. This occurs because the abattoir operators are trying to maintain throughput, and since there is excess

slaughtering capacity in the UK at present they are prepared to pay to ensure a known, regular supply of animals to their abbattoir.

(5) Cash flow is not as good as the live auction in most cases and normal terms of trade exist where payment may be delayed for between four and six weeks.

To summarise

It is noticeable that live market prices fluctuate more than deadweight ones. However, deadweight prices are quoted forward based upon the present liveweight prices. Any changes in live prices are reflected in deadweight prices about seven to ten days later. Thus on a rising trend in market prices liveweight will provide a better price, while on a falling trend in market prices deadweight is better (Fig. 17.2).

As well as there being a difference between live and dead prices through time, it is also vital to note that prices between abattoirs vary significantly. This is a result of several factors, among which will be capital costs, operating costs and, notably, market segmentations, i.e. each abattoir has identified an outlet for which it requires a particular quality animal. Since this is the case, the producer *must* shop around to find the abattoir (or live market for that matter) which will pay the best prices for his particular breed, size and quality of animal. Figure 17.3 illustrates this variation.

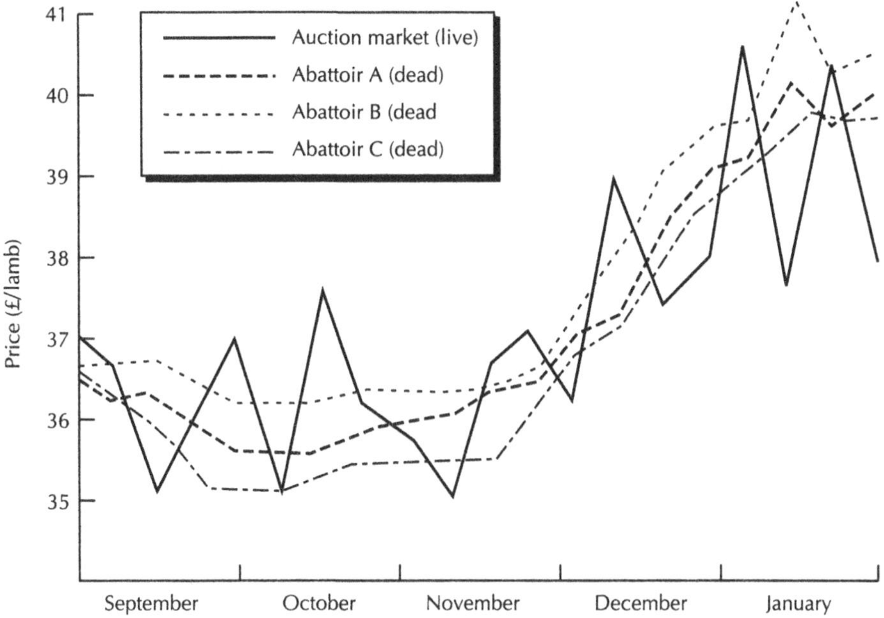

Fig. 17.2 Price fluctuations – liveweight and deadweight.

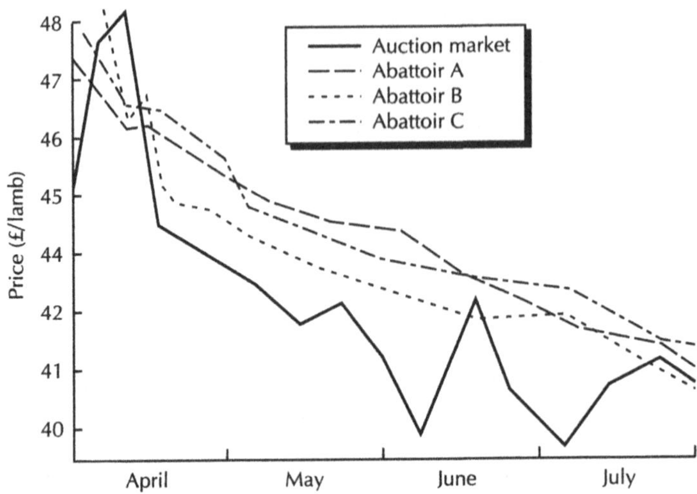

Fig. 17.3 Price variations between abattoirs and auction market.

Electronic auction

The electronic auction system is a computer-based method of selling livestock without assembling the livestock, the buyers and the sellers at one place, which happens at conventional open outcry auctions.

Benefits

Electronic auction offers the benefits of rapid transactions, instant administration and few disputes because the transaction process is transparent. Apart from saving time for buyers and sellers who can carry out the transactions from home, there is equal opportunity for buyers to obtain the animals they want at a price which they feel is justified. Transport costs are reduced, as is the stress on animals involved since they do not have to go to the slaughter house via a live auction market. The market price is made by buyers bidding for animals which are described using a standard classification, and immediate information is therefore available for all concerned. Since the buyers, sellers and lots are connected electronically it is theoretically possible for more people to take part, thus enlarging the market and providing a better indication of a 'fair' market price. Payment is normally received within two weeks.

Sellers can put a reserve price on their animals and usually transactions are on a no sale no fee basis. The value of the animal is adjusted, from a base standard, after the carcass quality is known following inspection in the abattoir.

Problems

In practice not many people are connected to the electronic auction system and only about 13% of livestock are sold by electronic auction in the UK each year. To get involved the 'players' need to have a PC, a modem and telephone link, and a backup power system, all of which involve a capital outlay which may be beyond the perceived cost effectiveness of electronic auctions.

Prices can be very volatile and will obviously depend upon the numbers of animals and buyers involved on any particular day. It is also possible for two buyers to bid the same price for a particular lot, although this happens infrequently and can be resolved relatively easily.

With few companies offering electronic auction facilities it may be that there is a vested interest in preventing change within the current systems and practices adopted.

Operation

The livestock owner usually invites the auctioneer's field officer to advise on which stock are suitable for auction. The animals will then be described on the classification system for conformation and fatness and entered onto the computer. Each buyer is linked to the auctioneer's computer via telephone line, modem and PC. Normally a catalogue of animals for sale is available for access and print out prior to the sale taking place. In the usual way each lot has a number, the vendor's name and address, breed description, grade and weight range of the animal, feeding regime and any other relevant information.

Buyers log on to the system and receive a number for that particular auction, which is known only to the auctioneer, thus preventing the growth of cartels or bidding rings.

A start price for the lot is displayed and automatically reduced by a set amount every three to five seconds, until someone makes a bid. Normally, the price is then increased by about 0.5 p/kg and the time counter restarts. If no-one else bids within a given time (eight to ten seconds) and the price is above the reserve, then the lot is sold.

If no-one bids a price for the lot, or the bid price is below the reserve, the animals can be re-entered at the end of the sale or may be offered for sale by private arrangement, via the telephone, to the buyer making the highest electronic bid.

The auctioneer informs the owner, who has purchased the livestock and between themselves they arrange for collection and subsequent slaughter of the animals.

The quality of the carcass

The quality of the carcass will be affected by:

- ❑ Breed or cross breed of animal
- ❑ Feed ingredients and method

- ❑ Pre-slaughter treatment of animals
- ❑ Post-slaughter treatment of animals

Breed

The conformation of the different breeds and their crosses is well documented and the producer has a wealth of information to aid in the selection of the right breeds to combine for his particular production purpose.

It is not intended to provide an exhaustive review of breed characteristics here, suffice to say that the main factors concerned in deciding which breed combination to use will be:

(1) Conformation:
   (a) the percentage of saleable meat in the high-value cuts
   (b) meat to bone ratio
   (c) killing out percentage
   (d) fat levels, both subcutaneous and inter-muscular
(2) Performance:
   (a) growth rates
   (b) feed conversion efficiency

Feeding

The basic (and complex) principles of nutrition are well documented and will not be examined here. Generally, feeding will vary according to the purpose for which the animals are being prepared and in relation to the time available to produce a finished animal. Put very simply, animals require enough feed each day to sustain bone and muscle growth. If they are fed more than this amount, then fat is laid down and continued overfeeding will produce the result shown in Fig. 17.4.

Animals which are the required weight and have reached it very quickly may tend to be too fat and therefore suffer poor grading results and a low price. This is especially true where concentrate feeding occurs on intensive systems, with pigs being particularly susceptible to this problem. So too are the smaller-framed breeds of cattle such as Aberdeen Angus and Hereford crosses, which will finish at lower weights than, say, pure Friesians or Charolais.

Figure 17.4 illustrates these points in pig production where the best grading results are for pigs maturing at 180 days, whereas those reaching bacon weight at 160 days are too fat and get down graded. Thus throughput is increased at the expense of quality and therefore price.

To try to prevent overfeeding, and since animals are different in their response to feed, some stylised weight-for-age curves have been used to monitor and control feed, in order that the best results are achieved. This simple principle is illustrated in Fig. 17.5. The basic principle is to get the correct weight and finish as soon as possible, by staying close to the curve throughout the life of the animal.

Despite attempts to get feeding right, it is very difficult in practice to be absolutely accurate with stylised theoretical boundaries. There is no substitute for experience in this matter, while discussions on improved feed and feeding with the slaughterers

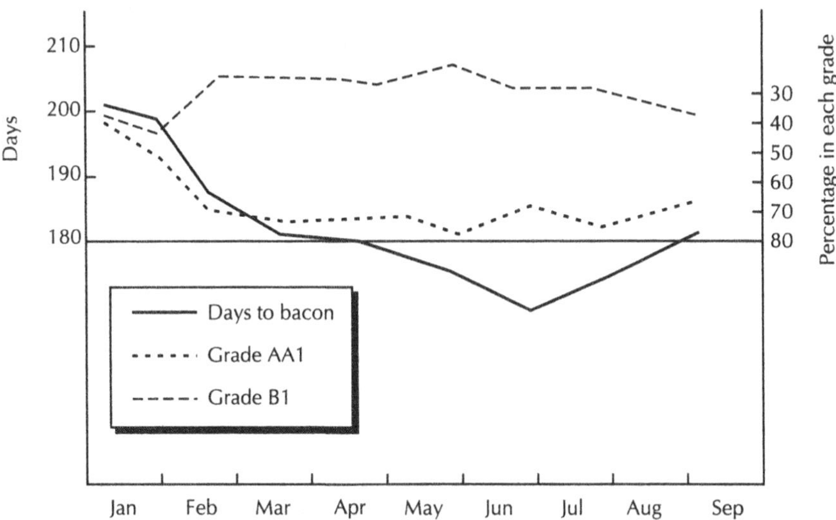

Fig. 17.4 Days to bacon and grading results.

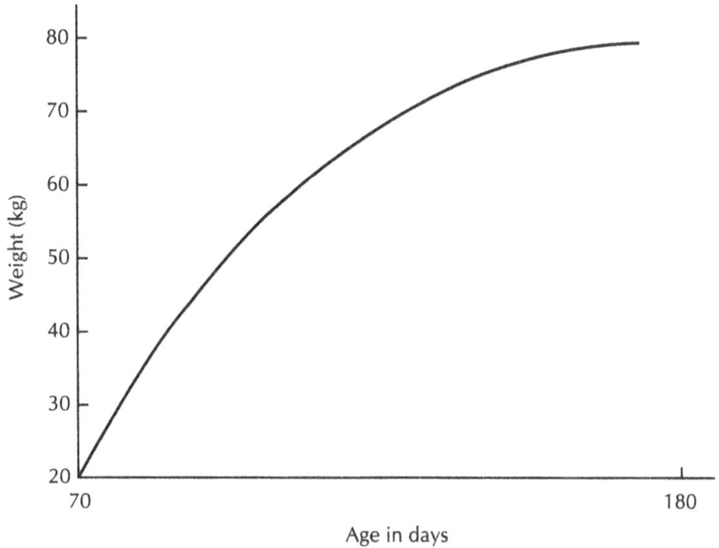

Fig. 17.5 Weight for age relationship in bacon pigs.

and feed manufacturers in relation to carcass conformation are 'a must' if continued progress is to be made towards the ultimate goal of fulfilling market requirements.

A further vital point to make is that the importance of selecting stock for market and abattoir regularly is much more important now than it has ever been. Gone are the days when it was acceptable to wait until a lorry could be filled with lambs or cattle as, by that time, some will be too fat.

Pre-slaughter treatment

This concerns:

(1) *Stress to the animal.* Stress is primarily caused by mixing groups of bulls together, either on a farm or during transport or prior to stunning. Some breeds of animal are particularly susceptible to stress and react in such a way that the meat is badly affected, e.g. pigs, particularly Pietrain, produce watery muscle (pale soft exudative – PSE) while dark, firm and dry pigments can also result from stress. Young bulls of all breeds can produce dark cutting meat (DCB) when subjected to stress, although it is by no means restricted to them. Generally speaking, the dark cutting beef is caused by high pH levels in the meat and the more active and dominant bulls in a group appear to be more susceptible. Low pH levels in bull's meat results in pale muscle in the carcass. If bulls are mixed, they require two days resting with feed to avoid DCB. Heifers and steers are not so susceptible and do not have this reaction to stress.

(2) *Fighting.* Fighting is far the biggest problem in pre-slaughter damage to pigs. In some surveys, 44% of carcasses are damaged and the situation appears to be worsening as more entire (and therefore belligerent) pigs are produced.

(3) *Weight loss during transportation and lairage.* Obviously, the distance from the abattoir will be vital. Sheep and cattle normally lose 2–3% of liveweight during transport and can lose up to 7% on long journeys. Most of the weight loss is incurred between market and abattoir (certainly a factor of importance to consider in avoiding liveweight sale since direct marketing to abbattoir cuts down weight loss) with farm to market being second, while the effects of fasting for 12 hours can result in a further 2% loss in sheep and cattle and up to 5% in pigs. Fasting for longer periods will cause even greater yield losses. The effect of fasting on meat quality is also important and the liver appears to be especially susceptible.

    It is therefore important for the producer to determine how far animals will travel under what conditions, when deciding on which market outlet to use, so that the cost–price relationship of each outlet can be determined and the best one selected.

Post-slaughter treatment

EU regulations on chilling mean that carcass meat must be cooled to 7°C within 24 hours of slaughter. This prevents the natural conditioning occurring and can lead to cold shortening of the meat, causing it to become tough when eaten. To fully condition meat, it should be allowed to cool naturally and then hang for 30 days. However, most of the conditioning occurs within the first seven days of hanging. The EU regulations, shortage of space and the need to increase turnover in the meat industry mean that the majority of meat is not conditioned fully by traditional hanging, and electrical stimulation of carcasses has been used to prevent cold shortening while allowing the advantages of fast chilling to be gained. Indeed most supermarkets demand that meat is conditioned for 10–14 days before they take delivery, and this is written into their contract with the meat plant. If the process of conditioning is carried out in vacuum then the tendency of the carcass to lose weight during conditioning is altered.

Some guidelines for avoiding stress and (therefore) weight and quality loss in animals

(1) Keep animals in social groups wherever possible. If they must be mixed, do so at loading as this will minimise fighting and physical damage.

(2) Avoid use of sticks or electric goads. Slap-marks for pigs must be used carefully and without too much force. sheep should not be moved using the fleece and must be lifted carefully, if lifting is necessary.

(3) Handle and move the animals quietly and calmly. Avoid sharp turns in races and, ideally, constructions should be solid without mesh or sharp points, particularly for sheep and pigs.

(4) Keep animals clean by fresh bedding, avoiding water spillage and overcrowding. Sheep should be moved from roots or muddy grazing two to three days prior to sale, onto short clean grazing. This will ensure that fleeces and hides/skins are clean and presentable and avoid contamination of the meat during the slaughtering process.

(5) Do not feed animals for slaughter on the day of dispatch. Those for next day's killing should be given a light feed with water available.

(6) The most obvious damage to animals is caused by injection abscesses, especially in sheep if they are injected through a dirty fleece or using a contaminated needle. Barbed wire fencing is another danger, particularly for cattle. High tensile wire or electric fencing is better. In feeding areas, steel or timber partitions are best.

(7) During transportation, consideration should be given to the following:
   (a) Provision of adequate ventilation on vehicles and in holding areas.
   (b) Reduction of loading density on hot days and for long journeys.
   (c) The slope of the loading ramp should be less than 20°.
   (d) Sufficient clearance on lower decks for cattle.
   (e) Proper partitions to prevent overcrowding, restrict movement and minimise mixing of groups.

(f) Holding pens for pigs should be long and narrow to restrict fighting.
(g) Smooth driving without sudden stops and starts.

(8) At the auction market and lairage, staff should be made aware of the consequences of stress and mishandling on the subsequent value of the meat, in order that animals are moved as smoothly and as comfortably as possible.

Timing of sales

For beef cattle and lambs the timing of sales to benefit from seasonal price fluctuations is more relevant nowadays because of the EU schemes which provide a headage payment and no guaranteed price.

There is no guaranteed price for pigs set by the EU or UK governments and so producers are subject to the vagaries of the market. For this reason most pigs are sold deadweight on contract so that a known price formula is negotiated and price failure is averted. With declining support in most products from the EU funds it is likely that producers of all red meats will be much more subject to open market prices in future and may be able to benefit by scrutinising the pattern of prices. These are shown in Figs 17.6–17.8.

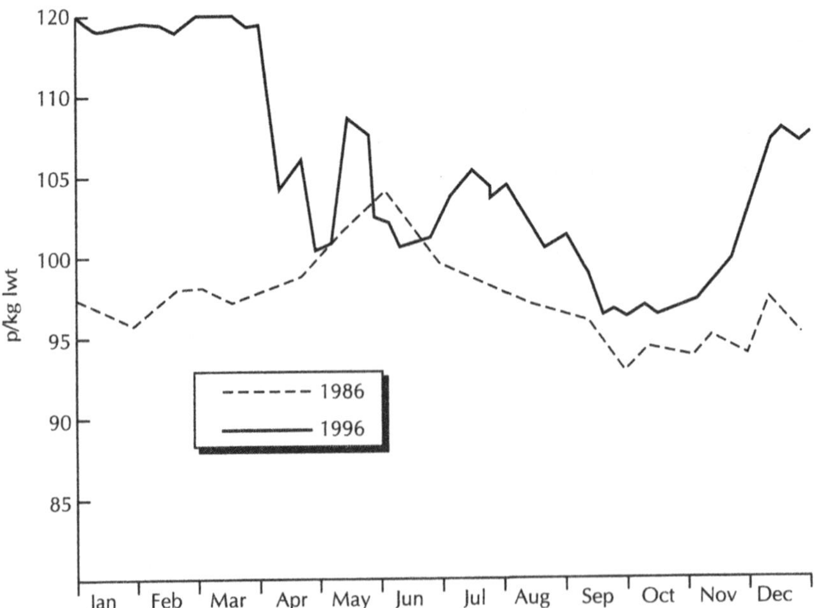

Fig. 17.6 Finished cattle prices.

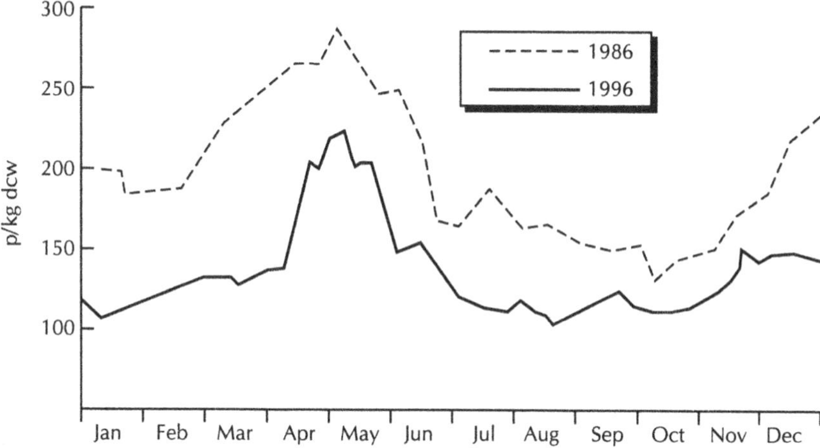

Fig. 17.7 Lamb prices.

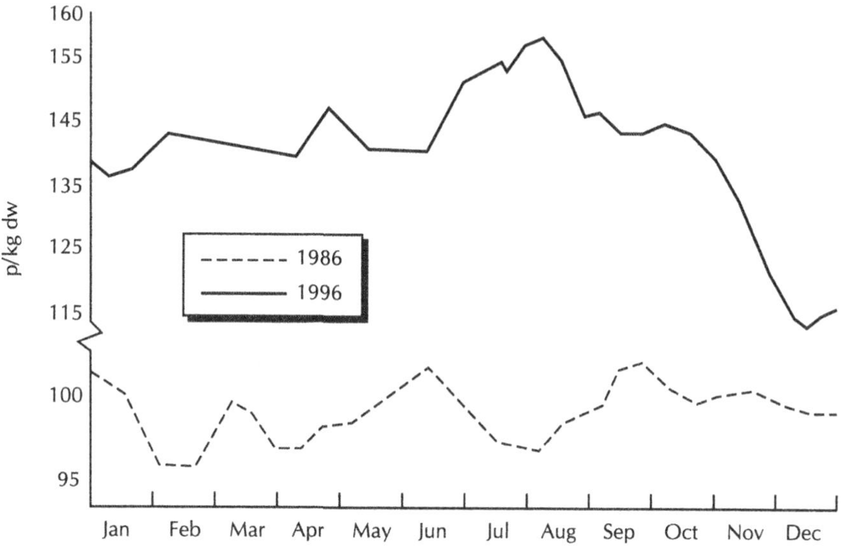

Fig. 17.8 Pig prices.

Premiums available for livestock

The sheep scheme (prior to 1993)

The sheep scheme was introduced in 1980. The aim at that time was to bring sheep market returns into line with French levels by 1984.

Prices in the UK were increased considerably as a result of the market support given, which amounted to around £15 per lamb per year, plus the ewe headage payment ranging between £2 and £6 per ewe per year. The principles of the scheme are illustrated in Fig. 17.9. When the UK market price was below the weekly guide price quoted by the Commission the difference was made up, on eligible lambs, by a variable premium. To be eligible lambs had to reach specified fatness and conformation standards. There was a maximum weight of 21 kg for lambs or 24.5 kg for hoggets upon which a premium was paid. The difference between the guide and the basic price over the year, from January to January, was made up by an annual payment divided between ewes in registered flocks, that is a headage payment which flock owners have to claim. The scheme increased profitability (and therefore the popularity of sheep) by between 25% and 30%. However, exporting sheep meat to other EU countries was difficult because of the *clawback* arrangements.

Sheep meat exported from the UK to an EU country had any variable premium payments clawed back. This meant that the exporter paid back the premium and therefore found it difficult to export lamb and remain competitive in the markets of Belgium, France and Germany. Attempts to extend clawback arrangements to sheep meat sold outside the EU were resisted. Other major producers in the EU, France and Eire, operated an intervention buying system to support their sheep meat prices, and were not therefore subject to clawback arrangements.

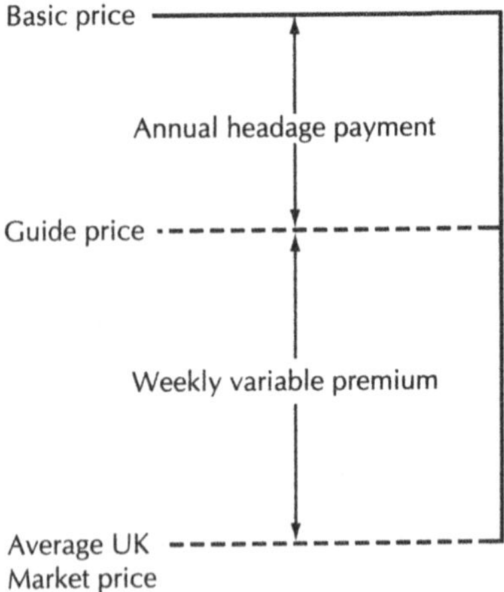

Fig. 17.9 The lamb support scheme prior to 1993.

Since the MacSharry reforms of 1992 the Sheep Scheme is completely different; there are no variable premiums, no clawback arrangements and support is given by a headage payment. The scheme is described in Chapter 14.

The beef scheme (prior to 1993)

Before 1 April 1989, the EU system for beef and veal was exceedingly complicated. The relationship between the reference price and the guide price determined whether any intervention measures were required at all. The reference price was determined by calculating the average of prices from selected markets and when this fell below 91% of the guide price the intervention buying of beef was triggered in the continental EU countries. If the reference was above the guide (reflecting a shortage of beef) then import levies would be reduced in order to allow cheaper meat into the EU. The principles of the old beef support scheme are illustrated in Fig. 17.10.

In the UK a target price was quoted which was the minimum weekly market price for eligible cattle. Any producer who sold cattle in a week when the average market price fell below the target price had the deficit made up by a variable premium, provided that the cattle were certified as meeting the minimum standards of fatness, conformation and weight by an MLC fatstock officer.

There was a maximum variable premium available of 10.04 p per kilogram (unlike the sheep scheme where no maximum applied) and clawback arrangements for beef exports were introduced in the 1985–1986 marketing year. The scheme meant the maintenance of reasonable margins on beef, the premium being worth around £35 on each animal, and helped to stabilise cattle numbers. The MacSharry reforms of 1992 altered the beef support scheme and it is described in full in Chapter 14.

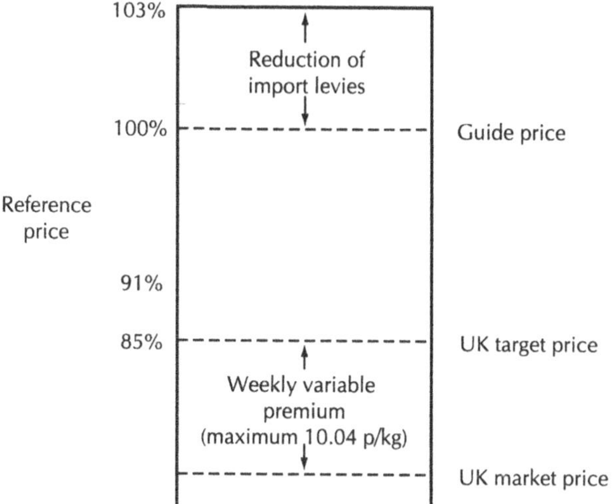

Fig. 17.10 The beef support scheme prior to 1993.

The pig meat scheme

The scheme runs from 1 July to 30 June each year. Introduced in 1973, the scheme offers no direct support to producers in the UK. Measures are available, however, to allow private storage of pig meat so that surpluses can be removed from the market if necessary. Buying into public storage has not been applied since 1971. Up to 30 June 1995 producers were protected from cheaper pig meat coming into the EU. Imports were controlled by fixing a sluicegate price below which no imported pig meat could enter the EU. A levy was added to the sluicegate which reflected the high cost of feed in EU countries (caused by high grain prices) and this raised the import threshold to a level which attempted to reflect the cost of producing pig meat in the EU. To this was added a further 7% of the sluicegate to maintain the competitiveness of EU-produced pigs. The principles of the pig meat support scheme are illustrated in Fig. 17.11.

As a result of the GATT agreement the sluicegate price was abolished on 30 June 1995. Protection for producers is now offered through a customs duty which will be reduced by 6% annually to 64% of the average levy operating in 1986–1988. Supplementary duties can, however, be imposed if imports rise too high. Minimum access quotas will have to be allowed (see Chapter 14, GATT/WTO) for which customs duty is limited to 32%. Exports will benefit from refunds which will be progressively reduced over 6 years to 64% (financially) and 79% (volume) of the average refunds in 1986–1990, i.e. 491 000 tonnes in year 1.

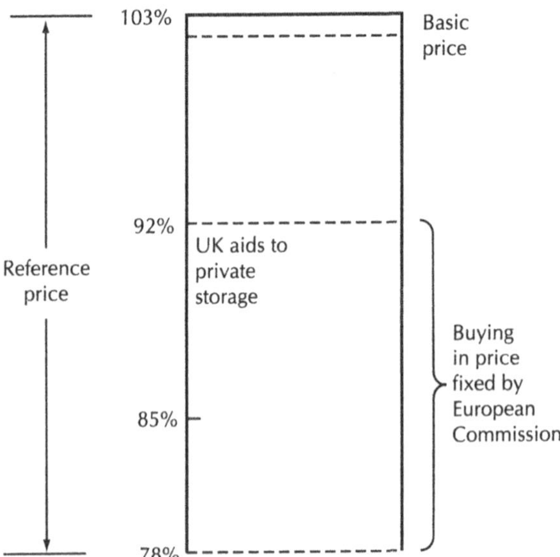

Fig. 17.11 The pig meat support scheme.

Conclusion on livestock marketing

The producer must take more account of the requirements of his customer (the butcher) and the consumer. The trend away from consuming red meat on health grounds is stabilising and there are signs that the consuming public is now aware that diet is only one of many factors (sex, age and inherited characteristics being more important) which need to be considered. The use of hormones and antibiotics to stimulate better performance will not be tolerated by the consumer and producers will also have to take greater care to improve their public image by making 'acceptable' changes to the environment within which animals are kept. For the most part this will not cause any undue difficulty as good husbandry encapsulates these practices anyway and to remain in the top 25% of producers, which is desirable, good husbandry is a paramount objective. It will, however, add cost to production and efficiencies will have to be sought.

Genetic modifications, under the general biotechnology banner, are suspiciously received by consumers as a result of events surrounding BSE. Consumers do not appear to be convinced by assurances of safety regarding genetic modifications and generally seem to feel that more time should be taken to test for the long-term effects of these innovations. A great deal has yet to be done to restore consumer confidence in the meat sector in particular and the food chain in general.

References

Ford, B.J. (1996) *BSE: The Facts*. Institute of Biology/Corgi.

Appendix Discount Factors

Present value of a single 'payment' due in the future

Future years	Present value factors at interest rates														
0	1%	2%	3%	4%	5%	6%	7%	8%	9%	10%	11%	12%	13%	14%	15%
1	0.990	0.980	0.971	0.962	0.952	0.943	0.935	0.926	0.917	0.909	0.901	0.893	0.885	0.877	0.870
2	0.980	0.961	0.943	0.925	0.907	0.890	0.873	0.857	0.842	0.826	0.812	0.797	0.783	0.769	0.756
3	0.971	0.942	0.915	0.889	0.864	0.840	0.816	0.794	0.772	0.751	0.731	0.712	0.693	0.675	0.658
4	0.961	0.924	0.888	0.855	0.823	0.792	0.763	0.735	0.708	0.683	0.659	0.636	0.613	0.592	0.572
5	0.951	0.906	0.863	0.822	0.784	0.747	0.713	0.681	0.650	0.621	0.593	0.567	0.543	0.519	0.497
6	0.942	0.888	0.838	0.790	0.746	0.705	0.666	0.630	0.596	0.565	0.535	0.507	0.480	0.456	0.432
7	0.933	0.871	0.813	0.760	0.711	0.665	0.623	0.584	0.547	0.513	0.482	0.452	0.425	0.400	0.376
8	0.924	0.854	0.789	0.731	0.677	0.627	0.582	0.540	0.502	0.467	0.434	0.404	0.376	0.351	0.327
9	0.914	0.837	0.766	0.703	0.645	0.592	0.544	0.500	0.460	0.424	0.391	0.361	0.333	0.308	0.284
10	0.905	0.820	0.744	0.676	0.614	0.588	0.508	0.463	0.422	0.386	0.352	0.322	0.295	0.270	0.247

Future years	Present value factors at interest rates														
0	16%	17%	18%	19%	20%	21%	22%	23%	24%	25%	26%	27%	28%	29%	30%
1	0.862	0.855	0.847	0.840	0.833	0.826	0.820	0.813	0.806	0.800	0.794	0.787	0.781	0.775	0.769
2	0.743	0.731	0.718	0.706	0.694	0.683	0.672	0.661	0.650	0.640	0.630	0.620	0.610	0.601	0.592
3	0.641	0.624	0.609	0.593	0.579	0.564	0.551	0.537	0.524	0.512	0.500	0.488	0.477	0.466	0.455
4	0.552	0.534	0.516	0.499	0.482	0.467	0.451	0.437	0.423	0.410	0.397	0.384	0.373	0.361	0.350
5	0.476	0.456	0.437	0.419	0.402	0.386	0.370	0.355	0.341	0.328	0.315	0.303	0.301	0.280	0.269
6	0.410	0.390	0.370	0.352	0.335	0.319	0.303	0.289	0.275	0.262	0.250	0.238	0.227	0.217	0.207
7	0.354	0.333	0.314	0.296	0.279	0.263	0.249	0.235	0.222	0.210	0.198	0.188	0.178	0.168	0.159
8	0.305	0.285	0.266	0.249	0.233	0.218	0.204	0.191	0.179	0.168	0.157	0.148	0.139	0.130	0.123
9	0.263	0.243	0.226	0.209	0.194	0.180	0.167	0.155	0.144	0.134	0.125	0.116	0.108	0.101	0.094
10	0.227	0.208	0.191	0.176	0.162	0.149	0.137	0.126	0.116	0.107	0.099	0.092	0.085	0.078	0.073

Index

9 780632 036035